AGENT CULTURE

*Human–Agent Interaction
in a Multicultural World*

AGENT CULTURE

*Human–Agent Interaction
in a Multicultural World*

Edited by

Sabine Payr and Robert Trappl
*Austrian Research Institute for
Artificial Intelligence and
University of Vienna*

LAWRENCE ERLBAUM ASSOCIATES, PUBLISHERS
2004 Mahwah, New Jersey   London

Copyright © 2004 by Lawrence Erlbaum Associates, Inc.
All rights reserved. No part of this book may be reproduced in
any form, by photostat, microform, retrieval system, or any other
means, without the prior written permission of the publisher.

Lawrence Erlbaum Associates, Inc., Publishers
10 Industrial Avenue
Mahwah, New Jersey 07430

Cover design by Sean Trane Sciarrone

Library of Congress Cataloging-in-Publication Data

Agent culture : human-agent interaction in a multicultural world / edited by Sabine Payr and Robert Trappl.
  p. cm.
Includes bibliographical references and index.
ISBN 0-8058-4808-8 (alk. paper)
 1. Human-computer interaction. 2. Intelligent agents (Computer software). I. Payr, Sabine, 1956– II. Trappl, Robert.

QA76.9.H85A49 2004
005.1—dc22   2003064220
                                                           CIP

Books published by Lawrence Erlbaum Associates are printed on acid-free paper,
and their bindings are chosen for strength and durability.

Printed in the United States of America
10 9 8 7 6 5 4 3 2 1

For Valentin

For Catherine

Contents

Preface ix

Introduction xiii
Sabine Payr

PART I. CULTURE(S) AND AGENT TECHNOLOGY 1

1 The Agents of McDonaldization 3
  Phoebe Sengers

2 Designing Technology, Designing Culture 21
  Lorna Heaton

3 Socially Intelligent Agents in Human Primate Culture 45
  Kerstin Dautenhahn

PART II. DESIGN FOR CROSS-CULTURAL BELIEVABILITY 73

4 Transcultural Believability in Embodied Agents:
  A Matter of Consistent Adaptation 75
  Fiorella de Rosis, Catherine Pelachaud, and Isabella Poggi

5	Creating Embodied Agents With Cultural Context Jan M. Allbeck and Norman I. Badler	107
6	Enculturating Agents With Expressive Role Behavior David R. Heise	127
7	Toward Cross-Cultural Believability in Character Design Heidy Maldonado and Barbara Hayes-Roth	143
8	Recruiting a Virtual Employee: Adaptive and Personalized Agents in Corporate Communication Benoît Morel	177
9	Lifelike Agents for the Internet: A Cross-Cultural Case Study Brigitte Krenn, Barbara Neumayr, Erich Gstrein, and Martine Grice	197

PART III. AGENTS FOR INTERCULTURAL COMMUNICATION   **231**

10	Building Bridges Through the Unspoken: Embodied Agents to Facilitate Intercultural Communication Katherine Isbister	233
11	Designing a Social Agent for Virtual Meeting Space Hideyuki Nakanishi, Toru Ishida, Katherine Isbister, and Clifford Nass	245
12	Designing Intercultural Agents for Multicultural Interactions Elaine M. Raybourn	267

Suggested Readings	287
Contributors	293
Author Index	301
Subject Index	307

Preface

Computer animation and computer simulation have been making such rapid progress during the last years that we are facing an increasing number of computer-generated, realistic, and believable actors in different roles and different media (e.g., in computer games and even in movies).

Whereas scientific research and technical development have been focusing mainly on the (individual) personality of synthetic actors, we wanted to investigate the role of the synthetic being as part of a heterogeneous society of real and virtual persons. We furthermore tried to investigate in which cultural context synthetic actors are developed and used. In doing so, we considered them as actors also in the sense of their becoming social beings through interactions with their users.

We only saw a chance for a success of our endeavor by initiating an interdisciplinary discourse among researchers in cultural and technical areas. However, this is an often attempted and rarely successful effort. We therefore looked for questions that seemed to be of current relevance also for technical developers. We found that several projects were modeling nonverbal communication of agents, especially by mimics, gestures, and postures. Clearly, those aspects are strongly culture-dependent, and ignoring these dependencies can lead to a breakdown in (intercultural) communication. This turned out to be one of the major starting points for the discourse we envisaged.

We therefore arranged an interdisciplinary international workshop on "Agent Culture" to be held at the premises of the Austrian Research Institute for Artificial Intelligence in Vienna. After searching the literature for potential participants, we found, to our own—positive—surprise, that nearly all scientists who considered this aspect in their research were women, therefore the vast majority of the participants were female. A second, also positive, surprise was the fact that in both the presentation and discussion, the main field of research of the participants, whether artificial intelligence, media design, communication, language, or cultural sciences or any combination of them, remained in the background, thus enabling vivid and truly cross-disciplinary discussions.

Among the many aspects we found during the discussions, only a few can be mentioned as a small appetizer for the book:

- Believability of an actor is based less on detailed personality or emotionality models, but on functioning social interaction.
- The cultural background of the synthetic actor is usually not modeled explicitly, but is taken for granted (i.e., taken implicitly from the culture of the developer).
- The more a synthetic actor is humanlike, the higher the expectancies of the users on its sociocultural capabilities.
- Nonhumanlike synthetic actors can escape these expectancies, but their acceptance by the users is considerably smaller in "serious" roles (e.g., as medical advisors).
- Synthetic actors can play an important role in the improvement of intercultural communication.

Many more facets and examples of this fascinating topic are found in the chapters of this book.

Transforming workshop results into a book would not be possible without cooperating participants. We foremost want to thank the participants who took great pains to enhance their original position papers to book chapters by including new material and considering the comments in and outside the discussions, and those authors who contributed additional chapters to ensure a broad coverage of the issues. Furthermore, we want to thank our colleagues at the Austrian Research Institute for Artificial Intelligence for their support—especially Isabella Ghobrial-Willmann for efficiently organizing the travel and hotels for the participants, and Tanya Decmar for her hard work preparing transcripts from the tape recordings. Kristin Duch of Lawrence Erlbaum Associates was an ideal partner in our endeavor.

Finally, we want to thank the Austrian Taxpayers whose money enabled us to pay for the travel and hotels of the participants at the workshop. We are especially grateful to Dr. Ilse Koenig, head of the department of Social

Sciences, and her collaborators, Dr. Christina Lutter and Karin Harrasser, M.A., all at the Austrian Federal Ministry for Education, Science, and Culture, who channeled this money into a project of which this workshop formed an integral part. We hope you enjoy studying this book.

—Sabine Payr
—Robert Trappl
Vienna, Austria, March 2004

Introduction

Sabine Payr

This book is about *embodied agents* or *animated agents*. Both terms are used for interface agents with a virtual body or parts of it that are capable of some kind of motion. The talking head demonstrating facial expressions falls under this definition as well as the fully animated cartoon character, be it an object, a plant, an animal, or a humanlike or fantasy character. More specifically, this book is about embodied agents as cultural objects and subjects. They are cultural objects in that they are artifacts created by humans embedded in their culture, and they are cultural subjects in that they interact socially with humans, thus relying on and (re)creating the cultural background present in any social interaction.

Most research on embodied agents has, so far, implicitly assumed a cultural background shared by the agent and its human interaction partner. If we adopt the vision of agent researchers in which animated agents populate numerous online environments, and thus will have to interact with users from many different cultures, it is time to unravel these implicit assumptions and face this question: What is culture-specific in any given embodied agent, and how will agents fare in a multicultural world?

WHAT IS AN AGENT?

Software agents are programs to which one can delegate (aspects of) a task. They differ from traditional software in that they are personalized, continuously running, and (semi-)autonomous. Their salient features are autonomy and adaptiveness.

Autonomy

A basic distinction in user–computer interaction can be described by the metaphor of direct versus indirect management. Direct management requires the user to initiate all actions explicitly and to monitor all events. Nothing happens unless a person gives commands from a keyboard, mouse, or touch screen. The computer remains passive and provides little help for complex tasks or for carrying out actions. In the indirect management metaphor, the user accomplishes those tasks in a collaborative manner with the computer. Both user and computer can initiate actions and monitor events. Computer software that can provide this type of interaction is called an *autonomous agent*. Agents "know" users' interests and can act autonomously on their behalf.

The concept of proactiveness is closely related to the concept of autonomy. It means that agents do not simply act in response to their environment, but exhibit goal-directed behavior by taking the initiative (Feldman & Yu, 1999). The concept of mobility as the ability of agents to migrate in a self-directed way from one host to another on the network can also be considered as an extension of the autonomy concept.

An important aspect of autonomy is the fact that the user defines a goal but not the means to achieve it. The agent is in some sense free to choose the means depending on the behavior of the environment. This is one of the big advantages of agents because the behavior of the environment is in some ways always unpredictable for the user. The choices of the agent, however, are limited by the utility the agent has to maximize or, vice versa, the cost it has to minimize, which is defined by the user. Such costs could be time, money, size of result, and so on. It clearly is not easy to develop agents that have this capacity.

Adaptiveness

Agents should be able to learn as they react to or interact with their external environment so that their performance improves over time. The external environment may include the physical world, users, other agents, or the Internet. The qualities necessary for adaptiveness have been defined as reactivity and social ability. *Reactivity* means that agents perceive their environment and respond in a timely fashion to changes that occur in it. *Social ability* means that agents interact with other agents, and possibly humans, via some kind of agent-communication language.

Whether a computer program can be considered an agent is defined functionally and not on the basis of a common design, programming language, or method. Any kind of information-processing, statistical operation, knowledge base, and artificial intelligence methods can be found and combined in a software agent. The concept is used widely and in different senses. Watt (1996), for example, tried to clarify the conceptual confusion by introducing two distinctions.

AGENTS AS METAPHOR AND AS IDEAL

First, he makes a difference between the metaphorical and the ideal use of the concept. In the metaphorical sense, an agent is a tool for thinking about software design—a paradigm. The term *agent-oriented programming* points to such a metaphorical use of the concept. Agent as an ideal, in contrast, focuses on the kind of agency with which humans are familiar. Research with this concept in mind attempts to create artifacts that are embedded in their environment, which is physical as well as social:

> Taken to the limit this would mean that agents have to be full-blown artificial intelligences, but this does not necessarily inhibit us from making good use of the technology in the meantime. We can still use the concept metaphorically, but explicitly as an interim step to keep our feet on the right path. (Watt, 1996)

THE INTERNAL AND THE EXTERNAL STANCE

As a second way to cut up the concept of an agent, Watt suggested distinguishing between the internal and external stances. In artificial intelligence (AI), agents are described principally in terms of the internal states—the desires, beliefs, and goals—over which the agent has control. Most workers in this field, however, sidestep the discussion of whether these notions, in turn, are to be understood in a psychological sense or again as metaphors. Just as a researcher (Cañamero, 2003) is free to call the function that drives the primitive creatures in her software toward food *hormones*, any piece of knowledge represented in a program could be claimed to be a *belief* and the specific ordering of rules a *goal*. The only problem is that, in the case of hormones, it is clear that the term is used metaphorically as our understanding of hormones includes their being a physical substance, whereas beliefs and desires are states, and we understand that both humans and machines can have internal states. Many a discussion between philosophers and AI researchers has been conducted about the sense in which these and other concepts—especially intelligence—may or may not be used for states of machines. They are often pointless, but not useless, in the sense that they may have led to more precise questions about psychological phenomena that had been taken for granted before.

The external stance is taken by human–computer interaction (HCI) research. Here the word *agent* is used for any active entity that will take on a user's goals and act on them. The HCI literature avoids describing what is going on inside an agent, falling back to an intuitive definition of agent as something that initiates and performs actions. Programs are agents in the social sense if they are capable of acting on their own or, rather, their users treat them as if they were capable of acting on their own—and this is what really matters. The external stance looks to nothing but the behavior of both agent and user.

There are numerous applications of software agents. Most frequent are agents for information retrieval and filtering, which should help the user cope with the information overload of, for example, the World Wide Web (www). There are agents for matching, comparing, analyzing, and categorizing information. An agent like "Firefly," which offers to e-commerce clients products on the basis of their previous searching and purchasing behavior (and of those of other clients), falls into this category. Other agents monitor computer user behavior to automate repeated tasks such as formatting operations, updating spreadsheets, opening applications, and so on.

Only a small fraction of agents is or will ever be visualized and animated. In the case of animated agents, we clearly use the external stance of HCI research to define what an agent is because we hardly look at the underlying technology of it: More often than not, the animated agent is a sort of visual front end to a whole bunch of software agents and/or other software modules (e.g., databases). We would even be tempted to speak of an animated agent when the crucial criteria—autonomy and adaptiveness—are insufficiently implemented. In effect the only criterion that matters is that the virtual character is perceived as an agent in the social sense so that the user is willing to enter into a social relationship with it.

HYBRID CULTURES

We could imagine a virtual world where all types of virtual characters exist at the same time: users represented by avatars, synthetic actors playing their scripted role(s), and autonomous agents with their own beliefs, desires, and intentions:

> We are faced with the interesting situation where relationships and meaningful interactions need not be restricted to interactions with a human other. In virtual environments, the other may or may not be human, and if human, may or may not be presented in a form easily recognized as humanoid. (Nowak & Biocca, 1999)

INTRODUCTION   xvii

What is more, any kind of visual representation can be freely exchanged among the characters so that a user may not be sure under what kind of control a virtual character acts at any given moment. For example, an avatar may be controlled by a human and suddenly be switched to a kind of automatic pilot. It now could interact and communicate with the (virtually) present humans without them perceiving the switch. At the same time, it could continue communicating with its owner (cf. Gerhard & Moore, 1998). The owner may leave for whatever reason and allow the same virtual body to be inhabited and animated by a program and turned from an avatar into an agent. However, a human user could choose any kind of animal or object as her avatar—tree, box, dog, fly, and so on. Other users would have to find out that what they assumed to be an unintelligent character in the virtual world is, indeed, another person. Would the user just stop to care whether her interaction is with another human? Or would she run the Turing test with each character that she meets and again at any moment when she suspects a change in the control of the character?

In the book chapter "Being Real" (Donath, 2000), Judith Donath tried to answer these questions. She started out from two philosophical possibilities—the pragmatic stance and the skeptical stance. As an aside, a third position also exists—that of solipsism (i.e., one can never be certain that there is really a world ouside the "I"). Yet this position is easy to falsify, at least for an Austrian (and maybe also for other countries' citizens): No single "I" is capable of inventing Austrian bureaucracy. The pragmatic stance is taken by those who assume that the other whom they meet is human as long as there is no striking evidence to the contrary. This is the stance that we take in everyday life when we—until now—meet other human beings. From their appearance, which is similar to ours, we conclude that their mental life, emotions, needs, desires, and so on are similar to ours. This assumption is at the basis of any social interaction because we can never know for certain whether a mutual understanding does indeed exist. We have to assume that it is so, and—in real life—this pragmatic stance is functional in such a vast majority of cases that it is justified. From the viewpoint of cognitive economy, it is far more useful to be pragmatic than to be skeptical.

Donath (2000) analyzed extensively the interaction between "Julia" (cf. Foner, 1999) and a chatroom participant who knows there are chat-bots in virtual worlds, but is nevertheless fooled by Julia for a long time. Even when the shortcomings in Julia's part of the dialogue become evident, she first is inclined to believe that she has met a somewhat disturbed, but still human, partner. Only when clearly nonhuman errors occur (e.g., verbatim repetitions or ignorance about Julia's own history), she discovers that she has interacted with a robot. What is more, she does not stop the dialogue at this point, but continues to chat with Julia much as if the chatter-bot were a human. However, she then concludes that the knowledge did make a difference: She was no more ready to feel what she called *empathy* with the chatter-bot. She would not expect that the chatter-bot would be able to become attached to her and would consider it a waste of energy to try to make it have feelings toward her. Donath concluded that there is a distinctive quality of being real that is at stake in the spread of virtual worlds with virtual inhabitants.

If a human user wants to reserve this qualtiy of being real to other humans, she would have to test each virtual character until she could determine beyond reasonable doubt whether it is an avatar of a human user or an autonomous agent. However, with the progress of agent development, with agents that are endowed with models of personality and emotions and with a made-up biography, this task becomes more tedious. Certainty can never be established definitely; there is always room for more skepticism. Therefore, it is also conceivable that human users do not grant the quality of being real to any virtual character and withhold empathy from the start. In a (virtual) world, where the chances to meet autonomous characters increase, this would be more economic. It would be the pragmatic stance once again, but now turned upside down.

However, we know that human beings tend to build emotional, sometimes even love, relationships with other human beings. What if they turn out not to be real human beings? The doll Olympia in Hoffmann's Tales is one of the first fictional examples. Can we risk the disappointment or even catastrophe if a person finds out that the other person is only virtual? When will the first smart (probably U.S.-American) lawyer sue a company for the pain her or his client suffered from falling in love with an agent? Should there be a law for agents to declare themselves virtual?

Nowak and Biocca (1999) suggested that humans, rather than just changing their attitude, will develop a continuum of relationships based on the perceived level of co-presence of the other intelligence when it becomes impossible to decide clearly and immediately whether the present relationship is interpersonal (in the sense of human to human). This would mean that we have to break with our assumption, supported by experience in the natural world, that we can safely treat anything that appears intelligent as humanlike and anything that looks and acts human as human.

> For example, it may become less meaningful to users whether an interactant is human or non-human for certain kinds of interactions in virtual environments: commercial transactions, telephone directory services, etc. It may only be meaningful to the user when the level of copresence needed for the interaction is very high and demands the full attention and intelligence of another human. . . . It may be enough in some interactions that the entity, like a dog or cat, is interactive, copresent, responsive, or makes us feel "less alone." The intelligence may not need to be human for many types of interaction. (Nowakk & Biocca, 1999)

All that we know about culture leads us to suppose that a hybrid culture would adopt a pragmatic stance in one or the other way. Culture, in a sense, is economy of the mind: We need not, very literally, reinvent the wheel, but can use it—even as it appears to our introspection unthinkingly. Likewise it is probable that we will develop ways to move and interact in a virtual world where the characters we meet might be agents or the avatars of human users. A passage from a science fiction novel may offer us a glimpse on how such a hybrid culture may work. In Tad Williams' (1998) tetralogy "Otherland," avatars are called "citizens," whereas agents are "puppets":

> "Just for your information, " she said, "it isn't really polite to ask if people are Citizens or not. . . ."
> "But I thought you said that it was the law. . . ."
> "It is the law. But it's a social thing—a little delicate. If you're talking to a Citizen and you ask them that question, then you're implying that they're . . . well, boring enough or mechanical enough to be artificial."
> "Ah. So one should only ask if one is reasonably certain that the person in question is a Puppet." (Williams, 1998, p. 61)

In this case, we could add, it does not really make sense anymore to ask. There is the law, but it is abstract, and there are ways to live with the law that have developed by use—in other words, a culture.[1]

The plot of "Otherland" revolves around the problem of virtual worlds that have become too realistic. Users are actually caught up in them and become unable to go offline. Their advanced virtual reality equipment makes them vulnerable to real pain and real death. Even today virtual worlds and the synthetic actors inhabiting them are becoming more and more lifelike (i.e., indistinguishable from the real world and real persons), all the more as the real world is made more and more simulationlike through digitization and mediatization. As a consequence, not far from today, if a person kills an actor in a computer game, there are three possibilities: (a) the person is killing an agent, (b) the person is killing an avatar that represents a human being, or (c) the person is looking into the real world via the cameras of a robot and is killing a human being.

AGENT CULTURE MEETS USER CULTURE

This visit to the hybrid culture of virtual worlds serves as a backdrop for the questions raised and discussed in this book. The virtual world, be it a game, learning environment, or social meeting space, appears to be the

---

[1] In this culture, for example, the "Puppet question" can be used to intentionally insult a citizen.

country of origin of the agent. In a world where everything is polygon-, code-, and text-based, the agent has almost the same possibilities of action and expression as the human user. What is more, we assist the emergence of genuine online cultures differently from the users' real-life cultures where virtual characters can fit in seamlessly.

By contrast, embodied interface agents mostly interact with users about some world that is external to the virtual space where the agent exists and is at home. The prototypical user that the authors of this volume have in mind is not the well-versed cyberspace traveler who feels at home in every MOO and chatroom, but a person who wants to get real-life tasks done using software and the Internet (e.g., shopping, learning, or getting advice). On the user's ground, the agent is confronted with the endless complexity and diversity of the world out there. One of the problems it has to face is the user's culture, which is probably different from its own.

Of course the interface agent is part and product of a particular culture. In its appearance, knowledge, and behavior, we can—possibly—detect its designer's explicit and implicit assumptions about the functioning of the world, the user's preferences and expectations, and so on. Only possibly, because our own culture remains invisible as long as it is not put into question and becomes the subject of our reflections and discourse. It functions as the unquestioned and unproblematic background to our actions (Habermas, 1999). Only when breakdowns of communication occur does it become obvious that the partners do not share a part of the background that is relevant in the current situation. Ideally, the misunderstanding should lead the partners to make their assumptions explicit and renegotiate the meanings of their utterances. There are at least two obstacles to this solution: First, partners may not even be aware that they have misunderstood each other. They may have interpreted the other's sign to their own satisfaction (i.e., integrated successfully into the complex sign of the utterance) without any of them realizing that the interpretation is not the one intended by the speaker. Second, the breakdown may be caused by culture-specific phenomena that are hardly accessible to conscious awareness, let alone verbal expression. Body rhythm is one example of such a phenomenon. Hall (1981) pointed out the importance of *being in sync* for successful communication: "Basically, people in interactions move together in a kind of dance, but they are not aware of their synchronous movement, and they do it without music or conscious orchestration" (p. 71). For Raffler-Engel (1988), acquisition of culture-specific rhythms starts before birth through the mother's movements and rhythms. That is why we are not aware of them and normally are not able to change them consciously. "The persistence of the problem throughout the interaction coupled with the inability of either party to identify its source make rhythmic dyssynchrony the most serious problem in cross-cultural communica-

INTRODUCTION   xxi

tion" (Raffler-Engel, 1988, p. 89). Communicative problems on such a level do not lead to misunderstandings, but to mistrust.

Agent designers want their virtual characters not only to be understood, but also trusted across cultures. They have to question and analyze their choices down to the deep levels of cultural phenomena they are unaware of in their everyday life as designers. We hope this book motivates these reflections and shows how to deal with cultural diversity in agent development.

OUTLINE OF THE BOOK

Part I of this volume is concerned with basic questions about agent technology and culture—or cultures. Phoebe Sengers, in *The Agents of McDonaldization* (chap. 1), argues that the concept of agency, as well as major trends in agent design today, are deeply rooted in Western, industrialized culture. She traces back the idea of agents to the principles of Taylorism like quantifying and rationalizing human behavior, reducing intelligent behavior to a set of independent, predictable, and interchangeable parts that live on in the engineering tradition in computer science. She reminds us that "agents are profoundly cultural" and challenges agent researchers to prove her wrong in her conclusion that "we are likely to see just-enough cosmetic adaptation in which agents appear inoffensive to members of other cultures, while maintaining a covert cultural campaign."

In *Designing Technology, Designing Culture* (chap. 2), Lorna Heaton presents two case studies on the design of computer-supported collaborative work (CSCW) software in Japan. She shows in detail how their understanding of the world around them guides the designers' choices: "As participants in their larger professional, organizational, and national cultures, individual designers link their creations with larger social or cultural values." Agent design is in no way different, and Heaton suggests means to reflect the cultural considerations that enter into interaction, personality, visual design, and context of embodied agents.

The first two chapters draw our attention to the fact of the cultural embeddedness of technology in general and agent design in particular, and they remind us that there cannot be an "agent without culture." In other words, culture is not some functionality that can be added to an agent architecture, but is always already present, introduced implicitly by the designers' cultural background. Chapter 3, then, turns to another aspect of agent culture. In *Socially Intelligent Agents in Human Primate Culture*, Kerstin Dautenhahn discusses her concept of socially intelligent agents (SIAs) in the context of human culture. She points out the risks of globalization and leveling of cultural diversity, but argues that SIAs could play a role in preserving this diversity. Agents as culturally situated artifacts could not only adapt to different cultures, but could support cultural diversity. With her conclusion that agent systems can be used to establish a shared understanding, but can also promote the diversity of understanding and identity, the author points to Part III of this volume.

Part II consists of chapters dealing with design concepts and reflections on cross-cultural believability. Fiorella de Rosis, Catherine Pelachaud, and Isabella Poggi open the section with their considerations on *Transcultural Believability in Embodied Agents: A Matter of Consistent Adaptation* (chap. 4). After presenting and discussing their ecologically oriented definition of *culture*, they identify two levels on which agents might deal with cultural diversity: a surface level of cultural sensitivity and a deeper level of cultural consistency. From their experiences with building a context-adapted animated agent, the authors suggest how an agent's behavior may be adapted to the cultural context of application while maintaining consistency.

They describe an agent that was developed in the framework of a European project and was applied to medical counseling on therapy, especially on taking medication. In the project, a study was made that compared drug explanation attitudes in Italy and the United Kingdom. The authors report cultural differences in doctors' behavior: "Italian doctors . . . tended, in particular, to minimize or hide the side effects of the prescribed drugs because they were convinced that this knowledge could affect patient compliance negatively." They suggest a design that allows the agent to adapt to these cultural differences so that the Italian agent puts information on side effects last in its list of conversation priorities. Inevitably, this decision invites a question from readers with other cultural backgrounds: Can and should mimicking current practices in human interaction in a given domain really always be the ultimate goal in designing an interface agent? Is the ethical problem of agents as "persuasive technology" (see chap. 3) solved satisfactorily if they "do what everyone (in that culture) does"?

In concluding their chapter, de Rosis et al. point to Hofstede's five dimensions of cultural variation as a possible model for building culturally consistent characters, in analogy to the "Big Five Factors" of personality. Such a model would indeed be needed to reduce the multitude of observations on culture-specific phenomena to a small set of dimensions that is able to represent different cultures.

Jan Allbeck and Norman Badler, in *Creating Embodied Agents With Cultural Context* (chap. 5), describe an approach based on the OCC model, which is widely known and used in the embodied agents research community (Ortony, Clore, & Collins 1988). Based on observations of cultural differences in nonverbal behavior and values, they suggest that "a different doctrine can be created for each culture and used as input to the OCC model." Such a doctrine would be a collection of settings for different dimensions of culture used as parameters for (re)action modulation. Without a doubt, it is the developer's dream to have just a few parameters with which to tune a system to a different culture. Cross-cultural psychology, in fact, has made considerable progress on relating a wide range of phenomena to Hofstede's "individualism–collectivism" dimension (Matsumoto, 2000). However, culture is the result of the highly complex interplay of historical, economical, religious, geographical, and political factors that may lead to specific expressions that are hard to capture by a few dimensions. It remains to be seen how far the approach of parametrizing culture can go.

Both chapters discussed previously are embedded in what we may call mainstream of current agent research, in that they are based on psychological modeling. Consequently, cultural variation may be brought into the model on the level of personality modeling. The point of departure of this research is the individual. As a sociologist, David Heise takes a different approach in *Enculturating Agents With Expressive Role Behavior* (chap. 6). He holds that affect control theory, an empirically based, mathematically elaborated perspective on microsociology, grown out of symbolic interactionism, can be incorporated into agents to give them a capacity for normative role behaviors and emotional displays. It allows the agent to be enculturated with a specific culture's interpretations of people and actions and its manner of processing events mentally. The focus here is on the social role of the agent, not so much on its deep personality—the assumption being that known roles are played in similar ways by social actors. The model relies on data about how people from different cultures judge people, relationships, and situations on the three scales of evaluation, potency, and activity. A culturally adapted agent that acts out its social role consistently (in the eyes of people from this culture) could be achieved by linking these different measures to behavior.

With *Toward Cross-Cultural Believability in Character Design* by Heidy Maldonado and Barbara Hayes-Roth (chap. 7), we pass on from more general considerations to the design and implementation of cross-cultural characters. Their example agent, Kyra, is a Web-based tutor for preteen children that has been localized to U.S., Venezuelan, and Brazilian audiences. The localization has been undertaken by native researchers of the respective cultures. Their work is firmly based on knowledge about cross-cultural communication and the key principles of character design developed by the research group. However, the results, illustrated by examples of behavior and dialogue, show such subtleties in cultural variation that we are led to doubt they could have been generated by any theory. What would a culture model have to capture to lead to a similarly fine-grained cultural adaptation of an agent?

Benoît Morel presents virtual character design from the perspective of the artist and practitioner. His company creates virtual characters for commercial Web sites—a process described in *Recruiting a Virtual Employee* (chap. 8). Corporate culture and audience culture(s) both guide the design choices, but as in the previous chapter the resulting culturally adapted agent is handcrafted. The case of the visualization of Microsoft's agent for Windows XP as a question mark illustrates a point also made by other authors in this volume: The price for a global character that offends no age, gender, social, or ethnic group may be too high, resulting instead in a semantically poor agent that is not able to transport personality or character.

Brigitte Krenn, Barbara Neumayr, Erich Gstrein, and Martine Grice describe an online entertainment application called "Flirtboat" and report cross-cultural user studies made in the United Kingdom, Austria, and Croatia in *Lifelike Agents for the Internet: A Cross-Cultural Case Study* (chap. 9). The team collected user (or, more precisely, avatar) data on gender, age groups, frequency of use, avatar choice, personality, and other aspects. The relevance of this case study lies not only in the results of the cross-cultural similarities and differences it reveals, but also in the methodology that allowed it to build such a rich collection of data. Flirtboat users can design their own avatar, among others via questionnaires, which result in detailed user profiles, and obviously they are willing to spend time and disclose information when the reward is adequate.

Part III documents a different approach to the issue of agent culture. Instead of asking how agents could adapt to the user's culture, these authors discuss the potential of agents as mediators in intercultural communication. The fact that the agent belongs to a different culture is seen as an asset in this perspective.

In *Building Bridges Through the Unspoken: Embodied Agents to Facilitate Intercultural Communication* (chap. 10), Katherine Isbister outlines the ways in which embodied agents are and could be used to coach the acquisition of intercultural communication skills. Agents can act as virtual conversation partners from the foreign culture with whom communication can be practiced repeatedly, without fear of failure and embarrassment, and with feedback behavior that is more characteristic and easier to detect and interpret than, possibly, with real interlocutors. Technology to track user behavior is being continually improved so that it becomes possible to extend intercultural agent-based training to the long-neglected, but important, nonverbal aspects of communication.

In *Designing a Social Agent for Virtual Meeting Space* (chap. 11), Hideyuki Nakanishi, Toru Ishida, Katherine Isbister, and Clifford Nass describe such an agent that plays the role of *party host* in a virtual space where people from different cultures meet. They conducted an experiment with students from the United States and Japan, which led to interesting results. The agent's intervention not only influenced the perception of the conversation partner from the other culture, but even stereotypes about the partner's nationality. The authors suggest that this finding could be used to intentionally mold intercultural communication. However, given that it seems to have been rather unexpected, this is a domain that should be approached with extreme caution.

On another level, the study points to a phenomenon that is easily overlooked: Culture is dynamic, and it changes in social interaction. Interaction with virtual characters being social in character, these characters, wherever they appear and whenever they interact with the user, take part in cultural change. In *Designing Intercultural Agents for Multicultural Interactions* (chap. 12), Elaine Raybourn addresses the need for a shared sense of community in large-scale collaboration systems for multinational organizations that transcends any one cultural orientation and that is truly multicultural. Designers of such environments should strive for computer-mediated communication that is capable of issuing rich cultural cues to create a living space for users and to support intercultural communication. *Multiculturalism* in virtual environments, defined as "the recognition that several different cultural orientations can co-exist in the same environment and benefit each other," finally leads to the emergence of a third culture. This emergent culture, co-created by multicultural users and intercultural agents, pertains to all parties involved and transcends any one single culture. After a long round trip, we come back to the hybrid culture mentioned earlier, but now from a culturally sensitized perspective. Raybourn concludes that "we should carefully consider our own cultural biases and those communicated by our agent-based software solutions." Her final remark is the best conclusion I can think of for this introduction and the whole volume:

> It is my hope that in this ensuing dialogue we designers, researchers, and consumers of these agents hold each other accountable for honing the sensibilities each of us needs to responsibly co-create intelligent, adaptive user interfaces and virtual characters that are truly equitable and multicultural.

REFERENCES

Cañamero, D. (2003). Designing emotions for activity selection in autonomous agents. In R. Trappl, P. Petta, & S. Payr (Eds.), *Emotions in humans and artifacts* (pp. 115–148). Cambridge, MA: MIT Press.

Donath, J. S. (2000). Being real. In K. Goldberg (Ed.), *The robot in the garden. Telerobotics and telepistemology in the age of the Internet.* Cambridge, MA: MIT Press. Also available at http://smg.media.mul.edu/papers/Donoth/BeingReal/BelingReal.html (last visited 11 Sept., 2003)

Feldman, S., & Yu, E. (1999). Intelligent agents: A primer. *Searcher, 7*(9). http://www.infotoday.com/searcher/oct99/feldmantyu.htm (last visited 11 Sept., 2003).

Foner, L. N. (1999). Are we having fun yet? Using social agents in social domains. In K. Dautenhahn (Ed.), *Human cognition and social agent technology* (pp. 323–348). Amsterdam/Philadelphia: John Benjamins.

Gerhard, M., & Moore, D. (1998). *User embodiment in educational CVEs: Towards continuous presence.* http://www.lmu.ac.uk/ies/conferences/ Gerhard.html (last visited 6 Feb, 2003).

Habermas, J. (1999). *Theorie des kommunikativen Handelns* (The theory of communicative action). Frankfurt a. M.: Suhrkamp. (Original publication 1981)

Hall, E. T. (1981). *Beyond culture.* New York: Anchor Books.

Matsumoto, D. (2000). *Culture and psychology* (2nd ed.). Belmont, CA: Wadsworth/Thomson Learning.

Nowak, K., & Biocca, F. (1999). "I think there is someone else here with me!": The role of the virtual body in the sensation of co-presene with other humans and artificial intelligences in advanced virtual environments. http://www.cogtech.org/CT99/Nowak.htm (last visited 06 May, 2003).

Ortony, A., Clore, G. L., & Collins, A. (1988). *The cognitive structure of emotions.* Cambridge, England: Cambridge University Press.

Raffler-Engel, W. (1988). The impact of covert factors in cross-cultural communication. In F. Poyatos (Ed.), *Cross-cultural perspectives in nonverbal communication* (pp. 71–104). Toronto, Lewiston/NY: Hogrefe.

Watt, S. (1996). Arificial societies and psychological agents. *BT Technology Journal, 14*(4). http://kmi.open.oc.uk/publications/techreports.html (last visited 11 Sept., 2003)

Williams, T. (1998). *Otherland: City of golden shadow* (Vol. 1). New York: Daw Books.

PART

I

CULTURE(S) AND AGENT TECHNOLOGY

CHAPTER

1

The Agents of McDonaldization

Phoebe Sengers
Cornell University

I used to build agents. In the old days, I built interactive computer characters, or animated creatures with idiosyncratic personality, who can express that personality in interaction with human users. To do so I used a programming framework that focused on expressing the agent author's vision of personality through an ad hoc and flexible organization of complex, hand-written behaviors (Loyall & Bates, 1991). This approach excited me because of its openness: It did not impose a rigid theory or structure on the personalities that could be expressed. This minimum-commitment model made possible the creation of arbitrarily complex and idiosyncratic personality. In this approach, agent design was thought of as craftwork: Authors could manipulate the materials of electronic code to express their unique personal vision of personality and life.

One evening I had a horrible nightmare. In my dream, 10 years had passed. I had graduated and the field of agents with personality had boomed. In this brave new world, I had found a job at a large fast-food corporation, working on the software assembly line building intelligent, franchised characters to be packaged with Happy Meals. My dream of intelligent characters who represented the full breadth, beauty, and complexity of life had been reduced to Ronald McDonald and the Hamburglar. I woke with a shudder. As dawn broke, I laughed at my imagination, went back to sleep, and forgot about the dream.

Fast forward 10 years. Agent-building has, in fact, exploded. My research group at the time has formed a start-up company, Zoesis, whose current product is intelligent, trademarked characters to support branding at company Web sites. Agents are being developed for use as virtual salespeople in virtual malls. Virtual people, like Lara Croft, are branded and marketed as mega-stars. Just as I had subconsciously feared, agents have been reincorporated to create the kind of personalities with which corporate America feels most comfortable: standardized, optimized, and under control.

In this chapter, I argue that this convergence between agenthood and American consumer culture is no accident. The concept of agency, as well as major trends in agent design today, is deeply rooted in Western, industrialized culture. Agents are an example of and propagate the trends to McDonaldization that Ritzer (1993) identified as emblematic of postindustrial consumer culture: efficiency, quantifiability, predictability, and control. Agents are profoundly cultural; their very definition incorporates assumptions of Western, industrial culture. Because of this, the notion of culturally flexible agents is an oxymoron.

To make this argument, I take you on a trip back in time. We go back to the beginning of Western industrialization to understand how the shift from craftwork to industrialized labor on the assembly line changed our understanding of work and, eventually, human behavior. We see how human behavior became something to be rationalized, quantified, and mechanized. As we move forward in time, we recognize how contemporary trends in consumer culture are rooted in the mechanisms and philosophies of the assembly line. All along the way, we find cultural shifts that are oddly familiar to those building agents; they make the notion of agenthood possible and shape the way we think of and build agents today.

AGENTS: THE VERY IDEA

Various definitions of *agent* have been put forward by the agent research community, none of which has acquired universality. Here I use the term *agent* in its broadest sense. By *agent* I mean a piece of soft- and/or hardware that is intended to represent a complete person, animal, or personality. An agent may be an animated figure, a robot that may or may not be human- or animallike in form, or a nonvisualized piece of code. By using the term *agent*, an agent's designers implicitly state that the code is best thought of as a self-contained, semi-intelligent unit corresponding in a vague sense to what we think of as a conscious being.

This definition of agent is intentionally quite broad, encompassing a huge range of potential technologies, applications, and motivations for agent-building. A wide range of agent applications is currently being used and developed. These include animated characters for video games; interface agents that provide a humanlike presence on Web sites, databases,

1. THE AGENTS OF McDONALDIZATION

and other programs; sales agents that engage in retail activity with human users; robotic toys such as the robotic dog Aibo and the hyperrealistic doll, My Real Baby; telephone-answering agents who can help users sort through the complexities of voice-mail systems; and so on.

What these agents generally share is a basic process of agent design. To build an agent, generally speaking, you begin with an idea of a kind of agent you wish to build. This agent may be a natural agent, such as an insect, dog, or person, or an imagined agent, such as a component of a large software application or a personality one wishes to embody in code. Once the concept of the agent is chosen, you break the behavior of the agent into units, dividing each of the units up into steps, implementing each of the steps and units as a mechanical process, and assembling them into the final agent. The kind of unit you choose varies depending on the kind of agent technology one is using and the aspect of the agent you are trying to model. Typically, agents are split into *functionalities*—problem-solving units that give agents specific capabilities, such as the ability to see, the ability to understand English-language texts, or the ability to understand basic rules of social interaction—and/or *behaviors*—integrated collections of functionalities that work together to engage in some coordinated real-world activity, such as hunting, finding and sorting trash, or searching for and greeting other agents. Although we have the technology to build subtle, complex, and powerful functionalities and behaviors, it is typically a great challenge to knit together the various functionalities or behaviors into an agent whose overall behavior is coordinated and coherent. What we see first is that this process of agent design has its roots in early industrialized restructuring of human labor.

INDUSTRIALIZATION AS AN EARLY FORM OF AI

> *In a sense, the mechanical intelligence provided by computers is the quintessential phenomenon of capitalism. To replace human judgement with mechanical judgement—to record and codify the logic by which rational, profit-maximizing decisions are made—manifests the process that distinguishes capitalism: the rationalization and mechanization of productive processes in the pursuit of profit.* (Kennedy, 1989, p. 6)

The history of the Industrial Revolution is often told as follows: In the beginning, there were craftspeople who owned their own tools, manufactured articles in their own idiosyncratic ways, and whose work was largely integrated with their way of living. As the Industrial Revolution begins, these workers are collected into factories, where they work together using the owner's tools. This owner, in an attempt to make work more efficient, begins to streamline the production process. Instead of having each worker build a piece from beginning to end, the production line is developed, where each worker works on some small part of the final piece. Work is broken up into stages, each of which is accomplished by a single worker; each stage is standardized so that articles can move from stage to stage without breaking work rhythm. Once work is divided up into standard stages, some of the steps can be done by a machine. Instead of building an article from beginning to end, workers now tend machines, which are each doing small steps of the article's production.

At each stage, work becomes more rationalized, predictable, and efficient. Workers on the assembly line can generate more articles, and the articles lack the idiosyncratic variation of normal craftwork. Instead of doing whatever he or she wants in a haphazard order, a worker has a fixed set of steps in which he or she engages. The intelligence of the worker, which he or she previously needed to monitor what he or she was doing and make active decisions about how work should proceed, is now embodied in the structure of the assembly line. Workers no longer need to think; the factory machinery does the thinking for them. Even before computers, industrialization takes the first baby steps of AI.

These traces of industrialization can be seen in the way we build agents today. The AI researcher building an agent follows the same basic line as the factory manager designing new production processes. Just as the factory manager attempts to design and reproduce a preexisting work process, the AI researcher would like to copy a natural process—an agent or idea of an agent. The factory manager breaks this process into logical steps, figuring out which steps should happen and in which order. Similarly, AI researchers analyze the agent's behavior to categorize its activity into typical behaviors, enumerate the conditions under which those behaviors are appropriate, or identify functionalities and develop well-defined algorithms by which those functionalities can be implemented.

Just as the factory manager embodies each step in machinery that can run with a minimum of human supervision, agent designers implement a mechanical version of each behavior, hooking them together so that they largely reproduce the imagined or real behavioral dynamics of the original creature. The early industrialist and the AI researcher are engaged in the same project: We analyze, rationalize, and reproduce natural behavior.

If agent design is like factory management, the nature of agent behavior may share characteristics with industrialized labor. In fact postindustrial work is radically different from preindustrial work, in both the qualities of the articles produced and the human experience of engaging in that work. The act of embodying work in the production line changes the nature of work. Work becomes more rationalized and less personal; workers are more dependable and more bored; the articles produced become more standardized and less individual. Several special qualities of postindustrial work and life have been identified by cultural theorists and industrial historians.

Reification. Things that were once thought of as ineffable or abstract become thought of as concrete. Labor, for example, which was once not strictly separated from the rest of life, becomes something that is measured and sold per piece or per hour. Once things are reified, they can be sold, becoming *commodified*, to be exchanged for particular sums of money.

In agent design, we too reify qualities of our agents. We sort ineffable behavior into well-defined categories, defining them cleanly and embodying them in code. We create rigid, formal definitions of categories of life, which are then said to drive behavior. In emotional modeling, for example, we define small sets of well-defined emotions, which are supposed to represent the rich complexity of human experience. Similarly, interaction between people and agents is reified; all too often interaction with agents is, at heart, simply a choice among some predefined options, although it may be masked by the appearance of more complex behavior.

Specialization/Atomization. Workers no longer engage in the entire work process. Rather, they each perform some small function within the process as a whole. Without an overview of the process, workers no longer need or are able to adapt to one another; each part takes place without reference to the others. Without feedback among the pieces, each piece is built in isolation, the whole being merely the sum of each individual, separately designed atomic part.

The production process, which was once a wholistic attribute of individual workers, is broken into rationalized parts, each of which is embodied in pieces of machinery or in production rules that regulate how they interact. Workers, who were once thought of as individual humans deeply embedded in the context of their daily life, now become interchangeable parts of the production process, whose time is to be sold to the highest bidder. They move from factory to factory, no longer connected to their home place or even a particular manufacturer. Workers see themselves as free and atomic individuals bound by no human ties.

Agents' functionalities or behaviors are similarly specialized, developed separately from one another and difficult to integrate. Agents are specialized for particular uses. Rather than representing the full breadth of human behavior, they are optimized for their use in particular situations. Interaction with them can be unnatural precisely because of their goal-directed, specialized design; unlike people, agents cannot handle behavior outside of their mandate. Like the real estate-selling agent REA's small talk (Bickmore & Cassell, 1999), focused on making the sale, agents interpret all behavior within the limited confines of their purpose.

Standardization. The idiosyncrasies of craftwork mean that one can never be sure what the produced goods will be like. The factory owner, however, who consolidates craftwork and has promised broader distribution networks goods of a particular kind and quality, wants to have some guarantees that the factory will produce similar goods no matter which workers are present on a particular day. The idiosyncrasies of personal work are no longer valuable. Instead the owner introduces steps of production control to ensure that the output is always similar. The qualitative, human, individual dimension of work is eliminated and replaced by efficient, controlled, and standardized work processes.

Similarly, standardization is one of the goals of agent architecture: to provide one single technology that can support multitudinous agents, each variations on a theme. Some architectures, like Loyall and Bates' (1991) Hap, allow for idiosyncratic variation according to the taste of the designer, but by and large such labor-intensive variations are considered a liability by the agent-building community. Variations in which one can alter agent behavior simply by changing the levels of a few parameters are preferred. A strong theme in agent research is the search for general rules behind behavior, which allow one to see behavior variations as parameter setting. In this way of thinking, all agents are fundamentally alike. Interactions between agents and people are standardized; the same technology is used in all cultures.

Formalization. The individual, material properties of workers and the material they operate on is ignored except insofar as it impinges on the production process. As the production process becomes more and more efficient, extrinsic considerations—whether social, spiritual, or physical—are left out. The production manager thinks of the production process in terms of abstract steps without reference to the particular identity of the worker or chunk of material involved. The factory is set up to enforce this abstract, impersonal view, which then seems to be an accurate reflection of the real. Individual differences become noise, unvalued and only reflected on to control their effects.

The individual, material properties of humans are also degraded in agent research. Generally speaking, agents are immaterial, presented as images on a screen. At best their physicality is simulated. There is some work on robotic agents, yet their physicality has little in common with human physicality, especially the aspects of physicality that are degraded or problematic in our culture. Few robotic agents bleed when cut and fewer still menstruate.

More generally, all aspects of being human that are considered inessential to the task at hand—whether this is a specific task the agent must fulfill or the more general problem of modeling an abstractly defined intelligence—are ignored. Individual differences are parameters in a general model, nothing fundamental—we are modeling intelligence, not a particular person at a particular time. Formal models of being human are developed and seen as explanations of the complex, difficult to define real.

With agents, human interactions are formalized and instantiated as code. Unlike telemarketers, agents can never break out of the script. For us this means unformalized interactions with them are no longer possible.

Mechanization. To maintain standard production, workers are given less and less leeway in decisions about their jobs. Rather than relying on the worker's judgment, the factory manager uses standardized production rules to ascertain that the product is always made the same way. As the steps of the production process are more and more formalized, the worker's intelligence becomes less and less pertinent. Once the worker's intelligence is no longer needed, the worker can be replaced by a machine. This is precisely what happens with agents: Work previously done by humans is now done by agents. Work is fully automated, but is given a false, *human* face.

Agenthood Under Industrialization: Shifts in Consciousness

Under industrialization, work has changed, and our understanding of what it means to be human is altered. For humans, industrialization is often an experience of being more and more dominated by systems of machinery, of both the technical and bureaucratic kinds. This is certainly the case for craftworkers, whose work historically consisted of skilled tinkering in the workshops of their houses, but presently generally involves the monitoring of raw materials as they are fed through massive machinery. Rather than applying their intelligence and skill to an ever-renewed activity, taking pride in the result of their handiwork, workers go through repetitive and mindless motions that are stipulated from beginning to end by the production handbook to create finished products they will never see. The experience of being a worker was once work. Now it is being an appendage to a machine.

In a sense, it is workers who have become mechanized. Lukács (1971) showed that the mechanization of the work process does not stop with production: The workers are progressively designed and controlled as machines.

> If we follow the path taken by labour in its development from the handicraft via co-operation and manufacture to machine industry we can see a continuous trend towards greater rationalisation, the progressive elimination of the qualitative, human and individual attributes of the worker. On the one hand,

the process of labour is progressively broken down into abstract, rational, specialized operations so that the worker loses contact with the finished product and his work is reduced to the mechanical repetition of a specialised set of actions. On the other hand, the period of time necessary for work to be accomplished (which forms the basis of rational calculation) is converted, as mechanisation and rationalisation are intensified, from a merely empirical average figure to an objectively calculable work-stint that confronts the worker as a fixed and established reality. With the modern "psychological" analysis of the work-process (in Taylorism) this rational mechanisation extends right into the worker's "soul": even his psychological attributes are separated from his total personality and placed in opposition to it so as to facilitate their integration into specialised rational systems and their reduction to statistically viable concepts. (Lukács, 1971, p. 88)

Taylorism, or scientific management, is the apogee of this view of worker as machine. The goal of Taylor's scientific management is to increase the efficiency of work processes by analyzing and optimizing not only the machinery, but also the way in which the worker uses the machines. Through time and motion studies, the worker's motions are examined; all extraneous motions are forbidden. Rather than letting workers interact with machinery in any way that they saw fit, Taylorists determine the "one best way"—the most efficient possible use of the machinery. On Taylorization, workers are generally given detailed instructions of every movement they should use to accomplish their job. Nothing is left to chance, nothing is left to worker ingenuity, and nothing ever varies. With Taylorism, the rationalization of the work process, having extended into the worker's mind, is complete.

Despite the radical successes of Taylorism in improving the efficiency of industrial work, it also has some unexpected negative effects. Taylor thought that workers would be happy to work more efficiently and make more money. Instead unions object to Taylorist techniques because they reduced workers to mindless objects, ignoring the expertise of skilled workers in favor of scientific analyses by outside experts. Workers find the absolute banalization of the work process that Taylorism implies unbearable; Taylorist work is both repetitive and mindless—on the one hand wearing out workers' bodies with repetitive stress injuries, on the other hand boring them senseless (Doray, 1988).

Ironically, whereas Taylorism leaves something to be desired for its original goals, it is extremely well suited as a basis for AI. Although workers cannot handle these repetitive, mindless activities, they are perfect for robots. Numerous scholars have pointed out that Taylorism is the last intellectual stop before AI: As soon as work is reduced to mindless, rote movement, idiosyncratic and moody human workers can be replaced by controllable and indefatigable robots, removing the last unpredictable part of the production process.

1. THE AGENTS OF McDONALDIZATION

The principles of Taylor—quantifying and rationalizing human behavior, reducing intelligent behavior to a set of independent, predictable, and interchangeable parts, removing all traces of human idiosyncrasy, creativity, and craftwork—are now suspect in management, but live on in the engineering tradition in computer science. Michael Mahoney (1997), a historian of science, pointed out with some surprise that in software engineering the arguments are not about whether the principles of Taylor are correct, but about how to apply them. This observation extends to AI—in many ways, AI is simply the late 20th-century reincarnation of turn-of-the-century traditions of human engineering and control.

Despite its problems, Taylorism's focus on efficiency over quality of experience has become an essential part of Western, especially American, consciousness. These trends in industrial culture are rooted in factory work, but they did not stay confined to the factory for long. Workers, who spend a large portion of their waking hours interacting with machinery on the production line, take home the values that that system has ground into their bodies. Production line designers, factory owners, and managers, spending their days designing machinery and optimal control of the human–machine interface, do not always forget their machinic view on life on their days off.

More insidiously, the drive of the assembly line, powered by the money its efficiencies can bring, spreads into other intellectual fields: Factory owners hire inventors, scientists, and engineers to design machinery and the production processes they support; they hire social scientists and management experts to design worker compliance (total quality management [TQM] is born). Each of these fields, applying itself within the context of the assembly line, starts to find more and more ways to generate interesting results within the factory work framework, slowly and mostly unconsciously taking over the assumptions of formalization, atomization, and so on that that framework presupposes. Factory owners lobby for laws that support and reflect their point of view on factory work. Both private and public institutions are set up to explicitly market these practices. For example, the military played a large hand in encouraging development of standardized manufacturing (Delanda, 1991). Other businesses, which are not strictly factory-oriented, envy the efficiency and rationality of factory work and begin to apply some of their ideas to improve their own processes. Soon the countryside is dotted with identical, standardized, efficient fast-food restaurants with its teenage automatons taking on the role of factory machinery. No one living in these cultures can escape the force of industrialization, even on their lunch break.

Artificial intelligence (AI) is no exception. From planning and scheduling of shop activities to robots for the assembly line, to reinforcement learning for process control, to automatic translation of manuals for equipment assembly, AI works on the problems of industrialization and, in turn, imbibes its values. They are reflected in the fundamental hope of AI: that most if not all of human behavior can be rationally analyzed, quantified, and reduced to algorithms reproducible in the machine.

THE McDONALDIZATION OF CULTURE

All this talk of workers in the factory and the assembly line may come across as antiquated today. How many of us are still factory workers on the assembly line? When *us* means the readers of this chapter, the answer must be very few. After all, we early 21st-century Westerners are no longer in the industrial era; we are brave new citizens of the Information Age!

Yet the forces of industrialization, far from having disappeared, have become so ingrained in our daily lives that they are taken for granted. If you live in the West, and especially if you are American, industrialization is the air you breathe and the prepackaged food you eat. Your life becomes more and more mechanized as your bank teller is replaced by an automated teller machine (ATM), your receptionist is replaced by a voice-mail program, and the telephone solicitor who has interrupted your dinner every night for the last 6 years is replaced by an autodialer with a cheery robotic recording. The last bastions of your craftwork slowly give way as universities become digital diploma mills, offering impersonal and standardized distance learning to students who are no longer limited by the bonds of location or social interaction (Noble, 1998). When in Paris, you eat your standardized lunch at McDonald's knowing that, although it may not be very good, it will not expose you to the idiosyncrasies of local cuisine—any calf brains will be ground beyond recognition into your Big Mac. You reify yourself as you sell 4 hours a day to each of three part-time jobs, trying to still maintain a full sanctified hour of quality time with your youngster—go ahead, sell yourself until you can afford to buy yourself back! You have specialized yourself, become the world expert on polynomial kernel support vector machine with fractional degree, unable to discuss your work with more than three or four colleagues because it is so hopelessly obscure (but nevertheless breathtakingly important). You atomize yourself, cut off from your extended family, perhaps even from your spouse and children, moving every 7 years in an evanescent search for the better life. How many times a day do you formalize yourself, jacking into cyberspace to become blissfully unaware of the constraints of your undeniably material, geographically located, and mortal flesh—at least until your carpal tunnel syndrome kicks in?

Industrial culture is not confined to the 19th-century factory; it continues to live through us on a day-to-day basis. Industrial culture is not just an attribute of a now marginal work-life; it colors nearly every aspect of late 20th-

1. THE AGENTS OF McDONALDIZATION

century Western existence (Strasser, 1982). It is not just a way of producing goods, but a new and not always positive way of being. We are postindustrial humanity: reified, specialized, atomized, standardized, formalized, and mechanized. We are nonstandard flesh, the weak link in a network of machines.

Sociologist Ritzer (1993) studied the extent to which themes from assembly line work, Taylorism, and bureaucracy have infiltrated our daily lives. This McDonaldization of society is based on the growing cultural importance of formal rationality, or systems under which people try to find the best ways to achieve pregiven and unquestioned goals—not according to their personal feeling for how it should be done, but by reference to formal systems of rules and regulations. Under formal rationality, idiosyncratic humans are replaced by nonhuman technologies, from protocols of behavior (like the *script* that McDonald's workers must follow) to the assembly line to computer software. This kind of rationality is interested, like Taylorism, in "the one best way" to do things, and is suspicious of the ability of individual people to judge things for themselves. Instead of leaving decisions up to the people involved, technical, legal, and bureaucratic systems are set up so that the best way to do things is also the natural way. In industrialized society, *best* is judged along four axes:

Efficiency. The production line maximizes the efficiency of craftwork; everything extraneous to optimal performance is removed, including personal idiosyncrasy and the joy of handiwork. For industrial culture, the number one goal is to satisfy needs quickly and without waste. Rather than lingering over a satisfying meal, the goal is to get customers in and out as quickly as possible. Why waste valuable time cooking a meal from scratch when a frozen prepackaged pot pie is cheap and so easy?

Quantifiability. To maximize efficiency, engineers calculate as much of the work process as possible: piece rates, material usage, worker movements, labor costs, and recidivism rates. Things that cannot be calculated are devalued. Cost–benefit analyses weigh quality of life against cold, hard, calculable cash. The soul cannot be weighed, so it must not exist. Quantifiability implies that more is better. The chain with the most stores must be the best. We buy not the best-tasting sandwich, but the one with the largest pile of unidentifiable ground meat. Airlines advertise not "We have the most pleasant flights," but "We fly to the most cities." The more you buy, the more you save!

Predictability. One of the major advantages of assembly line work is that the output is predictable. The phalanx of cars come marching off the assembly line, each exactly identical, with interchangeable parts, each with the same new car smell, the same ride, the same fluffy upholstery, and the same engineering mistakes. Predictability substitutes for familiarity:

Wherever we go, the Days Inn is exactly the same, with the same cheerful desk clerks telling you to have a nice day and the same style of insipid sit com grinding out its laugh track from the satellite TV in your room. On your bus tour of Europe, there are no unpleasant surprises: 1 day per city, carefully sanitized local color, and English-speaking natives.

Control. Life (and, in particular, human behavior) is in many ways not inherently predictable, so the holy grail of predictability is only achievable through the hefty use of controls. Unpredictable humans are replaced and controlled by technology and bureaucracy: The ATM never miscounts, the computerized tram keeps its doors open for exactly 5.3 seconds, and the fast-food worker does not get a chance to misspeak while regurgitating the manual's "Fries with that?" The customer can remain cost-effectively always right when he or she only has a choice of five menu items, and the high-intensity fluorescent lighting chases him or her out of the restaurant before he or she becomes an economic liability.

Each of these values certainly has its place. Inefficiency, incalculability, unpredictability, and lack of control are clearly not particularly preferable to their opposites. Yet Ritzer (1993) pointed out that under formal rationality each of these values is elevated to an absolute. When rationality is pushed too far, the result is, paradoxically, irrationality. A cheap fast-food meal with huge portions, wolfed down in 5 minutes, is not necessarily better than a less efficient home-cooked meal with quality ingredients. A packaged group tour with all activities carefully homogenized and isolated is safer, but not necessarily better, than a vacation requiring true contact with alien cultures and experiences. America has certainly pushed the envelope of homogenized, commodity-laden, safe, and predictable existence. Yet whether we truly maintain quality of life is an open question in a nation of the obese who fuel the purchase of all the latest high-tech fantasies by road raging across miles of asphalt from the suburbs to a 10-hour work day, then crawl back home to frozen dinners consumed silently in front of the stereo, big-screen, high-definition TV. Even Taylorism, which subsumes all human values to the goal of efficiency, is inefficient in the sense that, by reducing work to repetitive motion, it wastes the worker's talents and judgment.

AGENTS AS McDONALDIZATION

The connection between technology development and efficiency, predictability, calculability, and control is trivially clear—the development of technology in general is oriented to making processes and, by extension, our lives more efficient, predictable, calculable, and controllable. Yet there are more subtle implications for how this ideology of technology infiltrates the details of agent research. Efficiency, quantifiability, predictability, control: Ritzer's values of industrialization are also the values of AI. They can be seen in the view of intelligence in AI, so different from most people's day-to-day experience of existence: the calls for rational, goal-seeking, provably correct agents working efficiently to solve problems. More specifically, they can be seen in the research strategies that are considered appropriate for agent research.

Efficiency. One of the major standards for agent technology is that agents work efficiently. Fundamental problems in agent research are couched in terms of efficiency. For example, the subfield of planning research focuses on how agents can efficiently achieve goals; sources of inefficiency are considered problems to be solved. The action-selection problem, central to behavior-based agent research, focuses on how agents can choose the most appropriate actions for their environment, where *appropriate* is defined as a kind of problem-solving behavior to be optimized.

> [I]t favors actions ... that contribute to several goals at once, ... it exploits opportunities and is highly adaptive to unpredictable and changing situations, it favors actions that contribute to the ongoing goal/plan (unless another action rates a lot better), ... [,] it looks ahead (or "plans"), in particular to avoid hazardous situations and handle interacting and conflicting goals, it is robust ..., and it is reactive and fast. (Maes, 1989, p. 1)

The language of agent research is a language of goals to be achieved, preferably as directly, efficiently, and quickly as possible.

Agents are chosen for efficiency—in the Western context, agents are a form of cheap labor, which can replace more expensive human labor. In other cultures, where human labor is cheap, plentiful, and in desperate need to be subsidized, replacing human with mechanical labor may be considered inappropriate.

Quantifiability. Quantifiability (or calculability) is essential for any computational system. One can only create code to the extent that one has reified and found calculable mechanisms for the activity under consideration. Yet there are ways to structure agents that are more or less calculable. Quantifiability is not only a precursor for agent creation, but the process of agent creation on the computer can encourage greater quantifiability.

For example, consider the problem of behavior selection, or how a behavior-based agent chooses which behavior to engage in at any particular point in time. Under Loyall and Bates' (1991) Hap, for instance, behaviors are fundamentally symbolic units among which agents choose using mostly symbolic mechanisms. Under Blumberg's (1996) common-currency system, however, numbers are assigned to each behavior, and the behavior with the highest value wins. Under such a system, the goodness of any behavior is considered to be independently calculable and reducible to a number, ironically mimicking critiques of consumer culture that complain that all values are reduced to the value of products in dollar amounts.

Technically speaking, Blumberg's system lacks the flexibility to write arbitrarily complex, context-dependent arbitration rules, and behavior writing under a common currency system involves a substantial amount of tuning of not particularly intuitive numeric parameters. Yet there are clear advantages to a common currency system—similarly to cost–benefit analysis, it makes it easy to combine multiple considerations in a single format, enabling apples-and-oranges comparisons that may make no sense using more subtle mechanisms, yet still need to be made for the system to work. Although the common currency system is a worse match to the intuition of the programmer, it is a better match to the underlying demands of the computational system, under which everything, in the end, is numeric anyway.

More generally, AI represents the dream that everything about human behavior is fundamentally calculable. Nothing that is pertinent to or important about human behavior will escape the calculating engines of AI. There is no room for the enigmatic, the vague; everything is information, not meaning. Even the ultimate of irrationality, our emotions, can be calculated and quantified by affective computing.

Predictability. As a technology developed by humans, autonomous agents need to be predictable. We cannot know whether the technology has succeeded if we do not know what it will probably do. Although it is not always proper for agents to seem predictable to users, because it destroys the impression of life-likeness, in these cases they need to be predictably unpredictable: We need to be guaranteed that their surface behavior is not so predictable that users will find them implausible.

A subtle twist on predictability is presented by the notion of emergence. Behavior is said to be emergent when it is not programmed in a priori, but instead appears as a consequence of the interactions between more simple, programmed rules. Emergence may be touted as a form of unpredictability because the behavior of the agent exceeds what it was directly programmed to do. In practice, however, emergence is engineered in agent programming—the simpler rules and behaviors are tweaked until, as if by magic, the desired behavior emerges.

Control. Predictability is closely related to control. Although one may think of an agent as something that is autonomous and has its own, independent existence, in many cases agents are explicitly intended to be a tool controlled by their users. Even when an agent *is* thought of as autonomous, it is clear that a major goal for agent *programmers* in making agents is to be able to control their behavior (i.e., to build them to have particular aspects and not to have others).

Compared with humans engaged in the same activity, programmed agents can be (or people want them to be) fully controlled. As Siebel (n.d.) rapturously described,

> These ... agents will follow simple directions and tirelessly scan highly structured sources of data such as product and competitive information. People, of course, could do the same thing, but agents will be able to do it better and faster. They'll never get bored by repetitive operations, and they'll never stop for lunch.

Agents can be the perfect workers: They do anything you want, they never complain, and they never go on strike. Agents will be your perfect servant (Wardrip-Fruin, 1999; Wise, 1998).

If agents embody the values of consumer culture, and if those values make them useful for us, then where is the problem in McDonaldization? Ritzer (1993) argued that the consequence of heavy reliance on efficiency, quantifiability, predictability, and control is, paradoxically, irrationality. If we push the values of rationality too far, we end up in an irrational state, in which our interactions stop making sense, even when measured by the standards of McDonaldization; standing in line for 20 minutes for a McDonald's meal is no longer efficient, at least not for the consumer.

For agents, the paradox of McDonaldization is rooted in the fact that we begin building them because we value human behavior over that of machines; otherwise there would be no need to build agents that appear to be human. Yet agents can be more efficient, quantifiable, predictable, and controllable than people. In the values that matter to postindustrial consumer culture, we come up short in the comparison. In the process, human behavior and experience is devalued (Lanier, 1996). In a culture that uses computational metaphors to understand the world, the way we think about agents also becomes the way we think about people. As Agre and Chapman (1989) argued, agents carry with them a philosophy of human experience, which is exported along with our agents. The agents we build are the agents of McDonaldization, spreading the values of mass production along with the fruits of our labor.

CONCLUSION

> *In milling and baking, bread is deprived of any taste whatever and of all vitamins. Some of the vitamins are then added again (taste is provided by advertising). Quite similarly with all mass-produced articles. They can no more express the indi-*

> *vidual taste of producers than that of consumers. They become impersonal objects, however pseudo-personalized. Producers and consumers go through the mass production mill to come out homogenized and de-characterized—only it does not seem possible to reinject the individualities which have been ground out, the way the vitamins are added to enrich bread.* (Haag, 1962, p. 183)

The same culture that has replaced local stores in thriving downtowns with identical suburban mega-malls across the United States is now generating one-size-fits-all virtual personalities for domestic and international consumption. These personalities are superficially adaptable and personalizable, but under the hood are deeply embedded in Western assumptions and values that are not so easily changed. Because of the culture-boundedness of agent research, naïve approaches to intercultural agents, in which, for example, one is said to change the culture of an agent by switching a few personality parameters, offer a superficial culture-friendliness while masking a more deeply rooted cultural imperialism. Truly intercultural agents involve a negotiation of the terms under which agency is understood based on a deep respect for the aspects of local culture that may be incompatible with the assumptions of agent technology.

In my own work, I have tried to reflect on the meaning of agents and develop alternatives based on the notion of an agent as narrative (Sengers, 2000, 2003). In this way of thinking, agents can be thought of as stories about a character being told through the technology by programmers to users. This conception of agents allows one to be aware of one's own role as a culturally located agent creator, as well as flexible and thoughtful about the cultural background and values of the intended users.

What would an agent mean to a culture in which physicality plays a much more central role than in Western culture? How would its rule-driven behavior be adapted to a culture that values intuition, empathy, and wisdom over formal laws and rules? How could an efficient, predictable, task-oriented agent be made appropriate for a culture that has not adopted the Tayloristic tendencies of postindustrial consumer culture? In these cases, agents cannot simply be altered—they need to be fundamentally rethought. Agents may not even be appropriate to use at all for cases in which there are substantial cultural differences from mainstream Western industrialized culture—including subcultures within the West, from Alaskan subsistence hunter-gatherers to all of us who are tired of depersonalized interactions.

In my opinion, such rethinking is unlikely to happen. Agents are a perfect fit to McDonaldized culture—a culture whose *raison d'être* is unthinking growth. Why should corporations market culturally adaptive agents when agents that are embedded in consumer culture prepare the ground for other forms of globalization? Instead we are likely to see just enough cosmetic adaptation, in which agents appear inoffensive to members of other cultures while maintaining a covert cultural campaign: be more efficient and see how much you can save. Engage in pseudohuman interaction while enjoying the safety of controlled predictability! I hope I am proved wrong.

REFERENCES

Agre, P. E., & Chapman, D. (1989). *What are plans for?* Cambridge, MA: MIT AI Lab Memo 1050.
Bickmore, T., & Cassell, J. (1999). Small talk and conversational storytelling in embodied interface agents. In M. Mateas & P. Sengers (Eds.), *Proceedings of the AAAI fall symposium on Narrative Intelligence* (pp. 87–92). Cape Cod, MA.
Blumberg, B. (1996). *Old tricks, new dogs: Ethology and interactive creatures.* Cambridge, MA: MIT Media Laboratory PhD Thesis.
DeLanda, M. (1991). *War in the age of intelligent machines.* New York: Zone Books.
Doray, B. (1988). *From Taylorism to Fordism: A rational madness* (D. Macey, Trans.). London: Free Association.
Haag, E. van den. (1962). Of happiness and despair we have no measure. In E. & M. Josephson (Eds.), *Man alone: Alienation in modern society* (pp. 180–199). New York: Dell.
Kennedy, N. (1989). *The industrialization of intelligence: Mind and machine in the modern age.* Boston: Unwin Hyman.
Lanier, J. (1996). My problem with agents. *Wired, 4*(11).
Loyall, A. B., & Bates, J. (1991). *Hap: A reactive, adaptive architecture for agents.* Pittsburgh: Carnegie Mellon University (Technical Report CMU-CS-91-147).
Lukács, G. (1971). *Reification and the consciousness of the proletariat. History and class consciousness: Studies in Marxist dialectics.* Cambridge, MA: MIT Press.
Maes, P. (1989). *How to do the right thing.* Cambridge, MA: MIT AI Lab Memo No. 1180.
Mahoney, M. S. (1997). *Software and the assembly line.* Carnegie Mellon Software Engineering Institute Invited Talk.
Noble, D. F. (1998). Digital diploma mills: The automation of higher education. *First Monday, 3,* 1.
Ritzer, G. (1993). *The McDonaldization of American society: An investigation into the changing character of contemporary social life.* Newbury Park, CA: Pine Forge Press.
Sengers, P. (2000). Narrative intelligence. In K. Dautenhahn (Ed.), *Human cognition and social agent technology. Advances in Consciousness series* (pp. 1–26). Amsterdam: John Benjamins.
Sengers, P. (2003). Narrative and schizophrenia in artificial agents. In M. Mateas & P. Sengers (Eds.), *Narrative intelligence. Advances in Consciousness series* (pp. 259–278). Amsterdam: John Benjamins.
Siebel, T. (n.d.). Software sales agents. *Siebel Magazine, 1,* 3.
Strasser, S. (1982). *Never done: A history of American housework.* New York: Pantheon.
Wardrip-Fruin, N. (1999). Hypermedia, eternal life, and the impermanence agent. *Critical Essays, SigGraph '99 Art Gallery* (http://www.siggraph.org/artdesign/gallery/S99/essays/noahfull.html; Last visited Jan. 16, 2003).
Wise, J. M. (1998). Intelligent agency. *Cultural Studies, 12,* 3.

CHAPTER

2

Designing Technology, Designing Culture

Lorna Heaton
Université de Montréal

This chapter does not explicitly discuss embodied agents. Rather, it examines the social construction of another group of technologies—systems for computer supported cooperative work (CSCW). We describe the design of CSCW in Japan, with particular attention to the influence of culture on design choices. Two case studies are presented to illustrate the argument that culture is an important factor in technology design despite commonly held assumptions about the neutrality and objectivity of science and technology. We argue that, rather than think of the technology as merely located or situated socially, the social context must be constructed simultaneously. This leads us to question the necessity of a logical distinction between design and use.

The chapter goes on to discuss implications of our findings in the context of CSCW for the design of agents for multicultural situations. We suggest several avenues for reflection on the themes of the book: (a) what cultural considerations enter into interaction, personality, visual design, and context of embodied agents; (b) how agents might reflect and represent the social and cultural systems from which they come; and (c) the cross-cultural portability of embodied agents. Finally, we propose an A–B–X system as a formalization of a way to look at the relationship among agents (both human and artificial), the material context in which they evolve, and the ongoing construction of social context. Before moving into the body of the chapter, however, some background is in order.

BACKGROUND

Cultural attitudes toward technology and cultural dimensions in the implementation and use of technology are topics of increasing interest worldwide. The subject has become all the more significant with the proliferation of new communications technologies that facilitate globalization and intercultural contact. When we move into the domain of intelligent agents, the potential and challenges of communication not only between and across cultures, but between humans and nonhuman entities, and between humans with nonhuman agents as intermediaries, introduces new levels of complexity. The relative novelty of computer-mediated communication networks does not, however, mean that we must start from scratch in attempting to understand how people from different cultures will use them and how diverse cultural attitudes are likely to affect their use. Over the past 20 years, these questions have in fact been explored in the fields of both organizational and development communication.

In development communication, a turn-key approach to technology transfer has been rejected in favor of other models that accord substantial importance to culture. Among them, there has been considerable research on the importance of technological infrastructure and predisposition or competency as preconditions for technology transfer (Andrews & Miller, 1987), as well as various measures for increasing the likelihood of successful transfer: modification of imported technology by local engineers to make it more "appropriate" (DeLaet, 1992; Ito, 1986), a two-step flow in which new ideas or technology are first introduced to an opinion leader or technological gatekeeper who then persuades others to adopt them (Rogers & Shoemaker, 1971), or involving stakeholders in planning and decisions (Ackoff, 1981; Madu, 1992). All this work shares a concern for facilitating accommodation to changes produced in an environment with the introduction of new technology. In other words, making the technology fit its context of implementation and use has been found to considerably improve the chances of optimal use.

Understanding the reciprocal link between organizational practices and technologies has also been a key concern of organizational communication scholars, particularly with the advent of office automation and computerization. Many have drawn on Giddens' structuration work (Orlikowski, 1992; Orlikowski & Gash, 1994; Poole & DeSanctis, 1990) to explain how computerization may result in changes to organizational structure, whereas a diversity of approaches ranging from ethnography and workplace studies (Button, 1993; Luff, Heath, & Hindmarsh, 2000), to activity theory (Engeström & Middleton, 1996; Nardi, 1996) to social network analysis (Haythornthwaite & Wellman, 2001), to sociotechnical systems (Coakes, Willis, & Lloyd-Jones, 2000), to distributed cognition (Hutchins, 1995; Y. Rogers, 1992) have been

2. DESIGNING TECHNOLOGY, DESIGNING CULTURE

employed to study the evolution of social interaction in technological environments.

In short, studies in development and organizational communication over the past two decades have consistently pointed to three key factors in explaining successful information technology (IT) implementation: (a) existing technological infrastructure and predisposition—the context; (b) the process of implementation itself; and (c) the importance of viewing use as a process in which uses change over time.

It is now generally recognized in the social sciences that the path a given technology takes may not be inevitable and absolute. In recent years, numerous instances of how technical artifacts embody political, cultural, or economic positions have been identified (see e.g., the collections edited by Bijker, Hughes, & Pinch, 1987; Bijker & Law, 1992; Mackenzie & Wacjman, 1985; Winner, 1993). Increasingly, we seek to understand how technological artifacts are constructed and how the end result relates to their conditions of construction if we are to understand their implementation and use.

The challenge for social science, in our view, is to go a step further to examine *how* this process of social construction is accomplished and to determine which aspects of the black box called *technology* are more or less susceptible to social influences. By asking how ideas and circumstance affect action, we are in fact raising larger issues of the relationship between technology and context—a question of growing significance given increasing globalization and the increasing impact of technology in our lives (Hales, 1994; Jackson, 1996).

ON CULTURE

Before turning to our cases, we need to clarify what we mean by culture. Our interpretation of culture involves not just national culture, but organizational and professional cultures as well. In recent years, various forces have heightened awareness of the importance of the cultural factor, and a number of studies on work organization and work attitudes have consistently demonstrated significant differences across national cultures (Deans & Ricks, 1991; Erez & Earley, 1993; Raman & Watson, 1994). Among a number of typologies of cultures, the most widely cited and one of the most thorough is that of Geert Hofstede. In an attempt to identify cultural predispositions that Bourdieu (1983) called *habitus*,[1] Hofstede (1980) administered standardized questionnaires to some 116,000 people working for IBM in a

---

[1] Bourdieu's idea is that certain conditions of existence produce a habitus—a system of permanent and transferable dispositions. A habitus functions as the basis for practices and images that can be collectively orchestrated without an actual conductor.

variety of professions in over 50 countries in 1968 and again in 1972. On the basis of these data, Hofstede defined several *dimensions of culture*.[2] This and other similar studies clearly indicate that people from different cultures bring different attitudes to their work; this results in national differences in the way work is organized as well as different work practices.

Japan, for example, can be characterized as a group-oriented society with a long-term orientation, strong uncertainty avoidance, highly differentiated gender roles, and accepting of the unequal distribution of power. In contrast, North American society is highly individualistic and less tolerant of the unequal distribution of power, with a short-term orientation and medium degrees of uncertainty avoidance and gender role distinction. The four Scandinavian countries form a relatively homogeneous group, with few gender distinctions and generally low power distance, more group-oriented than North America, but less so than Japan.

Another body of literature has examined differences in attitudes, values, and practices among professions. A person's occupation or training undoubtedly has a major influence on how he or she approaches the world. For example, computer scientists likely draw on a similar pool of knowledge and techniques relative to systems development, which in turn calls for and constitutes a particular way of looking at the world. Similarly, social scientists may not always share common frames of reference, but most share certain elements of common knowledge.

Hannerz (1992) coined the term *transnational cultures*, which he defined as "structures of meaning carried by social networks which are not wholly based in any single territory" (p. 249). Many transnational cultures are occupational. Hannerz suggested that, although it makes sense to see them as a particular phenomenon, they must also be seen in their relationships with territorially based cultures, arguing that their real significance lies in their mediating possibilities. Although "transnational cultures are penetrable to various degrees by the local meanings carried in settings and by participants in particular situations" (p. 251), they also provide points of contact between different territorial cultures.

---

[2]The first dimension, that of *power distance*, refers not to the actual distribution of power, but to the extent to which the less powerful members of institutions and organizations within a country expect and accept that power is distributed unequally. This dimension has implications for hierarchy, centralization, privilege, and status symbols. The *individualism–collectivism* dimension identifies the strength of ties to and belonging in a group. One might expect this dimension to be correlated with loyalty, trust, shared resources, and even the relative importance of verbal or nonverbal communication. The *masculinity–femininity* dimension measures the clarity of gender role distinction, with masculine cultures having clearly defined gender and feminine cultures considerable overlap. Finally, the *uncertainty avoidance* dimension measures the tolerance (or intolerance) of ambiguity, the way in which people cope with uncertain or unknown situations. In the workplace, one might expect correlations with the way the environment is structured, rules, precision and punctuality, tolerance of new ideas, as well as motivation (achievement, security, esteem, belonging).

The important point here is that occupational culture need not be a subset of national culture. Rather, the two are distinct and interrelated. Those involved in developing intelligent agents may share a common agent culture, but they also reflect and interpret this professional culture within the framework of their territorial cultures, just as professional training and perspectives lead them to interpret elements of territorial culture in certain ways. A given situation can be understood in cultural terms as the product of what is unique (national culture) and what is shared by all (occupational culture). The resulting combination of the two will necessarily differ between cultures and even between systems in the same national culture because conditions can never be identical.

Finally, there is organizational culture, which is perhaps best understood as a metaphor, rooted in the premise that organization rests in shared systems of meaning, and hence in the shared interpretative schemes that create and re-create that meaning.

In summary, for the purposes of this research, we define *culture* as a dynamic mix of national/geographic, organizational, and professional or disciplinary variables in constant interaction with one another. Culture changes according to context and over time and should be understood not in terms of preexisting, fixed categories, but as resources, accumulations of actions, and patterns that constitute, reinforce, and transform social life. In short, we suggest that culture is continually constructed and reconstructed.

OPERATIONALIZING CULTURE: TECHNOLOGICAL FRAMES

The notion of technological frame provides an interesting way to approach culture from a constructivist perspective. Law and Bijker (1992) used the notion to "refer to the concepts, techniques and resources used in a community—any community. . . . It is thus a combination of explicit theory, tacit knowledge, general engineering practice, cultural values, prescribed testing procedures, devices, material networks, and systems used in a community" (p. 301). It is simultaneously technical and social, intrinsically heterogeneous. The related expression *frame of meaning*, as coined by Collins and Pinch (1982) and adopted by Carlson (1992) in his study of Edison and the development of motion pictures, translates the specific focus of this chapter on how cultural patterns and assumptions inform actions and shape choices most closely:

> . . . in any given culture there are many ways in which a technology may be successfully used. . . . To select from among these alternatives, individuals must make assumptions about who will use a technology and the meanings

users might assign to it. These assumptions constitute a frame of meaning inventors and entrepreneurs use to guide their efforts at designing, manufacturing, and marketing their technological artifacts. Such frames thus directly link the inventor's unique artifact with larger social or cultural values. (Carlson, 1992, p. 177)

THE STUDY: JAPANESE CSCW

This section provides concrete illustrations of the importance of culture as a variable in the technology design process. It highlights common research themes in Japanese CSCW design through a presentation of two research projects, with particular focus on the relationship between designers' justifications for their choices and how these choices are reflected in the design of machines and software. This is followed by a discussion of general trends and characteristics and relates them to cultural characteristics and beliefs. Here we seek an explanation for regional differences in CSCW not in institutional variables nor in strictly professional ones, but at a mid-level between micro and macro—in culture, which is both an individual attribute and a collective phenomenon.

We define CSCW broadly as: *work by multiple active subjects sharing a common object and supported by information technology*. The focus of computer-supported cooperative work, then, is less on working with computers than on working with each other *through* computers. This orientation opens the door to a real contribution from social scientists toward understanding the complex relationship between technology and its context of emergence and implementation. We believe that CSCW is a particularly appropriate object for this type of inquiry because it is a field that integrates a variety of perspectives, ranging from those of hard science (engineering) to social science and even philosophy. From the outset, the CSCW research community has asked questions about how the individual, and individual perspectives, fit in social context, and about how the whole may be greater than the sum of its parts (Bannon & Schmidt, 1991). In this respect, issues raised in CSCW are also crucial in the design of agent systems, particularly multiagent systems.

Our cases are drawn from a comparative study of the design of computer-supported cooperative work (CSCW) systems in two cultural contexts: Japan and Scandinavia (Heaton, forthcoming). We used a case study approach to capture the subtleties of the multitude of situational variables and their interaction. During 5 months of observation in various CSCW laboratories, the author conducted extensive interviews with over 20 software designers and engineers and took part in numerous informal conversations with others involved in CSCW research. Earlier typologies of cultures, particularly as they have been applied to the world of work, were used as a starting point and a general guide for observation, although no attempt was made to fit the data gathered into these classificatory schemes. Analysis of documents produced by the laboratories in question was also an important part of the process.

What did we find? Virtually all those involved in designing CSCW systems in Japan are engineers or computer scientists who identify strongly with their profession. Building a *good* system—that is, one that works, is reliable, state-of-the art, and original—is both the goal and measure of their capabilities as engineers. Design work is done exclusively in the labs, and any evaluation of prototypes takes the form of controlled laboratory experiments. Designers are not generally concerned with who will use their systems or how they will be implemented. With so technical a focus, it is not surprising that the main justifications for design choices are technical ones.

There is, however, another more social element to Japanese design choices: that of Japanese culture. Professional engineering or scientific culture notwithstanding, Japanese CSCW researchers, like most Japanese people, clearly believe that Japanese culture and the Japanese way of working are different from Western ways.[3] How to reflect or cope with this difference in designing technology is a constant *leitmotif* among Japanese CSCW researchers. Although most would prefer to believe that science and technology are culturally neutral or universal, they nevertheless recognize that, if use is a consideration, designing a groupware system cannot be approached in the same way as designing a TV.

The dean of groupware in Japan, Professor Matsushita, cites five principal specifically cultural reasons that groupware must be different if it is to be used in Japan: cultural differences in views on cooperation and competition, negotiation style, degree of context, importance of human relations, and relation of the individual to the group. Even those who deny specifically cultural aspects in the design of CSCW and groupware in Japan acknowledge cultural effects in implementation and use. Some major Japanese companies are now selling workflow systems developed by American companies, but this has been problematic. In the words of another leading researcher, the biggest challenge facing Japanese groupware is "attaining widespread use. Managers don't want to change the way they work. They

---

[3]Mouer and Sugimoto (1986) traced the long history of the theme of Japanese uniqueness and suggested that, although the ideology of Japanese uniqueness has been used in the service of many interests, the basic assumption that all Japanese people possess a common set of attitudes and share similar behavior patterns has remained largely unquestioned, particularly in English language publications. They concluded that the relationship between this ideology and views of Japanese society is maintained by a complex network of interpersonal and interinstitutional relationships. In other words, Japanology is a self-fulfilling prophecy—a social construction almost universally subscribed to.

want to be able to consult with people as they usually do." The following two examples illustrate how this desire to reflect cultural particularities plays out in practice.

TeamWorkStation/Clearboard
(NTT Human Interface Labs)

TeamWorkStation (TWS) is one of the earliest and most documented Japanese CSCW projects (Ishii, 1990). It has been widely cited within the CSCW community and has inspired considerable research within Japan around the concepts of seamlessness and gaze awareness. Ishii and his collaborators at NTT Human Interface Labs were not the first to develop the concept of a seamless work environment, however, nor were they the first to explore peripheral awareness. Both were borrowed from work done originally at Xerox PARC, but the Japanese way of dealing with these issues is unique.

With TWS, "a desktop real-time shared workspace," which integrates both computer and desktop workspaces, the team set out to design a system that would allow users to maintain their preferred work practices using their preferred computer applications or even working with pencil and paper within a shared virtual workspace. A second design requirement was a shared drawing surface. The research team chose video as the basic media of TWS for its ability to fuse traditionally incompatible media such as papers and computer files (Ishii & Miyake, 1991). Live video image synthesis was employed to capture individual workspaces (both computer screens and physical desktops) and to display them in separate layers on a computer monitor. The overlay function created with this technique allowed users to combine individual workspaces, and to point to and draw on the overlaid images simultaneously.

The three-member design team began to use the prototype on a daily basis in July 1989, and informal evaluations of its use pointed to the importance of gesture as a means to enforce the sense of shared space. They preferred hand gestures to pointing or marking with a mouse "because hand gestures are much more expressive, and because hand marking is generally quicker" (Ishii & Miyake, 1991, p. 45). Because the TWS prototype was designed without a formal floor control mechanism for passing the input control among collaborators, voice contact played an important role in preserving informal social protocol and coordinating action, especially the use of the limited workspace on the shared screen (Ishii & Miyake, 1991).

The faces of collaborators were displayed in separate windows beside the shared workspace in TWS. Yet spatial awareness was already a concern, and the design team pushed their explorations further in this direction with ClearFace and later ClearBoard. In contrast to screen layouts, which required users to shift their focus between the shared drawing space and the facial images and deal with them separately (tiling [i.e., laying them side by side] or overlapping windows), the ClearFace interface proposed translucent, movable, and resizable face windows that overlay the shared workspace window. The user could see the drawing space and his or her collaborators' faces in the same space and shift easily between the two. In use they observed that people hesitated to draw or write over people's faces, inciting them to make the face windows movable and resizable.

With ClearFace, the design team began to explore the dynamic relationship between elements in design meetings. Their attention shifted away from task—what workers are doing—to how they are relating to each other as they do it—from a focus on *shared workspaces* to the creation of *interpersonal spaces* (Ishii, Kobayashi, & Grudin, 1993):

> At the same time, in the discussion, the participants are speaking to and seeing each other, and using facial expressions and gestures to communicate. In the conversations it is essential to see the partner's face and body. The facial expressions and gestures provide a variety of non-verbal cues that are essential in human communication. The focus of a design session changes dramatically. When we discuss abstract concepts or design philosophy, we often see each other's face. When we discuss concrete system architectures, we intensively use a whiteboard by drawing diagrams on it. (Ishii & Arita, 1991, p. 165)

The effort to simulate as closely as possible the collaboration in front of a whiteboard was taken a step further in ClearBoard, the first prototype to refer explicitly to eye contact and gaze awareness (see Fig. 2.1). The design metaphor here was talking *through* and drawing *on* a transparent glass window. The system used colored markers on a glass board and video and a half-mirror technique to capture and orient the drawings. In this case, users recognized their partner as being *behind* a glass board, and they did not hesitate to draw over the facial image. The large size of the drawing board supported awareness of gesture and of the partner's surrounding environment, as well as of his visual focus. "*Gaze awareness* lets a user know what the partner is looking at.... If the partner is looking at you, you know it. If the partner is gazing at an object in the shared workspace, you can know what the object is. Eye contact can be seen as just a special case of *gaze awareness*" (Ishii & Kobayashi, 1992, pp. 530–531).

Gaze awareness allows participants to better situate the interaction within its context, providing a wider variety of cues for feedback and a richer awareness of the environment and others' activities. The emphasis on nonverbal cues and direction of gaze rather than eye contact is particularly significant coming from a culture in which eye contact is much less common than in Western culture and is in many cases considered rude.

Finally, the ClearBoard-2 design led to some reflections on interpersonal distance:

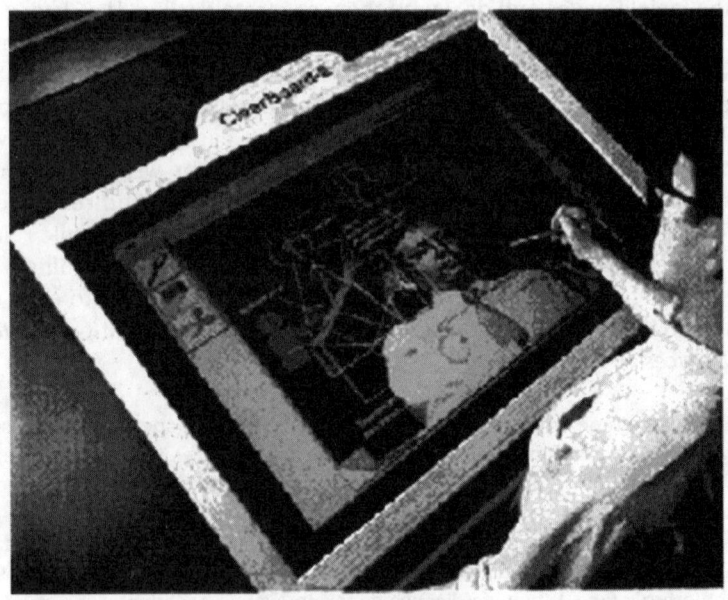

FIG. 2.1. ClearBoard.
© ACM, 1993, *TOIS*, **11** (4) Ishii, Kobayashi and Grudin

ClearBoard creates the impression of participants standing about *one meter apart*, because both sit (or stand) close enough to the screen to draw directly on its surface. This virtual distance belongs to the *personal distance* in Hall's classification. When people use ClearBoard with close friends or colleagues, this distance seems appropriate. However, for a formal meeting with a person of much higher rank, this virtual interpersonal distance might seem too small, and the participants might be uncomfortable. Therefore, we would like the media to provide users with some control over the virtual interpersonal distance. We are planning to provide an option of indirect drawing using a wireless tablet or pen-based personal computer for that purpose. (Ishii, Kobayashi, & Grudin, 1993, pp. 371–372)

This concern for interpersonal distance was picked up and further explored by another research group in our next case: MAJIC.

MAJIC (Matsushita Lab, Keio University)

Our second case is a system developed at the Matsushita Lab in the Instrumentation and Engineering Department of Keio University, a prestigious private university located near Tokyo. MAJIC illustrates many research themes characteristic of Japanese CSCW. It builds on earlier Japanese work at NTT on eye contact and gaze awareness, adding a multiparticipant di-

2. DESIGNING TECHNOLOGY, DESIGNING CULTURE

mension and a more explicit focus on the surrounding environment. The MAJIC team explained clearly why they feel this line of inquiry is important:

> When we have discussions in face-to-face situations and people approve of a statement, we can tell by their attitude, tone, eye movements, gestures and so forth, whether or not they approve wholeheartedly. It is difficult, on the other hand, to estimate how strongly they approve when we read only the minutes without attending a meeting. Hence, one of the purposes and/or advantages of face-to-face meetings is that all of the participants are *aware of the speaker's intent and the other listeners' reactions* based on both verbal and nonverbal communication. (Okada et al., 1994, p. 385)

As in TeamWorkStation, there are multiple references to the importance of context, orientation to the other (how what you say is being received), and a focus on interpretation of intention rather than surface meaning. The key design issues of MAJIC were defined as (a) support of multiway roundtable meetings and multiple eye contact; (b) maintenance of peripheral gaze awareness; (c) seamless presentation of life-size images of participants to achieve a sense of reality; and (d) a shared work space (Okada et al., 1994, p. 385).

The creation of a seamless environment and sense of presence in MAJIC relies extensively on nonverbal behavioral information, such as eye contact, gaze awareness, gesture and body language, and contextual cues such as image size, distance, and background. References to these elements are extremely specific. For example, the MAJIC team refers to symmetrical or asymmetrical postures and body orientations as important cues: "In this way we sense the atmosphere in the meeting room and the aura of the participants, and, consequently, we can understand the opinions of the participants clearly and make the meeting productive" (Okada et al., 1994, p. 386). They cite gaze as a means of controlling a meeting: "A chairperson sometimes gazes at participants to urge them to speak when there is silence in a meeting" (p. 386), and discuss the social uses of eye contact: "Of course eye contact is very important in communicating with one another, as mentioned above, but especially in Japan it is impolite to look into someone's eyes for a long time" (p. 387). In their observations of face-to-face meetings, the designers noted that participants most commonly averted their eyes by looking down at material on a table in front of them, and decided to provide such a table in their design.

Referring to Hall's (1976) classifications of appropriate distances for interactions, the MAJIC team discusses elements that may affect *virtual distance* (the sensed distance among participants): physical distance from the display, size and quality of video images, voice fidelity, backdrop, and so on. In fact, this has been the central focus of most of the MAJIC research. Starting with the assumption that image size of participants and background are the two important factors in achieving a sense of reality during videoconferencing, MAJIC I was designed to project life-size video images and simulate a virtual social distance of approximately 4 feet between participants.

The central element of MAJIC is a large (4 × 8 feet), curved, semitransparent screen. Each MAJIC unit also contains a workstation (with a recessed, tilted monitor), two video projectors, two video cameras, two directional microphones, and two loudspeakers. Video images of the participants are projected onto the screen and captured from behind it. Each participant sees the frontal view of the others, and the edges of the images overlap slightly (see Figs. 2.2 and 2.3).

The second factor deemed essential for "achieving a feeling of togetherness during videoconferencing" (p. 390) is the continuity of background images. In this interpretation of "seamlessness," if images run into each other, it is difficult to tell where one ends and the next begins; "if users are surrounded by other participants with a seamless background, they can feel as though they are together" (p. 386). In actual fact, the backgrounds must be matched at the seam. This is only a prototype, however: MAJIC proposes doing away with the actual background altogether and replacing it with an artificial one that can be chosen to create a desired mood, to relax, or to inspire. This would be done using a chromakey blue background.

A further extension of the idea of direct physical manipulation in MAJIC II is the Whisper Chair. By leaning right or left, the person sitting in this chair (equipped with sensors) can talk to one or the other persons on screen without the third party hearing. The rationale behind this development is that leaning is a more subtle, more natural way of confiding a secret than flipping a switch to turn off the audio channel.

FIG. 2.2. First draft of MAJIC. ©ACM, 1994 CSCW'94

FIG. 2.3. Gaze awareness in MAJIC. ©ACM, 1994 CSCW'94

DISCUSSION

The cases presented previously illustrate the close relationship between designers' preconceptions and frames of reference and the systems they design. Japanese CSCW researchers consistently invoke Japanese culture as a justification for decisions to focus on contextual awareness and nonverbal communication in Japanese CSCW systems. The preferred Japanese approach to CSCW design is to provide a channel for communication, which can be used to complement or supplement traditional ways of working. This channel should transmit as much information as possible (hence the widespread use of video and large displays), but should avoid specifying procedures or ways of doing things. The system is not a tool, but another element in the working environment that can offer important contextual information to enable coworkers to evaluate a situation and respond in accordance with existing social protocols.

Characterizations of Japan as a society in which human relations are all-important, relationships are dependent on positioning people on vertical (hierarchy) and horizontal (in- or outgroup) axes, and where communication is highly indexical or context-dependent have been widely discussed in the business and sociological literature on Japan (see e.g., Stewart, 1987; Barnlund, 1989, specifically on interpersonal communication in organizations). The extent of agreement in the literature suggests that they are firmly grounded in reality. In a society like Japan, where most behavior and the use of language is highly codified, the form is standard. It is important to look beneath the surface to interpret the meaning of an exchange, hence the importance of positioning and the emphasis on atmosphere. Much of the content of a message is implicit; interpretation is often based on intuition rather than facts, and relationships continually shift and are redefined.

Several common traits emerge in Japanese designers' attempts to deal with the particularities of their culture. First, fully conscious of the highly relativistic approach to relationships in their society, designers readily admit that there are limits to supporting the more subtle or situationally dependent aspects of work. Despite listing a shared workspace as one of the design issues and providing a workstation and table, no one has yet tried to work using MAJIC even in the laboratory. The NTT Software Labs team's research shifted its focus from shared workspace to interpersonal interaction during work.

A corollary of not trusting a computer system to model all instances of human communication or successfully translate the subtleties of day-to-day interaction is the focus of many Japanese CSCW systems on providing channels for communication rather than trying to specify content or process. This is clearly the case with MAJIC, in which research and evaluation have focused exclusively on the physical environment. In TWS/Clearboard, too, the focus is on providing an environment that simulates as closely as possible a face-to-face situation and does not in any way constrain potential use.

Another feature of Japanese CSCW systems is their provision of support for traditional, paper-based forms of working and ways of integrating paper and electronic information. TWS and Clearboard use video to capture texts or drawings on paper. The MAJIC system integrates a desk on which people can work. These systems also allow people to draw using pen or pen-based computing technology. This is all the more significant considering the transformations involved in converting keyboard input to Japanese ideograms or *kanji*. As one informant noted, "typing is not easy for us."

When language cannot convey all meaning, nonverbal communication becomes more important. Perhaps most significant, Japanese CSCW systems are also characterized by extensive emphasis on providing contextual cues so that Japanese people using these systems are able to orient their behavior appropriately. This emphasis on the contextual translates into research on spatial awareness, gaze awareness rather than eye contact, gesture, interpersonal distance, physical feedback, and large displays. One informant even went so far as to insist that physical feedback must be integrated into the interface design because he does not believe it is possible for Japanese people to have an entirely intellectual relationship with the computer.

Finally, considerable attention is paid to creating a sense of shared environment, and an environment that can be manipulated to influence the mood of the encounter. If a CSCW system is to be useful in Japan, it is considered important that a sense of atmosphere or feeling transpire through the system.

IMPLICATIONS—WHAT DOES THIS HAVE TO DO WITH AGENTS?

Clearly, the frames of meaning of Japanese CSCW researchers have a major impact on their design choices. These choices in turn guide the implementation and eventual use of these systems. Designers create artifacts to fit into cultural spaces as they understand them. New uses and new cultural meanings can only be developed after the fact. Design choices circumscribe a field of potential uses: some are built in, others are proscribed. Consequently, it is essential to consider design in studies of the implementation and use of technology. This raises the question of *which* cultural analyses and considerations enter into interaction, personality, visual design, and context of embodied agents.

My answer is that different cultures have different sensitivities to both how agents should look and to what they should be able to do. In terms of embodiment, we have seen that a realistic representation of one's coworkers is important in Japan. In Scandinavia, in contrast, the other case we are familiar with, simple iconic representations of collaborators (or agents) pose no problem. There are certainly cultural factors even in the choices of icons. For example, Microsoft's help agent is personified as a paper clip or a genius in the United States, but as a dolphin or an office lady in Asia.

WHAT SHOULD AGENTS LOOK LIKE?

If we strive for realism, there is the question of what makes an agent believable? The literature points to the ability to parse free speech, particularly intonation, and personality, often helped by body language and facial expression as key factors. Technically, with three-dimensional computer graphics and scanning technologies, we can make agents look as realistic as we want, or at least we will be able to shortly, and we can model their movements realistically. Think of the computer-generated film, *Final Fantasy*. Do we really want agents to look and sound human? The jury is still out on people's comfort levels with agent realism. The authors much prefer artificial-looking animations and agents, like Sony's artificial dog, AIBO. If our argument that culture is actualized in the present and gradually transformed holds, however, we may feel more accepting of human-looking agents in a few years, and our children may grow up expecting agents to look and behave like them. When compact disc (CD) audio recordings first came out, music critics complained that their sound was too "perfect" and that analogic recordings better captured the spirit of the music. Twenty years ago, hand-written letters were considered preferable to typed ones because one could more easily gauge the author's mood. Now we communicate by e-mail.

The explicit cultural sensitivity of Japanese CSCW work also points to a need for cultural sensitivity in the *design* of technological artifacts, this at a level that goes beyond ergonomics or changing surface details on an interface. In the case of Japan, the need for contextual information suggests that the use of language-based environments, even in Japanese, may be problematic. This difficulty goes far beyond the physical difficulty of inputting on a keyboard (although this is also a definite concern as reflected in the extensive research on pen-based computing, speech synthesis, and multimodal interfaces in Japan). Second, the assumed difficulty of fitting into a framework, or set way of doing things, suggests that organizing cooperative work as a series of procedures to be followed or channels to be taken may be inappropriate in Japan. In fact this is confirmed by the choice of Japan's leading workflow expert to focus on the use of resources rather than the paths they follow.

HOW SHOULD (CAN) AGENTS ACT?

This brings us to what we believe is the more important question from a cultural standpoint—not what agents should look like, but how they should behave (or not). In their study of robots playing humans in a multiplayer computer game, Huber and Hadley (1997) found that, despite their superior skill, the agents were quickly outwitted because they were too predictable and steadfastedly goal-minded when compared with human actors:

> The robotic agents are formidable dogfighters, reacting extremely quickly and effectively to attacks upon them while simultaneously effectively inflicting damage.... Few human players are competent enough dogfighters to be able to regularly defeat the robotic agents.... The robotic agents perform poorly in several aspects of the game, however. Foremost among these is the agents' predictability, lack of guile, and steadfastness to fulfill their missions. Human players quickly figure out the robotic agents' individual weaknesses and take advantage of them.... The robotic agents are also inferior to their human counterparts in coordinating their activities. (pp. 335–336)

When it comes to robots playing each other, as in the RoboCup football competitions, it seems that different teams develop different strategies and playing styles. In trying to play collaboratively, there is a trade-off between predictability and responsiveness. Balch's experiments (Balch, 1998; Balch & Arkin, 1995) have shown that a group reward system encourages diversity and cooperative soccer play, but researchers admit that human complexities such as motivation or jealousy are difficult, if not impossible, to reproduce in robotic systems.

Examples of soccer-playing, trash-collecting, and space-exploring robots reassure us that agents cannot yet do what comes quite easily to human beings. Yet why is that? The really hard problem—for agents—is the contextual knowledge required for them to be able to function in a natural environment. We call this *worldview*—the general knowledge about the objects, actions, distinctions, and relationships that matter in a particular domain. Our worldview helps us situate a request or comment and interpret its meaning. Having this bank of knowledge tells us what is likely to be important in a given situation; it helps us focus and interpret what something means in context. Human communication is multichannel—we draw on information from all over. Being able to focus helps us attend to several things at once—multitasking. Our worldview also helps us deal with ambiguity. It allows us to organize our actions and coordinate.

How do we acquire a worldview? Through past histories of interaction. We learn from past experience to judge what is foreground, what is background, and what is relevant in any given situation. If worldview is learned through past histories of interaction, it follows that they are intimately cultural. Hence, if Japanese are sensitive to context, and thus need a lot of nonverbal cues, the French or Germans may be particularly sensitive to hierarchy or bureaucracy and sometimes have trouble functioning if the rules are not clear. Canadians might be particularly sensitive to space?

The challenge, then, is to give agents appropriate worldviews. In developing agent-based systems, some worldview knowledge has to be expressed in the software. Groups of agents need to be given common views or the means to negotiate common understandings. For example, in manufacturing a pair of jeans, they need to agree on what constitutes a pair of jeans (e.g., denim only) and the sequence of processes used to produce them. This is relatively easy when something can be narrowly defined, say a single style of jeans, but when more flexibility is required the agents need to capture more and more of the underlying worldview.

If we are successful in endowing agents, embodied or not, with worldviews that correspond to their cultures of origin, we may reasonably hope that they will be faithful representatives when they go out and meet other agents. If we are able to synthesize information that is relevant across cultures, our agents will be portable cross-culturally. Yet as we saw earlier, agents are neither as flexible nor as unpredictable as human beings. How then can agents react to something new and still remain faithful to their owners? This leads to our final point: the individual is part of a collective, and the collective is constituted through actions in the environment, always with relation to others. The same must be true for agents.

LOCATING AGENTS IN THEIR COMMUNITIES: A–B–X

In concluding this chapter, we propose a formalization for examining the relationship between agents, the material context in which they evolve, and the ongoing construction of social context. Imagine an A–B–X system, where X is the physical and B the individual surface on which A (the community) finds its existence. Newcomb (1953) called this a *coorientation system*. The components of an A–B–X unit are defined by the structure of relationships and not the other way round. Objects, for example, are defined by the purposes of agents (they are the targets of intention). An agent, A or B, is defined by both the objects they are concerned with and for whom they act (or who acts for them). Figure 2.4 describes the A–B–X relationship.

The A–B dimension is that of communication. The relationship between any two actors (individual or collective), A and B, is always mediated by some object, X, to which they are cooriented. For example, if A is the software design community and B is the software designer, their relationship is mediated by X, the artifact produced by B (exemplified by a conference presentation or paper).

The B–X dimension is that of practice—an actor acting on some object, for example, or the construction of a software artifact by a designer. We term B an *agent* and X his or her *object*, where we understand the term *object* to have a double connotation: (a) that entity typically, but not necessarily, material that is affected by the action of B, and (b) that to which the intentionality or purpose of actor B is directed (as in "the object of this exercise is ..."). The term *agent*, however, also has a double meaning of (a) actor and (b) actor-*for*. The A–B relationship, it follows, is one of acting-for, where the A–B couple is cooriented to some object X.[4] All work is characterized by an agent (not the computer type)–object relationship, B–X, framed within an A–B matrix.

Unlike Newcomb, who visualized the A–B relationship as symmetrical, we argue instead that it is typically asymmetrical and complementary. Speech act theory, for example, points out that every act of communication, in addition to providing an exchange of information, has an illocutionary

---

[4]Note that for individual action the A–B distinction vanishes because acting (performing an operation on some object) and acting-for (realizing the intentionality of some person) are lodged in the same agent. One becomes one's own agent. The A–B link occurs when one actor acts for another to realize the latter's intention, and thus indexes the existence of a relationship of communication.

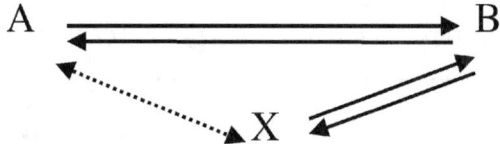

FIG. 2.4. The concept of a coorientation system.

point and force, which is what gives speech its effective capacity to channel human activities and thus generate organization (Austin, 1962; Searle, 1969; Winograd & Flores, 1986). In a similar vein, Giddens (1984) argued that all action, spoken or otherwise, exerts power. On a practical level, ordinary observation shows us to what extent organizational relations are expressed in asymmetrical communications: requests, orders, explanations, excuses, encouragements, and so on. The relationship of B to A is one of agent to principal. The existence of an A is the sanction for the action of B—a mandate to act, in effect. Thus, the A–B plane invariably indexes the existence of a community of discourse and practice.

Adopting a principle of structuration proposed by Giddens, we argue that it is the sanction of a community that legitimates the performance of work. The activities of an individual or group of individuals are embedded within some larger community (of discourse, constituency, country, community, corporation) that serves to legitimate the activities of its members (or at least a substantial proportion of them).

Let us try now to apply our A–B–X model to computer agents: Computer agents are the B, acting for and accountable to A (their community of reference or individual) in and on an object world, X. These agents may have autonomy, personality, and embodiment, but their activities are legitimated only when they report back and carry out the actions mandated by individual A, when they are the faithful representatives of A. This person or community, A, is in turn imbricated in a web of social relations in which he or she is the B and on whom he or she relies for sanction to continue performing. Thus, when agent systems are designed, the designer is B, working with technology, X, but he or she remains answerable to the agent design community, as well as to his or her culture (A). We believe that this view of the nature of communication between agents brings a new dimension to discussions of their embodiment and cross-cultural portability.

MOVING BETWEEN COMMUNITIES: THE NEXT LEVEL

One implication of our theory of action is the transformation of individual agency into group agency: Through communication, the group becomes an actor and is treated as such by others. The A whose agency is represented

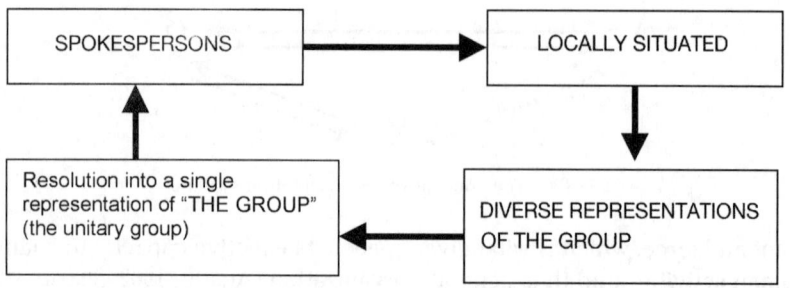

FIG. 2.5. Embedded levels of interaction.

by a series of Bs becomes a community. Yet how do different groups talk to each other? How can an agent that is representative of one culture interact with another agent or culture from outside its reference group?

We assume that the same model is applicable to more extended networks.[5] In short, we suggest that, just as an organization is composed of more than one group or unit of activity, interactions between distinct groups (such as interactions between agents across cultures) are joined to each other through shared circumstances. The result is the construction of units of coorientation and collective action involving, not just individuals, but groups and the culture itself. Because of the larger context, interfaces are created between groups, each locally situated, and across situations. In this context, conversational form can no longer be assumed to function only according to the rules of face-to-face interaction. The participants in this larger "conversation" speak for larger conversations (represent them), and frame knowledge becomes important.

An interface between two levels of communication is created. At one level, communication is a critical enabler in the performance of local circumstances. At the second level, communication is the linking up of all these locally situated interactions into a conversation that transcends the local. We visualize the embedding of different levels of interaction as a cycle in which the local is successively translated into the global and the global into the local (Fig. 2.5). Figure 2.5 thus describes a system with deep cultural roots. It is a historically situated artifact—a product of the collective undertakings of a particular society.

Grudin (1990) observed that the development of computer-assisted technology is the story of an enlarging interface with the user community: from a preoccupation with hardware (engineering issues and no direct contact with the user community), to software (the programmer as principal user), to the terminal interface (emphasis on ergonomic and human factors re-

---

[5]Taylor et al. (2001) propose several minor qualifications.

lated to individual use), to an interaction dialogue (mimicking features of normal conversation based on theories of human cognition and allowing for more creative uses), to the collaborative work setting (interfaces for collaborative working, learning, playing). Since Grudin's article was published, a further interface has become evident: the focus has shifted from enhancing connectivity (using groupware or CSCL environments) to designing environments. We have tended, historically, to see technology as an instrumentation, or set of tools, for conducting some activity in the transformation of a physical or symbolic environment. This is the B–X dimension. Our study of systems design in Japan suggests the necessity of seeing technology in its second manifestation, as the expression and concrete realization of an A–B environment. The new technology is no longer just about getting things done; it also has inscribed within its design a view of the social relationships involved in getting things done, including how it is managed and how it is sanctioned. Software-based systems not only have the capacity to represent organizational process, and thus render it transparent in a new way. More fundamentally, they have become active agents in constituting the part of the environment in which we conserve identity and adaptation. This new technology potentially instantiates a real, and a projected, A–B community. This is truly a new development.

CONCLUSION

We are only beginning to appreciate the complexity of the relationship between technology and its context and how changes in one inevitably affect the other. This chapter outlined how designers' views on Japanese culture find their way into the design rationale for CSCW systems: Japanese CSCW designers generally agree that Japan is unique and that designing for a Japanese context requires particular attention to a certain number of elements. Although it is not the only consideration in design, this attention to culture goes far beyond the stage of ideas to find concrete expression in the computer systems, as illustrated by our two examples. Based on how they understand the world around them, designers make assumptions that guide their design choices. The technologies they build contain some elements that are distinctive to their culture of origin without necessarily being unique to that context. They actualize their shared understandings of Japanese culture as they perform it in their daily design activities. As participants in their larger professional, organizational, and national cultures, individual designers link their creations with larger social or cultural values. By extension, we have argued that the design of agents is influenced by the cultural values and practices of their designers, and that an understanding of their *culture of origin* is invaluable in understanding the end result. In advancing a system of explanation that we call an A–B–X model, our motive is to draw attention to the interrelatedness of the B–X component (materially situated practical work) and the A–B component (the embeddedness of all work in a community of discourse and practice and its sanctions).

ACKNOWLEDGMENTS

We would like to thank the FCAR (Fonds pour la Formation des Chercheurs et l'Aide à la Recherche), whose financial support made this research possible. Thanks are also due to all those who participated in the study, as well as to James Taylor for his support and numerous discussions about A–B–X.

REFERENCES

Ackoff, R. (1981). *Creating the corporate future.* New York: Wiley.
Andrews, S. B., & Miller, H. G. (1987). Expanding market share: The role of American corporations in technical assistance. *International Journal of Manpower, 6,* 25–27.
Austin, J. L. (1962). *How to do things with words.* Oxford: Oxford University Press.
Balch, T. (1998). *Behavioral diversity in learning robot teams.* Unpublished PhD dissertation, College of Computing, Georgia Institute of Technology.
Balch, T., & Arkin, R. C. (1995). Communication in reactive multiagent robotic systems. *Autonomous Robots, 1*(1), 27–52.
Bannon, L., & Schmidt, K. (1991). CSCW: Four characters in search of a context. In J. Bowers & S. Benford (Eds.), *Studies in computer supported cooperative work* (pp. 3–16). Amsterdam: North-Holland.
Barnlund, D. C. (1989). *Communicative styles of Japanese and Americans: Images and realities.* Belmont, CA: Wadsworth.
Bijker, W. E., Hughes, T. P., & Pinch, T. J. (Eds.). (1987). *The social construction of technological systems: New directions in the sociology and history of technology.* Cambridge, MA: MIT Press.
Bijker, W. E., & Law, J. (Eds.). (1992). *Shaping technology/building society: Studies in sociotechnical change.* Cambridge, MA: MIT Press.
Bourdieu, P. (1983). *Le métier de sociologue: Préalables épistémologiques* (4th ed.). Berlin: Mouton.
Button, G. (Ed.). (1993). *Technology in working order.* London: Routledge.
Carlson, W. B. (1992). Artifacts and frames of meaning: The cultural construction of motion pictures. In W. Bijker & J. Law (Eds.), *Shaping technology, building society* (pp. 175–198). Cambridge, MA: MIT Press.
Coakes, E., Willis, D., & Lloyd-Jones, R. (Eds.). (2000). *The new sociotech: Graffiti on the long wall.* Berlin: Springer-Verlag.
Collins, H., & Pinch, T. (1982*). Frames of meaning: The social construction of extraordinary science.* Boston: Routledge & Kegan Paul.
Dahlbom, B., & Mathiasson, L. (1993). *Computers in context: The philosophy and practice of systems design.* Oxford: NCC Blackwell.
Deans, C. P., & Ricks, D. A. (1991). MIS Research: A model for incorporating the international dimension. *Journal of High Technology Management Review, 2*(1), 57–81.
DeLaet, C. (1992). *Des Outils pour un développement durable: Plaidoyer pour une réconfiguration.* Unpublished doctoral dissertation, Université de Montréal, Montréal.

Engeström, Y., & Middleton, D. (Eds.). (1996). *Cognition and communication at work*. Cambridge: Cambridge University Press.
Erez, M., & Earley, P. C. (1993). *Culture, self-identity, and work*. New York: Oxford University Press.
Giddens, A. (1984). *The constitution of society*. Cambridge, United Kingdom: Polity.
Grudin, J. (1990). The computer reaches out: The historical continuity of interface design. In *Proceedings of CHI'90* (pp. 261–268). New York: ACM.
Hales, M. (1994). Where are designers? Styles of design practice, objects of design and views of users in computer supported cooperative work. In D. Rosenberg & C. Hutchison (Eds.), *Design issues in CSCW* (pp. 151–177). London: Springer Verlag.
Hall, E. T. (1976). *Beyond culture*. Garden City, NY: Doubleday.
Hannerz, U. (1992). *Cultural complexity*. New York: Columbia University Press.
Haythornthwaite, C., & Wellman, B. (2001). The Internet in everyday life. *American Behavioral Scientist, 45*(3).
Heaton, L. (forthcoming). *Culture in design: The case of computer supported cooperative work*. Cresskill, NJ: Hampton Press.
Hofstede, G. (1980). *Culture's consequences: International differences in work-related values*. Beverly Hills, CA: Sage.
Huber, M. J., & Hadley, T. (1997). Multiple roles, multiple teams, dynamic environment: Autonomous Netrek agents. In *Autonomous Agents 97* (pp. 332–339). Marina Del Rey, CA: ACM.
Hutchins, E. (1995). *Cognition in the wild*. Cambridge, MA: MIT Press.
Ishii, H. (1990). TeamWorkStation: Towards a seamless shared workspace. In *Proceedings of CSCW'90* (pp. 13–26). New York: ACM.
Ishii, H., & Arita, K. (1991). ClearFace: Translucent multiuser interface for TeamWorkStation. In L. Bannon, M. Robinson, & K. Schmidt (Eds.), *ECSCW '91 Proceedings* (pp. 163–174). Amsterdam: Kluwer.
Ishii, H., & Kobayashi, M. (1992). ClearBoard: A seamless medium for shared drawing and conversation with eye contact. In *Proceedings CHI'92* (pp. 525–532). New York: ACM.
Ishii, H., Kobayashi, M., & Grudin, J. (1993). Integration of interpersonal space and shared workspace: Clearboard design and experiments. *ACM Transactions on Information Systems (TOIS), 11*(4), 349–375.
Ishii, H., & Miyake, N. (1991). Toward an open shared workspace: Computer video fusion approach of TeamWorkStation. *Communications of the ACM, 34*(12), 37–50.
Ito, S. (1986). Modifying imported technology by local engineers: Hypothesis and case study of India. *Developing Economies, 24*, 334–348.
Jackson, M. H. (1996). The meaning of "communication technology": The technology-context scheme. In B. R. Burleson (Ed.), *Communication yearbook 19* (pp. 229–267). Thousand Oaks, CA: Sage.
Law, J., & Bijker, W. (1992). Postscript: Technology, stability and social theory. In W. Bijker & J. Law (Eds.), *Shaping technology, building society* (pp. 290–300). Cambridge, MA: MIT Press.
Luff, P., Heath, C., & Hindmarsh, J. (2000). *Workplace studies: Recovering work practice and informing system design*. Cambridge: Cambridge University Press.
Mackenzie, D., & Wacjman, J. (1985). *The social shaping of technology: How the refrigerator got its hum*. Milton Keynes: Open University.
Madu, C. (1992). *Strategic planning in technology transfer to less developed countries*. New York: Quorum.
Mouer, R., & Sugimoto, Y. (1986). *Images of Japanese society: A study in the social construction of reality*. London: KPI Ltd.
Nardi, B. A. (Ed.). (1996). *Context and consciousness: Activity and computer–human interaction*. Cambridge, MA: MIT Press.
Newcomb, T. (1953). An approach to the study of communicative acts. *Psychological Review, 60*, 393–404.

Okada, K., Maeda, F., Ichikawaa, Y., & Matsushita, Y. (1994). Multiparty videoconferencing at virtual social distance: MAJIC design. In *Proceedings of CSCW'94* (pp. 385–394). New York: ACM.
Orlikowski, W. J. (1992). Learning from notes: Organizational issues in groupware implementation. In *Proceedings of CSCW'92* (pp. 362–369). New York: ACM.
Orlikowski, W. J., & Gash, D. C. (1994). Technological frames: Making sense of information technology in organizations. *ACM Transactions on Information Systems, 12*(2), 174–207.
Poole, M. S., & DeSanctis, G. (1990). Understanding the use of group decision support systems: The theory of adaptive structuration. In C. W. Steinfeld & J. Fulk (Eds.), *Organizations and communication technology* (pp. 173–193). Newbury Park: Sage.
Raman, K. S., & Watson, R. T. (1994). National culture, IS, and organizational implications. In C. P. Deans & K. R. Karwan (Eds.), *Global information systems and technology: Focus on the organization and its functional area* (pp. 493–513). Harrisburg, PA: Idea Group.
Rogers, E. M., & Shoemaker, F. (1971). *Communication of innovation: A cross-cultural approach.* New York: The Free Press.
Rogers, Y. (1992). Ghosts in the network: Distributed troubleshooting in a shared work environment. In *CSCW'92 Proceedings* (pp. 346–355). New York: ACM.
Searle, J. R. (1969). *Speech acts.* London: Cambridge University Press.
Stewart, E. C. (1987). The Japanese culture of organizational communication. In L. Thayer (Ed.), *Organization—communications: Emerging perspectives II* (pp. 136–182). Norwood, NJ: Ablex.
Taylor, J. R., Groleau, C., Heaton, L., & Van Every, E. (2001). *The computerization of work: A communication perspective.* Thousand Oaks, CA: Sage.
Winner, L. (1993). Upon opening the black box and finding it empty: Social constructivism and the philosophy of technology. *Science, Technology, & Human Values, 18*(3), 362–378.
Winograd, T., & Flores, F. (1986). *Understanding computers and cognition: A new foundation for design.* Norwood, NJ: Ablex.

CHAPTER

3

Socially Intelligent Agents in Human Primate Culture

Kerstin Dautenhahn
University of Hertfordshire

This chapter discusses socially intelligent agents (SIA) in the context of human culture. The first part of the chapter provides an introduction to SIA research and focuses on the issue of realism versus believability—namely, to what extent SIAs need to imitate behavior and appearance of humans. The *lifelike agents hypothesis* is analyzed in particular from the perspective of social robots. I give examples showing that with respect to realism less might be more. Mori's "uncanny valley" analysis of believable robots is explained. It is shown that the issue of believability is particularly crucial for technology whose purpose is to change the behavior, attitudes, beliefs (or, more generally, minds) of humans, as it is the case for SIAs. Therefore, SIAs fall under the category of technology called *persuasive technology*. The second part of the chapter addresses primate cultures, their phylogenetic origins, and how human culture is grounded in and shaped by cognitive capacities that the human species have evolved. I suggest that these cognitive capacities are important to consider in the design of SIAs. The chapter concludes by discussing the possible role of SIAs in the evolution of human cultures. I argue that SIAs can play a role in *preserving cultural diversity*—a huge technological challenge, but a possibility to make SIA technology more humane by acknowledging the individual and cultural roots of people, rather than treating them as an increasingly homogeneous group of users.

SOCIALLY INTELLIGENT AGENTS

Let us first introduce a few concepts: *Social agents* can be either biological social agents, such as cetaceans and primates (including humans), or they can be artifacts with artificial social intelligence (i.e., agents made of software or hardware). SIAs are agents that show aspects of human-style intelligence (Dautenhahn, 1998).

Humans are above all SIAs. It has become more and more recognized in the areas of cognitive science and artificial intelligence (AI) that human intelligence and cognition is fundamentally socially situated (cf. Aronson, 1994; Tomasello, 1999). In line with this paradigm shift in how we regard the phylogenetic (evolutionary) as well as ontogenetic (developmental) origins of human intelligence, an increasing number of robotics and AI researchers are designing and programming social agents—agents that people can relate to and that might relate to people as well (Edmonds, 2000).

To build agents that show aspects of human-style intelligence, researchers focus on issues such as how agents can learn from humans by demonstration or imitation (Dautenhahn & Nehaniv, 2002a), how facial emotional expressions might be used to regulate interactions of humans with a humanoid robot (a line of research comprehensively investigated e.g. by Breazeal & Scassellati [Breazeal, 2002; Breazeal & Scassellati, 2000]) or using social strategies and nonverbal means of communication for virtual agents that can engage humans in dialogues. The latter is intensively studied in the area of *embodied conversational agents* (Cassell et al., 2000). On a more fundamental level, issues such as embodiment and social situatedness are studied in this area (Dautenhahn et al., 2002a).

On the level of the individual human, interactions with SIAs can influence a human being's attitudes, behavior, and mind, and in this way empower as well as manipulate humans (see Nehaniv & Dautenhahn, 2000; Norman, 1997). B. J. Fogg (1999) discussed computers as *persuasive technologies*. In contrast to other nonpersuasive technologies, "persuasive computing technology is a computing system, device, or application intentionally designed to change a person's attitudes or behavior in a predetermined way." Furthermore, Fogg called the study of planned persuasive effects of computer technologies *captology*. Figure 3.1 shows the functional triad of computer persuasion. Following Fogg's terminology, SIAs might fall under the category of *social actors*, where agents can adopt animate characteristics, play animate roles, and follow social dynamics for the purpose of creating relationships with humans and invoke social responses. In this sense, SIAs are persuasive technologies, and therefore issues of design, credibility (Tseng & Fogg, 1999), and ethics of persuasive technology (Berdichevsky & Neunschwander, 1999) also apply to SIAs technology, in particular to the new generation of highly interactive social software and robotic agents.

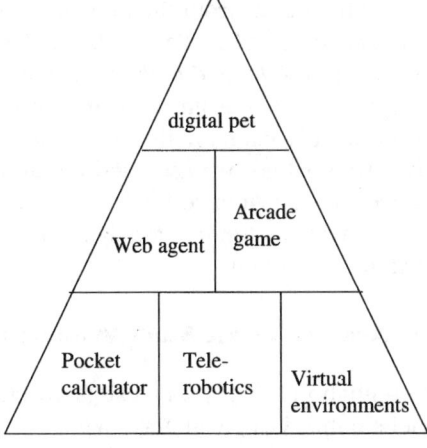

FIG. 3.1. The functional triad of computer persuasion, redrawn from Fogg (1999).

A popular strategy toward building SIAs that humans like to interact and engage with is to exploit the natural human tendency to anthropomorphize nature, including animals, robots, and anything that with some imaginative effort can be attributed agency, intentionality, and possibly even a mind (Dautenhahn, 1997). Such user-centered approaches focus on giving the illusion of life—namely, the *impression* of social intelligence (Bates, 1994; Persson et al., 2001)—rather than aiming to develop biologically plausible deep models of human intelligence (cf. Scassellati, 1999, 2001). Indeed it seems that humans cannot help but treat the world in social terms (Dautenhahn, 1995; Tomasello, 1999), and this tendency—which might be due to our evolutionary origins as a highly social species—ranges from socially interpreting the movements of points on a computer screen (Heider & Simmel, 1944) to treating computers as if they were people (Reeves & Nass, 1996). Similarly, robots are by most people spontaneously treated as intentional agents with personality, goals, and other mental and internal states. We are eagerly willing to suspend disbelief and pretend that a robot that looks and behaves doglike *is* a dog, and consequently we treat the robot like a dog. Entertainment products such as movies, literature, toy products, and so on routinely exploit this tendency. Even in situations where the setup is not deliberately playful, humans have been shown to behave toward computers as if they were machines. In extensive studies, Reeves and Nass (1996) demonstrated that humans tend to treat computers (and media in general) as people, applying social rules and heuristics from the domain of people to the domain of machines. This *media equation* (media equals real life) is particularly relevant for SIA research with the *human in the loop*—namely, where people interact with agents in the role of designers, users, observers, assistants, collaborators, competitors, customers, patients, or friends.

In a related case study where we assessed children's beliefs and fantasies about robots, we found evidence for the tendency to portray robots and interpret their behavior as fundamentally *social*. The following section describes this work in more detail.

Attitudes Toward Agents: A Case Study With Robots

In Bumby and Dautenhahn (1999), we investigated children's attitudes toward robots; a brief summary is given here. We were interested to find out how children interact and describe robots. Thirty-eight children (ages 7–11, 21 males, 17 females, BC1 socioeconomic category) were studied at St. Margaret's Junior School in Durham, United Kingdom. A number of working hypotheses were addressed with respect to how the children portrayed robots. In three studies, the children were asked to draw a picture of a robot and write a story about the robot they had drawn. These studies were observational. The third study had the format of an informal, guided, and filmed interview while the children were in a group interacting with two mobile robots (see Fig. 3.2) that were running in an environment with a lightsource. The robots were simple behavior-based Braitenberg vehicles (Braitenberg, 1984).

Results of Study a (pictures) showed, for example, that the children tend to give the robot humanoid faces. Figure 3.2 gives examples of a variety of drawings produced by 8- and 9-year-olds portraying robots. In Study b (stories), one result was that the children tend to put the robots in familiar settings, doing familiar tasks; see the following story told by a 9-year-old boy:

> Once there lived a robot called Rob, he was made by a mad professor called Brain-Box on Jan the 3rd 2,000,000 in Germany. Everybody thought it was brilliant and on Jan 30th the same year it ran away and it ran away to England and terrorised England people looked to see who or what had done it. The robot had killed someone and taken her brain they got the police out and they looked into it the police didn't know who or what had done it. One day someone called Tod saw Rob and called the police to come and get Rob but when the police cam Rob had gone. Tod got a fine for supposedly lying to the police. The next year the police had still not found Rob. Sometime in the last year Rob had fallen off something and he was found in pieces in a junk yard.

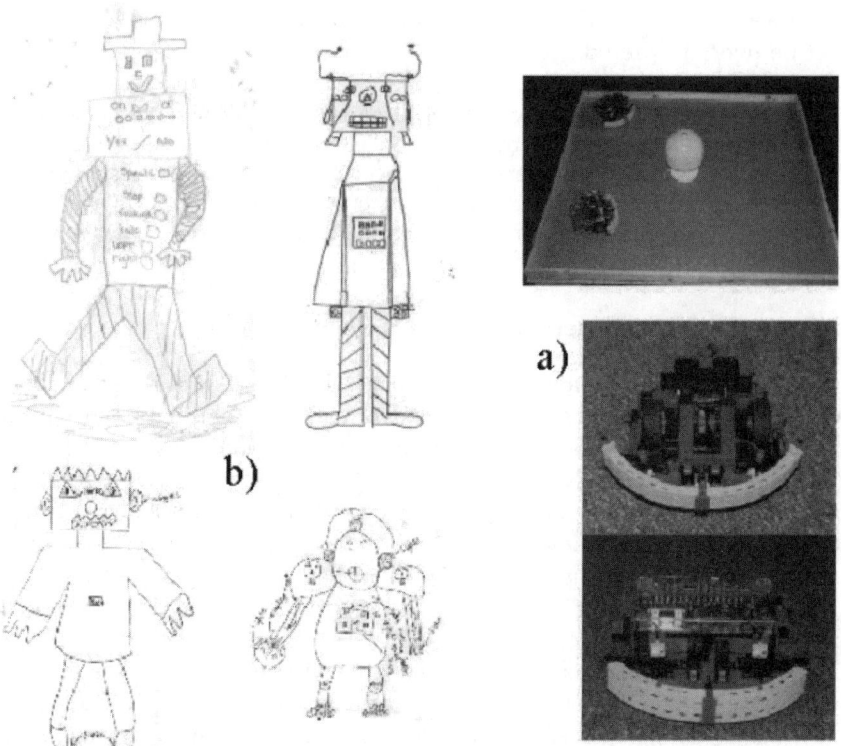

FIG. 3.2. (a) The experimental setup and the two autonomous, mobile *fischertechnik* robots; (b) drawings of 8- and 9-year-olds (Bumby & Dautenhahn, 1999).

Some people phoned the police the police came out and he had gone. The village dug a massive pit to try and catch Rob. Now Rob was as high as six houses he was eating a lot of junk metal. One day rob was walking along and he fell in the pit, the people found him in the pit and he was killed. (Based on a study summarized in Bumby & Dautenhahn, 1999)

The robots were significantly often put in a social context. Study c (interview) showed a clear tendency to anthropomorphize the robots (e.g., "I don't think it likes the light"). The children also often talked to the robots as if they were animals or small children. Other findings of this case study (e.g., with respect to attribution of gender or violence) are reported in more detail in Bumby and Dautenhahn (1999). Generally, the case study suggests that children tend to perceive robots as people. This single case study cannot answer the bigger question of how children of different ages and in different sociocultural contexts think about robots, but the results confirm findings along the lines of studies with computers (Reeves & Nass, 1996).

However, as pointed out by Erickson (1997), the media equation should not be overinterpreted:

> It is one thing for people to apply social heuristics to machines; it is quite another to assume that this accounts to social interaction, or to suggest that the ability to support social interaction between humans and machines is now within reach. Interaction is a two-way street: just as people act on and respond to computers, so computers act on and respond to people. Interaction is a partnership. (Erickson, 1997, p. 90)

Agents From the Observer Point of View

Given the immense influence of the observer point of view, is there any benefit in deep models of social intelligence for SIA researchers apart from those studying agents as models for understanding biological intelligence? This question is strongly reminiscent of discussions in the area of artificial life where researchers investigate systems that show lifelike properties (Langton, 1989). Even biologists who study life as we know it on Earth hesitate to agree on a complete and sufficient list of criteria of life. Thus, as lively reported in Levy (1992), at the first workshop that founded the area of artificial life, researchers jokingly discussed the *duck test* in honor of Jacques de Vaucanson's mechanical duck, which in 1738 became famous when exhibited in Paris because it showed a variety of duck behaviors such as drinking, eating, quacking, splashing around in water, and digesting food. The duck test goes as follows: "If it looks like a duck and quacks like a duck, it belongs in the class labeled ducks" (Levy, 1992, p. 117). Thus, if an agent looks like a human, behaves like a human, and talks like a human, should it be labeled *human*? Should it be treated like a human being with the same legal and human rights? Fortunately, at least today we are still far from having agents (robotic or in software) around that can fool us for longer than a few seconds or minutes about their true identity. In specialized domains (e.g., with limited communication channels as they exist in text-based multiuser dungeons [MUDs]), agents such as JULIA (Foner, 2000) might be able to fool us for a while as long as the discourse domain is limited. However, in unrestricted, repeated, long-term, face-to-face, agent–human communication, any discussion about agents that might be mistaken for humans is speculation for now.

In the recent movie *Artificial Intelligence*, the robot protagonist, which, central to storyline, is supposed to be the main object of the audience's sympathy and empathy, is played by a real human boy, a popular actor well known from various previous mainstream movies with emotional themes. Today's animation techniques are rapidly advancing, and entirely computer-generated movies such as *Antz* and *Toy Story* have been hugely successful, in addition to movies where animated characters supplement the human crew of actors (e.g., *Jurassic Park* or *The Matrix*). These movies were generated with new technology, but the characters and stories still followed the traditions of previous animation movies produced with little or no computer assistance such as *Bambi* or *101 Dalmatians*. In all of these cases, the stories and/or characters are clearly identifiable as fantastic, not from Earth, or not from the present day. The movies just mentioned are all in the realm of make-believe, let's pretend dinosaurs were still (or again) alive, what would happen if animals could talk?, what might the future look like?, and so on. These movies are grounded in the domain of imagination and fantasy, presented in the playful, consequence-free environment of movie entertainment. Yet what if movies were to look *really* realistic so that they could be mistaken for real life? *Final Fantasy* is a recent example pointing toward you-can-almost-taste-it computer-generated realism: a fantastic world populated by real, computer-generated people based on the latest state-of-the-art technology in computer generation, where even skin and hair look almost real. Despite huge expectations, it seems that *Final Fantasy* has not been the box office hit that was expected. Potential explanations proposed in the press (see *Los Angeles Times*, July 20, 2001[1]) identify a tension between realism and believability: Making animations more real does not necessarily make them more believable—an insight that is familiar to comic designers (McCloud, 1993).

Interestingly, the transition from comic-style to realistic agents might not be linear, as suggested by the roboticist Masahiro Mori (1982; cf. discussions in Reichardt, 1978; Bryant, 2002).[2] Mori's important contribution to this discussion is his proposal of the *uncanny valley* for lifelike robots (see Fig. 3.3). Here the prediction is that the more lifelike we make robots (i.e., the more similar they become to us), the more familiar (today we would use the term *believable*) they become until ultimately (100% similarity with healthy human beings) familiarity levels reach a maximum. However, the transition has a local minimum, characterized by a sharp drop in familiarity when robots are very lifelike and might be mistaken for real. In this case, they can create an uncanny and unpleasant feeling when slight but noticeable differences suddenly make it obvious that they are not real, thus violating our expectations. We might also call this the *Zombie effect*: moving corpses that are (strange looking but) similar to us until we touch their cold bodies. Not surprisingly, a lot of horror and science fiction stories are based on this effect, where familiar people that are *like us* are suddenly identified as aliens, zombies, or the like. Interestingly, in addition to the overall graph shown in Fig. 3.3(c), Mori distinguished two separate graphs reflecting two different components of similarity: movement (a), and ap-

---

[1] Many thanks to Chris Landauer for providing this article.
[2] Many thanks to Lola Cañamero for mentioning this work to me.

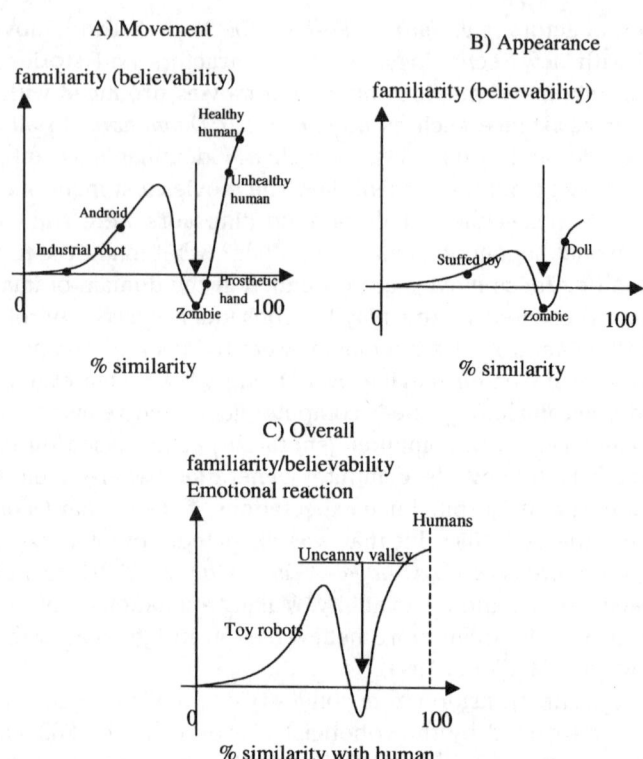

FIG. 3.3. How lifelike should a machine be? Masahiro Mori's uncanny valley, redrawn and modified from Reichardt (1978) and Bryant (2002).

pearance (b). For both criteria, he assumed curves of similar shape, although the movement curve is considered to be more dominant than the appearance curve. Indeed, Mori's suggestions are supported by psychological studies on anthropomorphism. Here it is often suggested that physical likeness, familiarity, phylogeny, and/or cultural stereotypes are important factors. Well known is the study by Eddy et al. (1993) who investigated people's tendency to anthropomorphize animals (see summary in Watt, 1997). The study identified two primary reasons that people attribute similar experiences or cognitive abilities to animals based on (a) the degree of physical similarity, and (b) the degree of an existing attachment bond (familiarity). For example, dogs and cats are more familiar to most people than frogs, and primates are physically (and behaviorally) more similar to humans than cats.

This study seems to support the *Life-Like Agents Hypothesis* discussed in more detail in the next section—namely, that humanoid agents that look like humans should be more believable and successful as social interaction partners for humans than nonhumanoid agents (assuming that humans mostly enjoy interacting with other humans). However, other evidence suggests that not physical similarity, but *behavior in context*, matters—a finding that can give support to Mori's relative emphasis on movements rather than appearance. Mitchell and Hamm (1997) provided undergraduate students with narratives depicting different mammalian agents (including humans) showing behavior that suggested jealousy or deception. The students were then asked to answer questions on particular psychological characterizations of the agents. The narratives varied according to species, context in which an agent's behavior occurred, and the degree of emphasis that the narrative was about a particular species of animal (or human). The behavior was constant in all narratives. Mitchell and Hamm found that variations in context influenced the psychological characterizations, but variations in species and emphasis did not (i.e., the psychological characterizations of all species were almost always similar):

> Nonscientists (and some scientists as well) apparently use a mammal's behavior-in-context (whether human or not) as evidence of its psychological nature, regardless of the mammal's physical similarity, familiarity, or phylogenetic closeness to humans, or the mammal's cultural stereotype; psychological terms are not used specifically for humans, but rather are depictive of behavior-in-context. (Mitchell & Hamm, 1997, pp. 173–174)

Interestingly, the notion that behavior matters more than appearance in ascribing intentionality is supported by an experimental study published in 1944 (Heider & Simmel, 1944) that convincingly demonstrated the effects of the intentional stance in an experiment with silent animations. Here human subjects created elaborate narratives about intentional agents when asked to describe movements of moving geometric shapes shown in a silent film. Figure 3.4 shows an example of a situation that can evoke a mentalizing response (e.g., chasing interpretation). Based on Heider and Simmel's studies more than 50 years ago, more recent studies have confirmed and further refined these results. Oatley and Yuill (1985) showed that cues which signal a social context increase people's tendency to use personal and mentalizing descriptions. Rimé et al. (1985) demonstrated that the replacement of geometric shapes by humanlike shapes lowered in subjects the intensity of attributing emotional states, showing that patterns of movements are more powerful in evoking anthropomorphic interpretations than the appearance of the characters. Children as young as 3 years can interpret simple patterns of movements as intentional, goal-directed behavior (Montgomery & Montgomery, 1999). Thus, the interpretation of patterns of movements as goal-directed and intentional, and the interpretation of interactions among

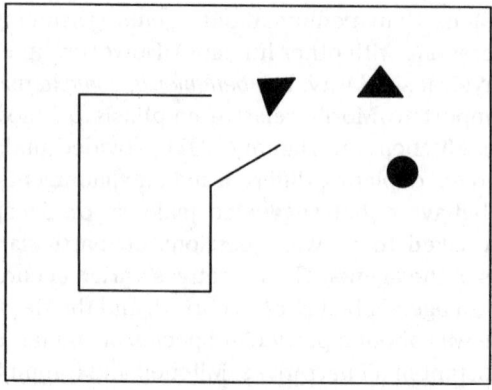

FIG. 3.4. A frame showing a scenario with simple geometrical shapes similar to the ones shown in Heider and Simmel's films, modified from Heider and Simmel (1944).

such moving objects as social or mentalistic, is fundamental to social intelligence in typically developing humans.[3]

Other future studies along research done by Heider and Simmel, Mitchell and Hamm, and others could confirm whether this also applies to nonmammalian animals. A particular challenge would be to include computational and robotic agents in such studies. I suggest that behavior reading might also apply to inanimate objects such as robots. Every robotics researcher who has ever given a demonstration of autonomous mobile robots to a general audience can confirm how readily humans anthropomorphize and view robots as people (cf. Braitenberg, 1984; Bumby & Dautenhahn, 1999; Dautenhahn, 1997).[4] The importance of behavior expression in agent building has also been recognized by Sengers (1998, 2000). Her argument is that *doing the right thing* (the classical approach of AI approaches to agent control) needs to be complemented by paying attention to *doing the thing right*, in particular creating believable transitions between agent behaviors.

Thus, going back to the topic of SIAs, how realistic do they need to be in appearance and behavior? How necessary is it that their architectures are psychologically and neurobiologically plausible models of social intelligence as it can be found in humans and other social animals? This debate of shallowness versus depth of modeling applies to all aspects of social intelligence, ranging from models of behavior generation, learning and memory,

---

[3]Children with autism have shown an impairment in using appropriate mental concepts when describing animations (Abell et al., 2000).

[4]For an excellent discussion of the many issues involved in anthropomorphism, see Mitchell et al. (1997).

communication and dialogue, to models of emotions (Cañamero, 2001). Can we rely on mechanisms of human fantasy or imagination that can virtually animate any object? Does the application domain matter (e.g., does it matter whether we face the agents in a consequence-free entertainment context, in a professional-office environment, in an e-commerce, or in a health care context)? It is likely that specific application areas require particular approaches toward SIA design and usage. Behavior that in a playful context might appear funny and entertaining might be judged annoying, pointless, or even insulting in a different context. Let us take the example of a person consulting a medical advice agent on possible choices for cancer treatment. It is unlikely that the person is making such inquiries for fun. Possibly it is due to a general or professional interest in the subject, but most likely advice is sought because the person or his or her friends/family are concerned and emotionally involved in this issue. How should such a medical advice agent look like and behave? For human beings, behaving appropriately in this situation does not necessarily come naturally, but requires medical and/or counseling training. Designing behavior and appearance of such agents (in addition to interface issues, etc.) provides a huge challenge. Few projects take up the challenge of using socially intelligent agents in therapy or personal and social education. Stacy Marsella's project "Carmen's Bright Ideas" is one exception. The project studies how animated agents can be useful in the area of providing counseling for mothers whose children undergo cancer treatment (Marsella et al., 2000). The European project Victec (Virtual ICT with Empathic Characters) is another example where empathy of users with virtual characters is a central concern (Victec, 2003). This project targets the area of personal and social education where virtual characters enact stories that school children can relate to in the context of bullying. The Aurora project gives an example of a project that investigates the use of social robots in autism therapy (Aurora, 2003).

Difficult as these projects are, they might provide important insights on what really matters in designing SIAs. Such scenarios are fundamentally different from entertainment contexts, where humans willingly accept impossible, absurd, insulting, or otherwise inappropriate or unrealistic behavior.

HOW REALISTIC? THE LIFELIKE AGENTS HYPOTHESIS

As discussed in the previous section, SIAs are often designed to imitate life. Based on what I previously called the *Lifelike agents hypothesis*, this approach can be characterized as follows:

> Artificial social agents (robotic or software) which are supposed to interact with humans are most successfully designed by imitating life, i.e. making the

agents mimic as closely as possible animals, in particular humans. This comprises both "shallow" approaches focusing on the presentation and believability of the agents, as well as "deep" architectures which attempt to model faithfully animal cognition and intelligence. (Dautenhahn, 1999, p. 359)

Generally it is argued that such lifelike agents are desirable because (a) the agents are supposed to act on behalf of or in collaboration with humans; they adopt roles and fulfill tasks normally done by humans, thus requiring human forms of (social) intelligence; (b) users prefer to interact ideally with other humans and less ideally with humanlike agents; thus, lifelike agents can naturally be integrated in human work and entertainment environments (e.g., as assistants or pets); or (c) such agents can serve as models for the scientific investigation of animal behavior and animal minds.

Argument (c) is certainly valid and need not be discussed here. However, Arguments (a) and (b) are not as straightforward as they seem. Lifelike agents that closely mimic human appearance or behavior can unnecessarily restrict and narrow the apparent and actual functionality of an agent. Similarly, imagine that mobile phones were designed so that they had the shape of old-fashioned dial-operated telephones. It then could be disturbing or at least puzzling for people to find out that the mobile phone might have more functionalities (e.g., sending and receiving e-mail, browsing the Web, etc.) than the original model. Thus, new designs not imitating any other previously existing object might better suit a piece of technology that is combining functionalities in a novel way or has new functionalities. To give another example, a social interface agent (e.g., in an e-commerce context) presented with humanoid appearance and behavior might evoke an initial feeling of familiarity in a human customer and elicit anthropomorphic tendencies. However, (a) human customers are then likely to expect the agent to show other human characteristics and functionalities, extensive social skills, personality and other traits of humans in general, and sales agents in particular (including that it understands jokes and possesses commonsense knowledge; Norman, 1997); and (b) new or different functionalities that the real agent does not possess need to be integrated in a plausible way in the agent's behavior without breaking the suspense of disbelief (Mateas, 1997; Nehaniv & Dautenhahn, 2000). When faced with a human in a department store, we might ask ourselves, "Who is this?" (customer? sales assistant? manager?). Yet we know clearly *what* the person is—namely, a member of the human species, which already allows us to make quite strong assumptions about his or her abilities, skills, and capacities. In contrast, a robot or software agent in the role of a sales assistant leaves us widely in the open about its skills and capacities. Can it talk? Can it understand English? Does it know what the color blue is? Thus, as suggested in Nehaniv and Dautenhahn (2000), it might be helpful if agents could clearly advertise what they are; what their skills, capacities, and limitations are; what they can and cannot do; and what their particular purpose is (e.g., who is the agent acting on behalf of? Is the agent legally liable?). Similar to how we interpret humans, using labels, batches, formal or informal clothing, dialects, or other signs might help, but often explicit introductions might work quicker and more efficiently.

Thus, what I suggest as a necessary requirement for progress in the field of SIAs is a systematic investigation of the following:

1. the extent to which agents should closely mimic human appearance and social behavior, rather than showing artificial appearance and behavior (cf. studies by McBreen & Jack [2001] on variations of how humanoid agents are presented and present themselves in e-commerce applications);

2. the extent to which lifelike agents need to possess explicit knowledge about social interaction and need architectures that model human social intelligence, as opposed to exploiting the natural human tendencies to anthropomorphize and animate the world;

3. possible mappings between the design space and niche space (Sloman, 1995) of SIA designs and architectures, on the one hand, and specific requirements in different application domains, on the other hand; and

4. the extent to which the sociocultural context or individual special needs of particular groups of users influence SIA design. Of vital importance are studies such as Takeuchi et al. (1998), which studied cultural differences in which U.S. and Japanese people reciprocate behaviors toward computers. They found that cultural differences regarding the emphasis on individuals as opposed to groups were reflected in the results. Such research shows that, instead of a general purpose SIA, we need to analyze particular requirements, and cultural differences are certainly among the most crucial.

In addition to important theoretical and conceptual work, empirical studies are vital to validate SIA theories and models, and to evaluate the performance of SIAs and how human users respond to them. I believe that comparative studies among different approaches, architectures, application domains, and models are particularly beneficial and necessary for systematically building up knowledge in the field for designing and understanding SIAs.

HOW SIMILAR TO US CAN AGENTS GET?

Will future humanoid SIAs be indistinguishable from us? What stages might lead to such agents with almost human social intelligence? The issue of how to distinguish naturally intelligent humans from artificially intelligent computer systems has a long tradition in AI. Particularly popular is the Turing Test (TT), by some perceived as the ultimate test for AI. The original formulation of the TT (Turing, 1950) was the following imitation game:

> It is played with three people, a man (A), a woman (B), and an interrogator (C) who may be of either sex. The interrogator stays in a room apart from the other two. The object of the game for the interrogator is to determine which of the other two is the man and which is the woman. He knows them by labels X and Y, and at the end of the game he says either "X is A and Y is B" or "X is B and Y is A." (p. 433)

To address the issue of machine intelligence, Turing then suggested a variation of this test—namely, having a machine taking the part of A in this game. The new question was then whether the interrogator will "decide wrongly as often when the game is played like this as he does when the game is played between a man and a woman?" (Turing, 1950, p. 433).

In subsequent years, the standard interpretation of the TT is to consider a scenario involving a human, a machine, and an interrogator, and the question of whether a machine could pass the test by communicating (traditionally in written format via typewriter or computer) with the interrogator indistinguishably from a human being. If, in a particular experimental setup over a limited period of time, the interrogator is not able to distinguish between the two candidates (machine and human), the machine is said to have passed the TT. The machine (computer program) is then either passing or failing the TT. Note that this scenario of text-based, symbolic communication, although not unrealistic (cf. pen-pals or e-mail-pals), substantially simplifies the process of natural human–human communication.

Interestingly, the TT for (artificial) intelligence is inherently a test for narrative intelligence—namely, presenting in an ongoing dialogue a believable story, a story that gives the appearance of being generated by a human. Although the TT can be dismissed as a trick, in the context of AI and intelligent machines, the TT can serve as an empirical criterion, setting the empirical goal to generate human-scale performance capacity (Harnad, 1992). Harnad (2001a, 2001b) extended the original TT scenario and proposed a TT hierarchy to discuss several degrees of indistinguishability instead of a yes/no evaluation. Note that each level subsumes the capacities shown at lower levels:

- T1: toy models of human total capacity
- T2: Total indistinguishability in symbolic (pen-pal) performance capacity (see standard interpretation of TT)
- T3: Total indistinguishability in robotic (including symbolic) performance capacity

- T4: Total indistinguishability in neural (including robotic) properties
- T5: Total physical indistinguishability

Surprisingly, this hierarchy is turning an evolutionary view on narrative intelligence upside-down (Dautenhahn, 2002): It implies that one might progress from level T1 to level T5—namely, by first building a convincing (narrative) pen-pal and then adding the physical matter or, to put it differently, by first building the (narrative) mind and then accounting for the situated, embodied, and experiential nature of the agent by building the body. One of the serious problems emerging from such a viewpoint is how meaning might arise, which is necessary to interpret, remember, and reconstruct stories in a socially adaptive sense.

Other candidates for a social intelligence test include the Prisoner's Dilemma (a simple two-person game originating in game theory), which is often used to demonstrate "social skills" of simulated agents. However, the simulations are often not addressing any important aspects relevant in real-life social interactions. An agent that is able to get a special discount while talking to a human on the phone and making a hotel room reservation might better address real-life social skills. An advanced version of such an agent, which can participate in a budget committee meeting and succeeds in getting the money, would be an example of a truly successful Machiavellian agent (cf. Sindermann, 1992).

More generally, Carley and Newell (1994) suggested a *Social Turing Test*:

> Construct a collection of artificial social agents according to the hypotheses about what makes agents social and put them in a social situation, as defined by the hypotheses. Then recognizably social behavior should result. Aspects not specified by the hypotheses, of which there will be many, can be determined at will. The behavior of the system can vary widely with such specification, but it should remain recognizably social. (p. 257)

Carley and Newell do not propose any concrete implementation of such a test, but offers a conceptual view of using such operational tests to investigate sets of hypotheses of what makes a social agent. Generally, it seems necessary to develop ways to measure, test, and assess social intelligence in agents.

WHO WE ARE: PRIMATE CULTURE

In this and the following sections, we look at some aspects of human cultures and their evolutionary origins. I argue that the evolution of human societies and cultures is interlinked with the evolution of cognitive capacities—capacities that put constraints on the way we can use and design SIAs and other interactive technologies.

Although millions of human beings live in large cities that (from an observer point of view) might resemble beehives or termite mounds, even (and in particular) under such circumstances individuality is at the heart of human nature. Humans are above all social beings who live in individualized societies.

The terms *anonymous* and *individualized* societies are used in biology to describe two different types of social organization (Dautenhahn, 1995). Social insects such as bees and termites are the most prominent example of anonymous (eusocial) societies, where group members usually do not recognize each other as individuals, but rather as group members. We do not observe bees or termites searching for missing members of their colony. Although individuals adopt specific roles in a colony, they do not show strong signs of individuality or personality.

The situation is quite different in individualized societies, among which primate societies (including humans) belong. Here we find complex recognition mechanisms of kin and group members. This gives rise to complex kinds of social interaction and the development of various forms of social relationships and networks. On the behavioral level, long-lasting social bonding, attachment, alliances, dynamic (not genetically determined) hierarchies, social learning, development of traditions, and so on are visible signs of individualized societies. In humans, the evolution of language, culture, and an elaborate cognitive system of mindreading and empathy are characteristics of human social intelligence in individualized societies (Dautenhahn, 1997). Humans are not only paying attention to other agents and their interactions individually (interactions between distinct personalities), but they use their mental capacities to reason about other agents and social interactions. It is at present unclear to what extent the social minds of members of other animal species—in particular, very social species like elephants, grey parrots, nonhuman apes, and cetaceans—are similar to or different from our own. Likewise the issues of cultural and memetic evolution is highly controversial. The concept of *memes*, first introduced by Dawkins (1976), comprises ideas, fashions, skills, and other components of *human* culture. Human culture and the memetic transmission of knowledge, ideas, and skills is often regarded as unique to human societies. According to Donald's (1993) discussion of the evolution of culture and cognition, modern humans have three systems of memory organization (mimetic skill, language, and external symbols) not available to our primate relatives, and these inventive capacities result in languages, gestures, social rituals, images, and so on. According to Tomasello (1999; Tomasello et al., 1993), cultural learning is a uniquely human form of social learning. Cultural learning requires three social-cognitive processes that emerge in human ontogeny: imitative learning,[5] instructed learning (teaching), and collaborative learning. Similarly, Blackmore (1999) argued that only sophisticated forms of imitation that are characteristic of humans, but not nonhuman primates, were a necessary prerequisite for memetic replication, which has led to human culture.

Others argue that culture as such is unlikely to be a feature unique to human societies and that the acquisition of novel behaviors in protocultures can be observed in nonhuman animals. To give an example, traditions have been observed among troops of Japanese macaque monkeys (Huffman, 1996): Japanese macaques or *Macaca fuscata* show several examples of the acquisition of innovative cultural behaviors (e.g., sweet potato washing and wheat washing was invented in 1953 by a young female and subsequently spread to older kin, siblings, playmates, and eventually other members of the troop). Other observed cultural behaviors are fish eating (as many newly acquired food sources initially spreading from peripheral males to adult females, then from older to younger individuals) and stone handling or stone play (initially spread only laterally among individuals of the same age). Subsequently, all these behaviors were passed down from older to younger individuals in successive generations (*tradition phase*). These examples clearly show the influence of social networks on the *transmission phase* of novel behavior: The nature of the behavior and social networks determine how the behaviors are initially transmitted depending on who is likely to be together in a certain context and therefore who is exposed to the novel behavior. Innovative behaviors of the kind described here have been independently observed at different sites. Various factors have been discussed that influence cultural transmission: environmental factors, gender, age, and other social and biological life history variables. For example, unlike potato or wheat washing, stone handling declines when individuals mature.

With respect to cultural transmission in nonhuman apes, recent evaluations of long-term field studies of chimpanzees or *Pan troglodytes* give compelling evidence for cultural behavioral variants (traditions) in different chimpanzee communities—data that cannot be explained by ecological differences of habitats alone. Evidence for chimpanzee culture comprises dozens of different behaviors, including tool usage, grooming, and courtship behaviors (Whiten et al., 1999; Whiten & Boesch, 2001).

Recent evidence about cultural evolution in cetaceans points toward animal cultures that parallel and might have evolved independently from human culture. Particularly convincing are studies on social learning in bottle-

---

[5]For an overview on imitation in animals and artifacts, see Nehaniv and Dautenhahn (2001) and Dautenhahn and Nehaniv (2002b).

nose dolphins (*Tursiops* spp.; Herman, 2002) and killer whales (*Orcinus orca*; Rendell & Whitehead, 2001).

Frans del Waal (1999) concluded his discussion on culture in primates: "The 'culture' label befits any species, such as the chimpanzee, in which one community can readily be distinguished from another by its unique suite of behavioral characteristics. Biologically speaking, humans have never been alone—now the same can be said of culture" (p. 636).

The striking similarity of cultural transmission of novel behavior exhibited by Japanese macaque monkeys or chimpanzees with human cultures questions the uniqueness of human societies. Note that culture has been observed in monkeys, who seem to be less skilled imitators than apes (Visalberghi & Fragaszy, 2002) and do not seem to possess higher level cognitive capacities necessary for complex social forms of primate politics shown by nonhuman apes and humans.

However, many nonhuman primates are good social learners (widely using nonimitative forms of social learning; e.g., stimulus enhancement or social facilitation). Reader and Laland (1999), therefore, argued that the meme concept can and should also be applied to cultural transmission among nonhuman animals. Animal societies can appear in various forms. Human societies, human culture, and human minds reflect in many ways their evolutionary origin in animal societies, animal culture, and animal minds. Considering human culture in an evolutionary context, linking it to precursors in nonhuman primate societies might help provide a better understanding of human culture. Thus, culture is not unique to the human species, although the kind of mechanisms that are necessary for and support culture in different animal species (e.g., cognitive mechanisms, language, imitation) might differ.

CULTURE AND THE BRAIN

Culture and other characteristics of human societies cannot be separated from specific environmental (including social) constraints and mental capacities, which evolved as adaptations for dealing with such constraints. Specific adaptations might have turned out to be prerequisites in the evolution of more sophisticated forms of primate societies and culture. New forms of media and technology seem to substantially expand the social life of humans. Yet the same mental capacities that constrained the evolution of the human social animal now pose cognitive limits on the complexity of our social life. Our primate social brain has a limit on the number of individuals with whom we can maintain direct social relationships—namely, relationships based on direct social knowledge (around 150) correlated with the relative size of the human neocortex (Barton & Dunbar, 1997; Dunbar,

1993). The figure can be identified consistently in various ancient and present human cultures. This number is significantly larger—namely, more than double the number observed in any population of nonhuman primates. It has also been suggested that language took on the new and more efficient role of social grouping (as opposed to physical grooming), being about 2.8 times more efficient. Not surprisingly, it can be shown that ideal human conversational groups more often than not consist of 3.8 participants (the speaker and two to three listeners).

As pointed out previously, biological evolution led to two distinctively different forms of social organization in animal societies (anonymous and individualized societies). It appears that individualized societies were a necessary (but not sufficient) prerequisite for the evolution of culture, providing a social environment that supported the evolution of complex forms of social learning (in particular, imitation). The capacity for phenotypic, cultural evolution seems correlated with particular mental capacities and social skills that facilitated the evolution of complex forms of primate societies and primate culture. Primate social behavior is well studied, but we know less about the social life and mental capacities of nonprimate species (crows, parrots, cetaceans, elephants, and others). However, when searching for animal culture, highly social animals in individualized societies are good candidates. Ants do not imitate, they do not learn from each other; primates do. Memes, as the replicators of culture, seem to require a social host, and memes are transmitted along social networks and depending on interactions in which its host is engaged. These seem to be the natural constraints under which culture is able to evolve in primate societies.

The "magic numbers" 150 and 3.8 discussed earlier indicate strong limitations and constraints for the future development of human societies. Systematic investigations that take these cognitive constraints into consideration could provide a basis for social agent technology that meets the cognitive demands of human primates.

We can speculate that unless drastic (technological) enhancements of human cognitive capacities are invented, these magic numbers of 150 and 3.8 pose cognitive limits and limitations on the way we can use agents or computers for supporting communication and interaction. These limits could possibly be exceeded by inventing new, more efficient ways of social grooming (exceeding the communicative capacities of language). However, these cognitive factors need to be considered in and during the design of SIAs.

The magic numbers have important implications for SIAs that are supposed to support and enhance human social intelligence. Different from our primate ancestors, for some humans whose real social networks are smaller than 150, the roles of friends and social partners might be filled by other partners, either human beings (e.g., actors in movies and soap operas, news presenters, or presenters of daily chat shows) or fictional characters such as Captain Picard in *Star Trek: The Next Generation* or Homer Simpson, including computer game characters such as Lara Croft in *Tomb Raider*. Although such social relationships are rather unidirectional (we bond with them, but they do not bond with us; they are not even aware of our existence), they might serve a similar role as real human networks (Dunbar, 1996). The boundaries between real and artificial are often nebulous or ambiguous (cf. interactions with chatbots or MUD robots in multi-user online environments—Foner, 2000; or a new generation of embodied conversational agents, e.g., software agents that might serve as estate agents—Bickmore & Cassell, 2000). Today interactive game software can create believable illusions that agents truly bond with their users (e.g., the Norns in the computer game *Creatures*; Grand et al., 1997; or robotic pets such as Furbies [Tiger Electronics, Ltd.] or Aibos [Sony Corporation] that are extending such acquaintances even to the physical level). However, these extensions of real social networks (Turkle, 1995) are not without limitations, constrained by the cognitive group size limit of 150 that characterizes human primates social networks. As Dunbar (1996) argued, modern information technology might change a number of characteristics of how, with whom, and with what speed we communicate, but not influence the size of social networks nor the necessities of direct personal contact that need to provide trust and credibility to social relationships. It often seems that language as a dominant means of communication in modern human societies can do remarkable things, "yet underlying it all are minds that are not infinitely flexible, whose cognitive predispositions are designed to handle the kinds of small-scale societies that have characterized all but the last minutes of our evolutionary history" (Dunbar, 1996, p. 207).

CONCLUSION: CHALLENGES AND CHANGES

One might question whether the future will look like many agent researchers tend to paint it—namely, a future with numerous and different types of robotic and software agents populating human-inhabited real and virtual spaces possibly indistinguishable from fellow humans. Such agents could be used in a variety of application areas, such as education, therapy, e-commerce, or entertainment (Dautenhahn et al., 2002). However, assuming that this might happen at some point in the future, one might ask: Is it a future that is desirable? Can and do we need to make an effort in shaping it?

First of all, aren't agents just another piece of technology and invention that will certainly change our lives, but that we are happy to accept, such as telephones, trains, cars, and so on? Is there anything special about agents? The argument presented in this chapter is saying that, indeed, this new emerging generation of agents is special. These agents are different from cars, trains, and typewriters because many of them have the primary, if not sole, purpose of being persuasive machines—machines that change our beliefs, attitudes, and behavior, agents that humans relate to and establish a relationship with. Pet robots and embodied conversational agents, just to give a few examples, embody representations of what we consider social or appropriate behavior—they convey meaning and values. Such agents are cultural artifacts, and they play a role in changing human society and culture as an active member of this society: They are social actors. It is because of this that they are different from tools or artworks that influence implicitly through usage and interpretation.

A possible danger of a future where most of our daily interactions might be with and mediated by artificial agents could be the loss of cultural diversity. Children in Tokyo watch the same cartoon movie and play with the same Aibo robot as children in Los Angeles, Sydney, and London (note that we use the *Aibo* as a metaphor for robot pets when in fact many owners of Aibos have been adults). Advantages of such a development are that children can in this way communicate cross-culturally about shared experiences. Disadvantages are that the diversity of traditional toys and patterns of play and interaction might get lost, being replaced by stereotypes that are being understood easily based on compromises and the lowest common denominator.

Biological evolution depends on the generation of diversity—the fact that offspring, no matter how similar they are to their parents, are usually not identical. Diverse populations are nature's pool for new ideas—new designs for body plans and adaptations. Similarly, I assume that cultural evolution depends on diversity too. A convergence of cultures and a mass extinction of local diversities is already underway (cf. the spread of Starbucks or McDonald's across the globe). If children everywhere in the world (or at least in many countries) wear the same clothes, eat the same food, listen to the same music, watch the same movies, use the same toys, learn with the same educational software, and play the same games, from where should new ideas come? Novelty is often not that novel in a strict sense; it is often a result of blending concepts and ideas from different domains. Similar to the mass extension of biological evolution that is underway and greatly due to human interference, a mass extinction of cultural diversity has already begun, and we might witness that agent technology might contribute to its acceleration.

How can any negative influences of SIAs on the evolution of human culture be counterbalanced? The answer is difficult, and naturally any suggestions that can be provided are speculative. However, a first useful step might be to identify the nature of agents that we would like to build. First, agents need to be explicitly perceived as culturally situated artifacts—as artifacts that convey meaning and values. Second, agent design as such needs to consider not only culturally adaptive agents, as suggested by O'Neill-Brown (1997), but agents that support cultural diversity—that embody meaning and values differently in different communities. Agent systems have already been used for people to express their values, facilitate communication, and establish a shared understanding (cf. Bers & Cassell, 2000). Yet it seems equally important for agent systems to support values and meaning that are specific to particular communities, supporting a diversity of understanding and identity rather than a shared understanding based on leveling out differences. Communication and shared understanding across cultures is important as long as these cultures still exist. Agents could play an important role in supporting cultural diversity. To give an example, mechanic toys produced for the mass market cannot adapt to local values and meaning. A computational or robotic agent that can learn and adapt its behavior and appearance, and that behaves autonomously as a persuasive machine, could embody values and meaning of particular communities and cultures. Different from the mechanical toy, it might actively contribute to the conservation of cultural diversity by addressing the cognitive and cultural needs of human primates.

EPILOGUE

I end this chapter with a *Gedanken experiment* resulting from discussions with Cynthia Breazeal in summer 1999[6]: We thought about a cross-cultural experiment with the humanoid robot, Kismet, mentioned earlier—a humanoid face robot developed at the MIT AI-Lab. Kismet uses a variety of vocal and facial expressions to engage humans in social interactions. We asked ourselves, what if Kismet can be culturally adaptive (i.e., can adapt the way it uses and expresses emotions) and can adapt its vocal intonations (if not vocabulary) to the particular sociocultural environment by which it is surrounded? What happens if four identical versions of Kismet are installed in public places (e.g., Tokyo, Edinburgh, San Francisco, and Hanover) equipped with software that can learn and adapt during interactions with people. Is a Californian Kismet more likely to be outgoing and enthusiastic in its emotional expressions? Is the British Kismet more likely to be reserved, but very polite? If a future Kismet possessed means to develop, then a variety of different Kismets could emerge. The different Kismets would act as a mirror, embodying and reflecting cultural characteristics acquired from their environments. The robots could even be sent to other

---

[6]Cynthia cannot be held responsible for the details of the picture portrayed in this chapter and that were not part of our discussions.

countries as robotic ambassadors (e.g., for children to playfully become aware of interesting cultural differences in interaction and communication).

Bringing these different robots together to a family gathering would certainly be interesting to watch. However, if the robots during this family gathering continue to adapt, they are likely to become identical in this melting pot. What is worse, if they are not able to go back in time and memorize, the acquired cultural adaptations are lost. However, if we freeze the robots' adaptation skills before the gathering, they might not be able to nicely interact or understand each other (as far as we can claim that any robot can show genuine understanding). What this chapter was concerned about is exactly this tension between (a) forces of convergence and adaptation, on the one hand (i.e., leveling out differences toward so-called *standards, favorites*, or *lowest common denominator*), and (b) the need for divergence—the creation and evolution of variations that can preserve (evolve, not freeze) cultural identities. Biology tells us that without variations populations are prone to extinction, in particular due to changes in the environment that might ask for novel characteristics that suddenly could turn out to be preadaptations. If humanity consisted of billions of clones of one single individual, extinction would occur rapidly probably due to a disease or any other kind of threat against which this particular individual (and all its copies) have no defensive mechanism. We can expect similar effects on culture if all human beings were biologically different, but culturally identical. A homogeneous culture is unlikely to create solutions to new problems, let alone its lack of evolvability (i.e., the potential for evolution).

I doubt that the agents of the future can be held responsible for future developments of human cultures, after all, there are many other factors that will impact the fate of human cultures on this planet (political, economical, geological, astronomical, etc.). However, if we do not recognize the potential impact of (socially intelligent) agents on culture, the way they could positively or negatively influence any such development, we might simply miss an opportunity to shape our future.

REFERENCES

Abell, F., Happe, F., & Frith, U. (2000). Do triangles play tricks? Attribution of mental states to animated shapes in normal and abnormal development. *Cognitive Development, 15,* 1–6.
Aronson, E. (1994). *The social animal.* New York: W.H. Freeman.
Aurora (2003). http://www.aurora-project.com, last accessed September 24, 2003.
Barton, R. A., & Dunbar, R. I. (1997). Evolution of the social brain. In A. Whiten & R. W. Byrne (Eds.), *Machiavellian intelligence II extensions and evaluations* (pp. 240–263). Cambridge: Cambridge University Press.
Bates, J. (1994). The role of emotion in believable agents. *Communications of the ACM, 37*(7), 122–125.

Berdichevsky, D., & Neunschwander, E. (1999). Toward and ethics of persuasive technology. *Communications of the ACM, 42*(5), 51–58.

Bers, M. U., & Cassell, J. (2000). Children as designers of interactive storytellers "let me tell you a story about myself." In K. Dautenhahn (Ed.), *Human cognition and social agent technology* (pp. 61–83). Amsterdam: John Benjamins.

Bickmore, T., & Cassell, J. (2000). "How about this weather?" Social dialog with embodied conversational agents (Technical report FS-00-04). In *Socially intelligent agents—The human in the loop* (pp. 4–8). Menlo Park, CA: AAAI Press.

Blackmore, S. (1999). *The meme machine.* Oxford: Oxford University Press.

Braitenberg, V. (1984). *Vehicles: Experiments in synthetic psychology.* Cambridge, MA: MIT Press.

Breazeal, C. (2002). *Designing sociable robots.* Cambridge, MA: MIT Press.

Breazeal, C., & Scassellati, B. (2000). Infant-like social interactions between a robot and a human caretaker. *Adaptive Behavior, 8*(1), 49–74.

Bryant, D. (2002). http://www.arclight.net/~pdb/glimpses/valley.html, last accessed September 24, 2003.

Bumby, K., & Dautenhahn, K. (1999, August). Investigating children's attitudes towards robots: A case study. In *Proc. CT99, The Third International Cognitive Technology Conference, San Francisco*, available at http://www.cogtech.org/CT99, pages 391–410, last accessed September 24, 2003.

Cañamero, L. C. (2001). Building emotional artifacts in social worlds: Challenges and perspectives (Technical report FS-01-02). In *Emotional and intelligent II: The tangled knot of social cognition* (pp. 22–30). AAAI Press.

Carley, K. M., & Newell, A. (1994). The nature of the social agent. *Journal of Mathematical Sociology, 19*(4), 221–262.

Cassell, J., Sullivan, J., Prevost, S., & Churchill, E. (Eds.). (2000). *Embodied conversational agents.* Cambridge, MA: MIT Press.

Dautenhahn, K. (1995). Getting to know each other—artificial social intelligence for autonomous robots. *Robotics and Autonomous Systems, 16*, 333–356.

Dautenhahn, K. (1997). I could be you—the phenomenological dimension of social understanding. *Cybernetics and Systems, 25*(8), 417–453.

Dautenhahn, K. (1998). The art of designing socially intelligent agents: Science, fiction and the human in the loop. *Applied Artificial Intelligence Journal, Special Issue on Socially Intelligent Agents, 12*(7–8), 573–617.

Dautenhahn, K. (1999, August). Robots as social actors: Aurora and the case of autism. In *Proc. CT99, The Third International Cognitive Technology Conference, San Francisco*, available at http://www.cogtech.org/CT99, pages 359–374.

Dautenhahn, K. (2002). The origins of narrative—in search for the transactional format of narratives in humans and other social animals. *International Journal of Cognition and Technology: Co-existence, Convergence, Co-evolution (IJCT), 1*(1), 97–123.

Dautenhahn, K., Bond, A., Cañamero, L. C., & Edmonds, B. (Eds.). (2002). *Socially intelligent agents—Creating relationships with computers and robots.* Norwell, MA: Kluwer Academic Publishers.

Dautenhahn, K., & Nehaniv, C. L. (2002a). An agent-based perspective on imitation. In K. Dautenhahn & C. L. Nehaniv (Eds.), *Imitation in animals and artifacts* (pp.). Cambridge, MA: MIT Press.

Dautenhahn, K., & Nehaniv, C. L. (Eds.). (2002b). *Imitation in animals and artifacts.* Cambridge, MA: MIT Press.

Dautenhahn, K., Ogden, B., & Quick, T. (2002). From embodied to socially embedded agents—Implications for interaction-aware robot. *Cognitive Systems Research, 3*(3), 397–428.

Dawkins, R. (1976). *The selfish gene.* Oxford: Oxford University Press.

de Waal, F. B. (1999). Cultural primatology comes of age. *Nature, 399*, 635–636.

Donald, M. (1993). Precis of origins of the modern mind: Three stages in the evolution of culture and cognition. *Behavioral and Brain Sciences, 16,* 737–791.
Dunbar, R. I. M. (1993). Coevolution of neocortical size, group size and language in humans. *Behavioral and Brain Sciences, 16,* 681–735.
Dunbar, R. I. M. (1996). *Grooming, gossip and the evolution of language.* London: Faber & Faber.
Eddy, T. J., Gallup, G. G., & Povinelli, D. J. (1993). Attribution of cognitive states to animals: Anthropomorphism in comparative perspective. *Journal of Social Issue, 49*(1), 87–101.
Edmonds, B. (2000). Developing agents who can relate to us—putting agents in the loop via situated self-creation (Technical report FS-00-04). In *Socially intelligent agents—The human in the loop* (pp. 55–60). Menlo Park, CA: AAAI Press.
Erickson, T. (1997). Designing agents as if people mattered. In J. M. Bradshaw (Ed.), *Software agents* (pp. 79–96). Menlo Park, CA: AAAI Press/The MIT Press.
Fogg, B. J. (1999). Introduction: Persuasive technologies. *Communications of the ACM, 42*(5), 27–29.
Foner, L. N. (2000). Are we having fun yet? Using social agents in social domains. In K. Dautenhahn (Ed.), *Human cognition and social agent technology* (pp. 323–348). Amsterdam: John Benjamins.
Grand, S., Cliff, D., & Malhotra, A. (1997, February). Creatures: Artificial life autonomous software agents for home entertainment. In *Proc. First International Conference on Autonomous Agents (Agents '97), Marina del Rey,* ACM.
Harnad, S. (1992, October). The turing test is not a trick: Turing indistinguishability is a scientific criterion. *SIGART Bulletin, 3*(4), 9–10.
Harnad, S. (2001a). Turing indistinguishability and the blind watchmaker. In J. Fetzer & G. Mulhauser (Eds.), *Evolving consciousness* (pp. 3–18). Amsterdam: John Benjamins.
Harnad, S. (2001b). Minds, machines and Turing: The indistinguishability of indistinguishables. *Journal of Logic, Language, and Information, 9*(4), 425–445.
Heider, F., & Simmel, M. (1944). An experimental study of apparent behavior. *American Journal of Psychology, 57,* 243–259.
Herman, L. M. (2002). Vocal, social, and self imitation by bottlenosed dolphins. In K. Dautenhahn & C. L. Nehaniv (Eds.), *Imitation in animals and artifacts* (pp. 63–108). Cambridge, MA: MIT Press.
Huffman, M. A. (1996). Acquisition of innovative cultural behaviors in nonhuman primates: A case study of stone handling, a socially transmitted behavior in japanese macaques. In C. M. Heyes (Ed.), *Social learning in animals* (pp. 267–289). New York: Academic Press.
Langton, C. G. (1989). Artificial life. In C. G. Langton (Ed.), *Proc. of an Interdisciplinary Workshop on the Synthesis and Simulation of Living Systems* (pp. 1–47). Los Alamos, New Mexico.
Levy, S. (1992). *Artificial life, the quest for a new creation.* London: Penguin.
Lieberman, H., & Shneiderman, B. (2001). *Your wish is my command: Programming by example.* San Francisco: Morgan Kaufmann.
Marsella, S. C., Johnson, W. L., & LaBore, C. (2000, June 3–7). Interactive pedagogical drama. In *Proceedings of the Fourth International Conference on Autonomous Agents* (pp. 301–308). Barcelona, Spain: ACM Press.
Mateas, M. (1997). Computational subjectivity in virtual world avatars (Technical report FS-97-02). In *Socially intelligent agents* (pp. 43–45). Menlo Park, CA: AAAI Press.
McBreen, H., & Jack, M. (2001). Evaluating humanoid synthetic agents in e-retail applications. *IEEE Transactions on Systems, Man, and Cybernetics—Part A: Systems and Humans, 31*(1), 394–405.
McCloud, S. (1993). *Understanding comics: The invisible art.* New York: HarperCollins.
Mitchell, R. W., & Hamm, M. (1997). The interpretation of animal psychology: Anthropomorphism or behavior reading? *Behavior, 134,* 173–204.
Mitchell, R. W., Thompson, N. S., & Miles, H. L. (Eds.). (1997). *Anthropomorphism, anecdotes, and animals.* Albany, NY: State University of New York Press.

Montgomery, D. E., & Montgomery, D. A. (1999). The influence of movement and outcome on young children's attribution of intention. *British Journal of Developmental Psychology, 17,* 245–261.

Mori, M. (1982). *The Buddha in the robot.* Boston, MA: Charles E. Tuttle.

Nehaniv, C. L., & Dautenhahn, K. (2000). Living with socially intelligent agents: A cognitive technology view. In K. Dautenhahn (Ed.), *Human cognition and social agent technology* (pp. 415–426). Amsterdam: John Benjamins.

Nehaniv, C. L., & Dautenhahn, K. (Eds.). (2001). *Cybernetics and systems, special issue on imitation in natural and artificial systems, 32*(1–2).

Norman, D. A. (1997). How might people interact with agents. In J. M. Bradshaw (Ed.), *Software agents* (pp. 49–55). Menlo Park, CA: AAAI Press/The MIT Press.

Oatley, K., & Yuill, N. (1985). Perception of personal and interpersonal action in a cartoon film. *British Journal of Social Psychology, 24,* 115–124.

O'Neill-Brown, P. (1997). Setting the stage for a culturally adaptive agent (Technical report FS-97-02). In *Socially intelligent agents* (pp. 93–97). Menlo Park, CA: AAAI Press.

Persson, P., Laaksolahti, J., & Lonnqvist, P. (2001). Learning and interacting in human-robot domains. *IEEE Transactions on Systems, Man, and Cybernetics, Part A, 31*(5), 349–360.

Reader, S. M., & Laland, K. N. (1999). Do animals have memes? *Journal of Memetics—Evolutionary Models of Information Transmission, 3*(2), http://jom-emit.cfpm.org/1999/vol3/reader_sm&laland_kn.html, last accessed September 24, 2003.

Reeves, B., & Nass, C. (1996). *The media equation.* Cambridge: Cambridge University Press.

Reichardt, J. (1978). *Robots: Fact, fiction and prediction.* London: Thames & Hudson.

Rendell, L., & Whitehead, H. (2001). Culture in whales and dolphins. *Behavioral and Brain Sciences, 24*(2), 309–382.

Rimé, B., Boulanger, B., Laubin, P., Richir, M., & Stroobants, K. (1985). The perception of interpersonal emotions originated by patterns of movement. *Motivation and Emotion, 9,* 241–260.

Scassellati, B. (1999). Imitation and mechanisms of joint attention: A developmental structure for building social skills. In C. L. Nehaniv (Ed.), *Computation for metaphors, analogy and agents* (pp. 176–195; Springer Lecture Notes in Artificial Intelligence, Volume 1562). Heidelberg: Springer.

Scassellati, B. (2001). *Foundations for a theory of mind for a humanoid robot.* Unpublished doctoral dissertation, MIT Department of Computer Science and Electrical Engineering.

Sengers, P. (1998, May 9–13). Do the thing right: An architecture for action-expression. In K. P. Sycara & M. Wooldrige (Eds.), *Proc. of the Second International Conference on Autonomous Agents* (pp. 24–31). Minneapolis/St. Paul: ACM Press.

Sengers, P. (2000). Narrative intelligence. In K. Dautenhahn (Ed.), *Human cognition and social agent technology.* Amsterdam: John Benjamins.

Sindermann, C. J. (1982). *Winning the games scientists play.* New York, London: Plenum.

Sloman, A. (1997). What sort of control system is able to have a personality. In R. Trappl & P. Petta (Eds.), *Creating personalities for synthetic actors.* Heidelberg: Springer.

Takeuchi, Y., Katagiri, Y., Nass, C. I., & Fogg, F. J. (1998). Social response and cultural dependency in human–computer interaction. In *Issues in cross cultural communication—Towards culturally situated agents* (pp. 114–123). Workshop Proc. PRICAI'98, National University of Singapore.

Tomasello, M. (1999). *The cultural origins of human cognition.* Cambridge, MA: Harvard University Press.

Tomasello, M., Kruger, A., & Ratner, H. (1993). Cultural learning. *Behavioral and Brain Sciences, 16*(3), 495–552.

Tseng, S., & Fogg, B. J. (1999). Credibility and computing technology. *Communications of the ACM, 42*(5), 39–44.

Turing, A. (1950). Computing machinery and intelligence. *Mind, 59*(236), 433–460.

Turkle, S. (1995). *Life on the screen, identity in the age of the internet.* New York: Simon & Schuster.

Victec (2003). http://www.victec.org, last accessed September 24, 2003.

Visalberghi, E., & Fragaszy, D. (2002). Do monkeys ape?—ten years after. In K. Dautenhahn & C. L. Nehaniv (Eds.), *Imitation in animals and artifacts* (pp. 471–499). Cambridge, MA: MIT Press.

Watt, S. N. K. (1997). *Seeing things as people: Anthropomorphism and common-sense psychology.* Unpublished doctoral dissertation, Department of Psychology, The Open University.

Whiten, A., & Boesch, C. (2001, January). The cultures of chimpanzees. *Scientific American*, pp. 61–67.

Whiten, A., Goodall, J., McGrew, W., Nishida, T., Reynolds, V., Sugiyama, Y., Tutin, C., Wrangham, R., & Boesch, C. (1999). Cultures in chimpanzees. *Nature, 399,* 682–685.

PART

II

DESIGN FOR CROSS-CULTURAL BELIEVABILITY

CHAPTER

4

Transcultural Believability in Embodied Agents: A Matter of Consistent Adaptation

Fiorella de Rosis
University of Bari

Catherine Pelachaud
University of Roma–"La Sapienza"

Isabella Poggi
University of Roma–Tre

In this chapter, we propose some reflections on how an embodied animated agent might be designed so as to adapt its behavior to the cultural context to which it applies. We start from a discussion of the meaning of the term *culture* to then analyze the literature findings about the way human beings' behavior (natural language expression, affect feeling and display, verbal and nonverbal components of their communication, etc.) varies according to the culture. The description of a context-adaptable embodied animated agent is a departure point to suggest how adaptation might be extended to cultural factors. Finally, the problem of how to ensure that the agent behavior does not lose in consistency while acquiring adaptation abilities is examined: Consistency is considered, at present, as an essential constituent of agent believability and should therefore guide the setting of values for adaptation parameters.

SOME BASIC QUESTIONS

What Do We Mean by Culture?

Several definitions of *culture* may be found in the literature. To Gudykunst and Kim (1992), culture is "a theory for interpreting the world and knowing how to behave." To Hall and Hall (1990), it is a term used by anthropologists to refer to "a system for creating, sending, storing and processing information developed by human beings, which differentiates them from other life forms." To Hofstede (1980), "culture is, to human collectivity, what personality is to the individual." This author analyzed how patterns of acting, feeling, and thinking are ingrained in people by late childhood and how differences in these cultural patterns are displayed in the choice of symbols, rituals, and values by a culture. After a large survey on 53 countries, he defined five dimensions of culture, which include:

- *Short-term versus long-term orientation* (emphasizing practical values vs. focusing on truth and certainty of beliefs);
- *Femininity versus masculinity* (blurring the lines between gender roles vs. displaying traditional differences in how age, gender, and family are viewed);
- *Power-distance* (emphasizing/deemphasizing equality among social and age groupings);
- *Collectivism versus individualism* (level of integration of people into strong groups that protect them in exchange for unbridled loyalty); and
- *Uncertainty avoidance* (level of tolerance for ambiguity).

To Brislin (1993), culture "consists of ideals, values and assumptions about life that are widely shared among people and that guide specific behaviors." To Brown and Nichols-English (1999), "cultural traditions are characterized by the values, beliefs, attitudes, practices, customs and behaviors of a group of people" (p. 61).

We would define *culture* as "a set of beliefs, shared by a population, regarding the environment in which the population lives and the best techniques to reach the biological terminal goals in that environment, given the means–end relations that hold in that physical environment and the accumulated set of beliefs." Culture also includes knowledge about the way beliefs may be gathered and organized and about the norms and values that establish how to best achieve those goals. Let us justify this definition.

Every action in our life is part of a plan aiming at some goal. For instance, Somali shepherds are daily in search of bush to get food, while Italian housewives go out and buy pasta and tomatoes to prepare lunch. The goals of our everyday plans are not ends in themselves; they all aim in turn at more general goals of humans: the biological goals of survival and reproduction, and some subgoals like physical well-being, safety, loving and being loved, self-realization, identity, and image and self-image. These are *terminal goals*, counting as ends in themselves; they are our most important goals, the ones with the highest weight. The goals of our everyday plans are *instrumental goals*, which directly or indirectly serve our terminal ones. For instance, searching the bush or buying pasta and tomatoes are means for survival, whereas cooking delicious spaghetti is a means for having a positive image as a good cook.

How do we choose the goals that are means to our terminal goals? This is the job of learning, tradition, and culture; this is how culture is linked to goals. Humans pursue their goals by using their internal and external resources. External resources are the objective conditions holding in the environment (presence of food, characteristics of the territory, climate conditions, etc.); internal resources are the human action capacities (physical strength, body agility, manual skill) and beliefs. Human beliefs and their processing ways are necessary in the pursuit of goals at different levels: in monitoring the presence–absence of favorable external conditions, ordering goals according to their values to choose which of them to pursue, and storing knowledge about the available or best means for achieving a given goal.

Of course in different environments, the physical conditions, the most easily available resources, and, consequently, the actions to get them are different. For instance, in the interior it is easier to get food by rearing sheep or cows, whereas on the coast fishing is the most direct way to feed. So in the interior the main beliefs to store and process and the main actions to learn concern sheep and cows rather than fish or shrimps, and people are more likely to become shepherds than fishermen. In summary, any population, given the environment in which it lives, accumulates a set of beliefs on the instrumental goals that most easily and economically serve the biological terminal goals in that environment. An instrumental goal then becomes more or less important in a culture depending on the strength of its link to a terminal goal. To the extent to which that instrumental goal is the only possible means to reach a terminal goal in that culture, that instrumental goal will receive a higher weight and alternative goals will be dropped. It then becomes a strategy of survival typical of that culture, and culture overall may be defined as a set of beliefs on the most typical techniques to pursue goals.

These beliefs are determined by beliefs about the environment. For example, as long as a population does not know the mechanisms of plant reproduction, agriculture cannot be chosen as an instrumental goal for survival. Therefore, culture entails beliefs about the external world. Because language is driven by beliefs and is, at the same time, their vehicle, culture typically shows up in it. Language is, on one side, an image of a population's beliefs; on the other side, it is a way to organize them—a set of rules on how to conceptualize and categorize information. Consequently, culture implies an outfit of typical communication techniques—that is, of settled instrumental goals stating how to convey information.

Culture also entails values and norms. *Values* are evaluations about what is good and bad, and therefore about what should be pursued as a goal. Because particular behaviors may be good or bad in a given environment, again according to the most effective survival techniques, different populations may hold different values. For example, in environments in which individualistic behavior proved to be advantageous, individualistic values develop, whereas in environments where collectivistic behavior is more fit, values centered on the family or group hold.

Norms are obligations that rule the relationships among people in a group. Again in a culture more centered on interdependency, a highly weighted goal and then a norm that will hold may prescribe to be cooperative even when this implies intruding in other persons' affairs. On the contrary, in a culture more centered on the individual's independence, keeping one's privacy is more weighted, and a norm holds of not intruding in others' affairs and contrasting others' intrusions.

Finally, both values and norms generate goals in people (the goal to pursue that value or respect that norm). If these goals are threatened, they provoke emotions. Violating one's own values may induce *shame*, whereas violating norms may cause a sense of *guilt*. Therefore, if two populations have different values and norms, they will also feel these emotions as a consequence of different events.

In the next part of this chapter, we discuss which differences in communication styles may be due to cultural differences. We try to answer the following questions: What is universal (biological) and what is culturally determined in the different aspects of communication that we simulate in a believable agent? What is universal and what is cultural, in particular, in the verbal communication, in the body language, and, specifically, in gesture, gaze, facial expression, body posture, and proxemic behavior? Is emotion expression more likely to be universally shared than information about the world?

How Does Culture Manifest Itself? Semantic Versus Interactional Rules in Communication Systems

The sociocultural context affects the behavior of individuals living in that context, and therefore also their expectations about how other individuals in the same context should or would behave. Although there is a common core of behaviors that respond to universal laws, several aspects are culture-dependent. According to Samovar and Porter (1972): "culture manifests itself both in patterns of language and thought and in forms of activity and behavior. These patterns become models for common adaptive acts and styles of expressive behaviors which enable people to live in a society within a given geographic environment at a given state of technical devel-

4. TRANSCULTURAL BELIEVABILITY 79

opment." Communication style is therefore one of the main aspects of behavior that are influenced by culture. To Condon and Yousef (1975): "we cannot separate culture from communication, for as soon as we start to talk about one we are almost inevitably talking about the other." If we wish our artificial agent to be believable for the particular cultural community in which it will be employed, we should design it so that both the reasoning style and the communication forms it adopts are an expression of the norms and values adopted by that community.

In every communication system, two kinds of rules are specified: semantic and interactional. *Semantic rules* define a correspondence between signals and meanings. For instance:

> if you want to communicate the meaning "I greet you," say "Hello"

A verbal language has several complex rules of this type: It is linguists' job to discover them. A communication system based on gesture or gaze also has its own lexical rules. For instance, in the system of gaze, the previous rule becomes:

> if you want to communicate the meaning "I greet you," raise your eyebrows

whereas in the system of gesture, it becomes:

> if you want to communicate the meaning "I greet you," wave your hand

In addition to semantic rules, we also have *interactional rules*, which do not state *how* some meaning should be conveyed, but *if* some meaning can, should, or should not be conveyed in a given situation. The following are two examples of these rules:

> if you meet a person you know, apply the rule for the meaning "I greet you"
>
> if you meet an unknown person, do not apply the rule for the meaning "I greet you"

In other words, a communication system tells you, on one side, what to do if you want to convey some meaning, and, on the other side, whether it is prescribed, accepted, or forbidden to convey such a meaning. The latter are the norms of use in a communication system, the so-called *norms of appropriateness*, which have been traditionally studied by sociolinguists.

We introduced this distinction because we think that, across communication systems, the two types of rules may or may not be subject to cultural variation. Typically, the facial expression of emotions might be universal from the point of view of semantic rules, but culturally determined on the side of interaction rules. In the verbal expression of emotions, both the semantic and interactional sides are subject to high cultural variation.

Yet how do the norms of use generally work? Some of these prescriptions are of a global type. For example, in some cultures, the use of gestures is sanctioned in general because they are considered a quite primitive or socially low-level way of communicating: The consequence is that gestures occur less frequently in communicative interaction and tend to be less varied and less conspicuous. For instance, British people gesticulate less than Neapolitan people, the repertoire of gestures they use is poorer, handshapes are less frequent, and the amplitude of gestures is narrower (Kendon, 1995).

In addition to these norms that globally influence the communication system, we may categorize the norms of use according to whether they define the meanings conveyed by signals or the physical performance of signals. In the first category, we may include norms of language interdiction: Using a particular signal may be forbidden because the meaning it conveys is tabooed. Hence, obscene words may be more tabooed in a traditional society than in one with fewer inhibitions. Therefore, as for obscene words, cultural variations hold not only on how a given meaning is conveyed (semantic rule), but also on whether that meaning may be conveyed (interactional rule—norm of use). In communication systems different from the verbal one, cultural variations hold, in our view, not so much about semantic rules as about interactional rules. Take gaze, for instance: We claim that a loving gaze, a seductive gaze, or a gaze looking down at someone are performed in the same way and have the same meanings all over the world. Yet, according to cultural norms, if looking down at someone is considered impolite or clearly showing love to someone is seen as incorrect or even obscene in some culture, then in that culture those types of gaze are tabooed, and they will not be performed easily.

What Is Universal and What Is Cultural in Emotion Feeling and Expression?

It is almost universally acknowledged that emotions have a social role. They are a means to coordinate social interactions and relationships and are dynamic processes that mediate the individual's relation to a continuously changing social environment (Keltner & Haidt, 1999). Research by anthropologists and psychologists has focused on how emotions help collections of interacting individuals who share common identities and goals to meet their shared goals or the superordinate goals of the group. As a consequence, emotions are claimed to play a critical role in the processes by which individuals assume cultural identities.

Some differences across cultures have been found on how emotions are manifested in language. For instance, affective lexicons differ across cultures. However, the absence of a name (and, possibly, of the concept) for some emotions in some cultures holds only for quite specific emotions, and not for the so-called *basic* emotions. Happiness, sadness, anger, fear, surprise, and disgust seem to be referred to in all cultures, and when they are felt a neural program for their facial expression is triggered; this program is wired in the human species and is therefore universal too (Ekman & Friesen, 1969). The fact that basic emotions seem to be universal does not necessarily imply that people in different cultures always show their emotions in the same way. For instance, Bagozzi et al. (1999) classified cultures as *independent* and *interdependent-based* (a factor similar to Hofstede's individualism vs. collectivism). They assigned Western cultures to the first class because these cultures "place considerable importance on self knowledge of one's attributes and inner psychological processes in general." Cultures in the second class (that they also call *Confucian-based*) tend to give importance, on the contrary, to things outside the individual. These authors described the result of an experimental study that shows people in the two classes of cultures experience positive and negative affect in different ways: Independent cultures tend to not associate negative and positive emotions, whereas interdependent-based cultures conceive of affect in characteristic dialectical ways and tend to associate positive and negative ones—to externalize them less and attach a lower social role to them.

Two cultural factors may intervene to produce differences in the emotion expression. The first one is that an emotion is triggered by the cognitive categorization of a situation on the part of the subject. A situation that in a culture—because of its beliefs, norms, and values—is categorized as a cause of sadness, in another culture—with different beliefs, norms, and values—might be categorized as a cause of happiness. For example, the death of a beloved person in a Catholic group may be seen as his getting back to Father; the death of a martyr in the Islamic culture may be greeted as a joyful event. Therefore, the intensity with which an emotion is shown by an individual in a given context is most likely influenced by the system of beliefs and goal weights established by people in that context. In different contexts, the same event can be interpreted in different ways and can elicit a different mixture of emotions, each with its own intensity. The second reason that the expression of emotions may vary culturally is that, in different cultures, the filter that decides whether and how to express the felt emotion is different. This filter may be represented in an agent by a set of factors, such as the cognitive and personality traits of the agent and the Interlocutor, their relationship, and the situation in which they interact. These factors also include the cultural norms about the expression/nonexpression of given emotions. In 1969, Ekman and Friesen introduced the term *display rules* to express how the display of facial expression was managed by sociocultural rules. They hypothesized that an emotion may be felt but not shown due to some circumstances, and they tested this idea in an experiment involving two cultures: American and Japanese. Subjects from both cultures were shown films inducing different emotions (stress or neutral) under two conditions: with or without an authority figure on their side while they watched the movies. When watching the movies alone, subjects from both cultures showed the same type of facial expressions. However, when an authoritative figure entered the room, Japanese subjects masked their negative expressions by positive ones (smiles); some American subjects did the same, but at a much lesser degree. This study proved that the difference in showing some expressions was not due to differences in emotion feeling, but rather in differences in their display due to cultural differences. Therefore, consistent evidence from different studies seems to support the idea that there is a universal meaning of some facial expressions, but there are cultural differences in the usage of these expressions (Knapp & Hall, 1997).

What Is Universal and What Is Cultural in Verbal Communication?

In verbal languages, words and syntactic rules are typically and obviously different across languages (Eibl-Eibesfeldt, 1974). Of course at a deep level the syntax is universal (as it is argued by the Chomskian universal grammar theory). However, specific syntactic rules exist in every language, and finding the path that links them to the underlying universal structure of grammars is not straightforward (Poggi, in press). Even if this issue is quite interesting from a theoretical viewpoint, it is not central for our task of simulating an agent's behavior. Some universally recurring mechanisms may be found in iconicity, which holds not only in single words (say, onomatopoetic words and phonosymbolism, beautifully simulated in Cassell's Gandalf), but also in some syntactic mechanisms that are employed in narratives to put events in a temporal sequence, like the "ordo naturalis." Apart from this, cultural variation across languages holds at various levels: from the more linear, hierarchical, or circular structure of planning to the importance given to politeness or rhetoric, to the rules that define what to talk about, how much to mention the self in communication, whether to convey new information or just how to make reference to shared knowledge, and so on.

An example in the medical domain follows. In an article aimed at educating the pharmacists about cultural issues associated with the provision of pharmaceutical care services, Brown and Nichols-English (1999) claimed that "All cultures have a system of health beliefs that explain the cause of illness, what to do about it and who should be involved" (p. 64). According to the experience of these authors, culture affects not only beliefs about illness etiology and treatment practices, but also the patients' acceptance and adherence to recommended treatments.[1] Therefore, the authors suggest that health care providers identify the important and salient beliefs and expectations of their patients to adapt those factors of communication that risk to impede the recommended behavior. We return to this particular example later to show how this type of adaptation might be implemented.

What Is Universal and What Is Cultural in Gesture?

Gesturing is considered inappropriate in some cultures. As a consequence, the degree and mix of verbal and nonverbal communication employed in a conversation are reflective of culture and language. Although some authors claim that the effect of differences in these constraints are not so strong as one might think (Cassell, 2000), some differences in the amount of gesturing have been found.

As far as the quality of gesturing is concerned, the issue of universally shared versus culture-specific signals is particularly tricky because different types of gestures exist. Gestures may be categorized according to their cognitive construction—that is, according to whether and how they are represented in memory (Magno & Poggi, 1995). Some gestures are "coded"—that is, they are represented in memory just as a lexicon, with a rule that links the signal to its meaning. Gestures of this kind are, for example, the "symbolic" ones (like the gesture for "ok"), but also the beats, which scan the syntactic structure of the sentence or emphasize parts of the discourse (Poggi, 2002). Other gestures are "creative": They are invented on the spot to illustrate the discourse; in this case, the correspondence between signal and meaning is not univocal. For these gestures, a small set of inference rules are represented in the people's mind, through which senders may create a gesture for a new meaning to convey, and addressees may understand the meaning of a gesture never seen before. Examples of creative gestures are the iconic and metaphoric gestures describing concrete and abstract objects and actions (Cassell, 2000; McNeill, 1992) and the inference rules to represent forms, actions, or locations of the referent. A "cello," for instance,

---

[1]As an example: "In the African-American community, expressions such as 'high-blood', 'low-blood' and 'hyper-tension' have all been connected to folk-based models of hypertension, and those who believe in these folk-based interpretations are more likely to treat their blood pressure with home or folk remedies and less likely to be compliant with prescribed antihypertensive medications" (Brown & Nichols-English, 1999, p. 64).

may be represented by depicting the form of a cello in the air or by moving hands as if a cello were being played.

Among coded gestures, some are culturally coded (e.g., the gesture for ok, the Churchill gesture for Victory). Others may be biologically coded (e.g., those that are a ritualization of physiological movements, like the gesture of raising fists up to show elation). In a culture-sensitive agent, the former has to be varied from one culture to another, whereas the latter might be the same regardless of the agent's culture. As far as creative gestures are concerned, as we mentioned, the only representations that are coded are general inference rules of similarity between a possible gesture and a represented referent. In our hypothesis, the majority of these rules are probably universal because they simply consist of imitating objects or actions. When a represented referent is typical of a culture or an action is performed in a typical way, the corresponding creative gestures are also culturally dependent. Further, the form of metaphoric gestures (such as a rolling gesture to indicate an ongoing process) differs among language communities (Cassell, 2000). Prosodic features of language and gestures (such as nods and gazes) vary in amount, duration, and timing (O'Neill-Brown, 1997). Pointing your finger in the air may be used to summon a waitperson in Vienna, whereas in Brussels it is impolite. Disagreement is indicated by shaking the head from side to side in Northern Italy, whereas nodding should be done up and down in Bulgaria and down to up in Southern Italy.

What Is Universal and What Is Cultural in Facial Expressions?

Facial expression and gaze are the parts of communication that are more likely to be universally shared; they are both devoted to communicate various types of information. First of all, information about the world: We may point at things with chin or gaze, and we may squeeze eyes to mean that something is little or difficult. Second, information on the speaker's beliefs, goals, and emotions: We raise eyes up to signal that we are trying to remember or make inferences; with different kinds of frowns, we may communicate that we are angry, we are paying attention, or we are giving an order. Finally, information on the speaker's identity: Faces convey important social information about who you are and what you are thinking through their structure, dynamics, and decorations. From the overall face aspect, viewers assess personality and make more or less approximate evaluations of age, gender, and ethnicity. Decorations such as eyeglasses, cosmetics, and hairstyle provide more or less subtle cultural cues. Based on the cues we see in a person's face, we may also make judgments about his or her character and personality (Donath, 2001). The literature on face perception demonstrates that the face forms the locus of many of our stereotypes, prejudices, and cultural values. Attractive faces tend to give the impression of social competence; babyfaced people are perceived to be more submissive, naive, honest, and warm; whereas male faces are more likely to be perceived as belonging to an expert (Branham, 2001).

Throughout his research, Ekman (1999) pursued the idea of universal facial expressions: He and his colleagues conducted several experiments to demonstrate this idea. Experiments consisted of showing photos of faces to subjects from a vast number of countries (Africa, South America, Europe, and Asia, including remote countries having no contact with other cultures). Subjects were asked to match photos with short stories. Prototype facial expressions were observed for six emotions (anger, disgust, fear, happiness, sadness, and surprise), the so-called *basic* emotions. Later on Keltner (1995) added the expression of embarrassment to this list. Ekman and his colleagues conducted further studies, in which the subjects could choose their own words to denote their emotions by selecting a term from a list of emotion names. Although they had the possibility of choosing their own words, the subjects nevertheless selected emotion terms similar to those that had been employed by researchers in the previous studies (Ekman, 1979).

If, therefore, at least the facial expression of basic emotions seems to be universal, cultural differences may be found in the amount of gaze allowed in social interaction. In some cultures (e.g., in Asia), it is impolite to stare at a person while talking to her; the opposite holds in Arabic countries. Differences may be found not only in the duration of gaze, but also in its direction (where to look, whom to look at; Argyle & Cook, 1976). In the United States and most Western cultures, looking directly into the eyes of the interlocutor is seen as a positive trait, a sign of confidence, and not looking directly is considered a sign of an evasive attitude. In the Japanese culture, on the contrary, looking directly at someone is considered negative, rude, and improper. Again a laugh in the former culture usually signals that one is happy, amused, or pleased, whereas in the latter it may signal nervousness or discomfort (O'Neill & Brown, 1997).

Despite these differences, some common patterns seem to exist as well: Too much gazing is usually a sign of anger or threat, whereas too little gazing might signal shyness but also carelessness (Knapp & Hall, 1997). Cassell (2000) claimed that, as with gestures, "there are more similarities in the use of the face than there are differences, at least with respect to the regulatory conversational functions of these behaviors."

Do Stereotypes Always Apply?

Not all persons of a particular ethnic group or nationality exhibit the broad patterns of thought and behavior that are generally attributed to their cultures. The extent to which members uphold these traditions depends on many factors, such as social class, education, language, family structure, and so on. So as always happens in user-adapted interaction, stereotypes have to be considered only as a default image of the application context, which is known to be imperfect and revisable. Mechanisms to revise stereotypes after a more accurate knowledge of the individual's behavior are commonly applied by humans and should be introduced in the design of any artificial agent.

HOW CULTURE INFLUENCES HUMAN–COMPUTER INTERACTION

If, as we saw in the previous section, cultural differences in norms, standards, and goals underlie cultural diversity, and these are reflected in differences in the way people reason, feel and display emotions, appear, and gesture, it is natural to wonder whether these differences affect human–computer interaction. In particular, should globalization be interpreted as going toward uniformity of interaction styles or should interaction with a machine be adapted to the cultural context to which it is applied, as it tends to be for other factors such as the user experience, preferences, interests, and so on?

Adaptation is usually justified by some empirical demonstration (or some well-grounded assumption) that it improves, in some sense, the usability of the application. So before investing efforts toward culture-adapted interaction (and, in particular, characters), some evidence should be given that this will improve the use of the systems to which it is applied.

In the scope of their computers as social actors long-term research plan, Nass and colleagues examined this question: Does the ethnicity of a computer agent have an effect on users' attitudes and behaviors? In a study comparing a group of Americans with a group of subjects from an ethnic minority (Koreans), they found that ethnic similarity had significant and consistent effects on the users' attitudes and behaviors. When the ethnicity of the subject was the same as that of the computer agent with whom the subject was interacting during the experiment, the agent was perceived to be more similar, more socially attractive, and more trustworthy. The agent's arguments were also perceived to be better and more convincing (Lee & Nass, 1998; Nass et al., 2000).

In lack of a strong background of empirical studies involving groups of different nationalities or ethnic origins, one may only speculate on the domains in which adaptation to culture might be justified. Lee and Nass (1998) proposed the domains of recommendation in general (e.g., training or medical advice), online shopping, and advertising. They claimed that users would be more willing to trust in the agents, take their suggestions, or give them their credit card number if these agents display a matching of values and norms with their own scale of values. Access to services in general and teaching are other domains in which expectations and communication forms are presumably informed by culture (O'Neill-Brown, 1997). In service systems and shopping, one of the differences between short- and long-term orientation cultures (or between masculinity and femininity) is in their being goal-directed versus being viewed as an opportunity for "living an experience" or "initiating casual conversations." Therefore, the kind of small-talk dialogues being developed in REA (Bickmore & Cassell, 2000) seems to be appropriate in the second type of culture, whereas in the first one helping the shopper get out of the shop more quickly might be preferable. Tutoring systems, on their side, reflect pedagogical approaches and viewpoints about teacher–student relationships: Tutoring agents have therefore to reflect these approaches and viewpoints in the culture in which they will be applied.

However, adaptation to culture did not yet influence HCI as it presumably should: Only a few cues on a tendency to act may be observed. For instance, although emoticons are ubiquitous, some cultural difference occurs even here: The Japanese emoticon for smile (.) depicts that women are not supposed to show their teeth when smiling, and the second most popular icon in Japan is the cold sweat (;), with no clear Western equivalent (Donath, 2001).

E-commerce recently increased the interest toward this research area (there is a Web site devoted to intercultural communication). To examine the influence of culture on Web site design, Sheridan (2001) compared various Web sites from the same companies. She described how the Mercedes' German site has a "clean, functional navigation with few choices to click," whereas the Japanese site requires more patience to achieve the navigation goals. This would reflect a difference in what Hofstede called long-term versus short-term orientation in the two cultures. On the contrary, the Web site of the Malaysian Association of Hotels would reflect a high power distance by featuring awards and authority figures.

What Happens With Embodied Animated Agents?

It is reasonable to expect that designers use the physical forms and the behavior of embodied agents in making sense of their actions, intentions, and emotions just as occurs with people (Bartneck, 2001). An agent's face may either clarify its intentions or complicate them: By altering the level of helpfulness or attractiveness, for instance, agents could subtly announce their willingness to listen and serve the user or engage in social activity (Branham, 2001). If the culture-dependent factors reviewed earlier are appropriately considered in the design phase, the agent should be conceived after a careful analysis of the situation in which it will be applied.

On the contrary, we are witnessing a situation in which existing embodied animated agents, be they stylized cartoonlike characters or more realistic creatures, reflect the culture of the environment in which they have been designed by mirroring the developer's reasoning style, communication modes, and so on. Even in tools in which the aspect and animations of the agent may be selected, the repertoire of available characters is culture-specific. For instance, by looking at the Web site of MS-Agents (http://www.msagent/ring.org/chars/), one notices that these agents may take the appearance of animals (the well-known Peedy, Oscar the Cat, Max the Dog, Claude or Milton the Bears, etc.), characters from traditions and novels (Merlin, Genie, Gar the Gargoyle, Santa Claus, etc.), or anthropomorphizations of technological objects (Alien, Spaceman, Robby, etc.). In all cases, they are strongly inspired from Western culture. When they are humanlike characters, they may represent a typical British waiter (James) or a blond- or brown-haired girl (Charlie or VRGirl) with again a typical Western appearance. The only three characters that seem to refer to non-Western cultures (Woo, Totem, and Miku) have a limited number of animations.

Therefore, what we may already see is a situation in which agents are "not able to communicate with people from cultures different from their designers or with agents developed by other developers" (O'Neill-Brown, 1997). A developer of a natural-language interface for a legal information system in Italy referred that they initially designed their character as an attractive young female assistant because they thought that the typical user of their system was going to be a male lawyer. However, after realizing that, in fact, the lawyer's (female) secretary was the one who most frequently used the system, they had to notice that the appearance and behavior of the character were disturbing to these users. Therefore, they had to design a new character with a more classical dressing, a more professional communication style, and so on.

The interest toward techniques for drawing multicultural characters is recent and was probably promoted by computer game designers. Agents might deal with cultural diversity at two different levels:

- at a surface level by showing cultural sensitivity—that is, by showing the ability to recognize the cultural manifestations of the targeted population;
- at a deeper level by showing cultural consistency—that is, by making their personal and programmatic activity (values, beliefs, attitudes, behavior, etc.) congruent or consistent with the cultural orientation and precepts of the targeted population. This attitude requires a deep understanding of the culture in which the agent is immersed and the ability to alter its communication behavior (verbal and nonverbal expressiveness), to take into account cultural issues that may affect the task of which the agent is in charge.

What Is It That Makes an Embodied Animated Agent Believable?

Believability has long been the main requirement in the design of embodied animated agents. After Bates (1994) first introduced this idea in the early 1990s, the main meaning of this (general) term was to give the illusion of life. Because the goal of showing a humanlike embodied character in the display was to contribute to moving the metaphor of human–computer interaction toward the idea of interacting with a friend rather than using a tool, the main requirement was that, when looking at an embodied animated agent, generic users should suspend their disbelief that what was in front of them was something different from a machine (Loyall & Bates, 1997). This would enable them to establish a new kind of communication with the technology, in which cooperation would be more effective and would be accompanied, if needed, by some entertainment or pleasure. The supporters of this new form of interaction found a strong background to their goals in the long series of Nass and colleagues' studies, which seemed to prove that humans tend to establish some kind of social relationship with conventional computers even when their aspect is nothing more than that of mere machines. Other authors (Sengers, 1999) have supported a lighter concept of believability, which they have named *comprehensibility*: "an agent is comprehensible if, based on the agent's observable actions, users can build an accurate interpretation of the agent's beliefs, reasoning, knowledge and so on." In this case, the agent should not necessarily give the illusion of life, but should at least show an interpretable behavior.

Which properties are successful in achieving believability? It is clear that a realistic, humanlike appearance is not always needed: Cartoons may be good and even more attractive in a number of applications, as proved in Benoît Morel's contribution to this book (see chap. 8), whereas more realistic agents may be useful in domains in which entertainment is not the main goal of interaction. It is also clear that, although a computer will never be able to "feel" emotions (Frijda & Swagerman, 1987), their ability at least to "display" emotions and "engage in social interactions" with the user is important, for instance, in application domains such as tutoring. However, what really makes an agent's behavior believable was not stated in clear terms for some time, and ad hoc solutions to the problem were proposed.

Thanks to Ortony's (2003) reflection of this subject, it recently became clear that consistency in the agent's behavior is an important ingredient of believability:

> What does it take to make an emotional agent, a believable emotional agent? If we take a broad view of believability—one that takes us beyond trying to induce an illusion of life . . . to the idea of generating behavior that is genuinely plausible—then we have to do more than just arrange for the coordination of,

for example, language and action. Rather, ... the behaviors to be generated—and the motivational states that subserve them—have to have some consistency, for consistency across similar situations is one of the most salient aspects of human behavior. ... But consistency is not sufficient for an agent to be believable. An agent's behavior also has to be coherent. In other words, believability entails not only that emotions, motivations and actions fit together in a meaningful and intelligent way at the local ... level, but also that they cohere at a more global level—across different kinds of situations and over quite long time periods. (Ortony, 2003)

Ortony distinguished between consistency across similar situations and coherence across different kinds of situations and over quite long time periods. Consistency is a general requirement of computer usability and is constantly mentioned in the guidelines about interface design (Shneiderman, 1998). The idea is that, if an interface is consistent, the user will apply rules learned by interacting with the system when performing previous tasks to envisage how new tasks will have to be performed. This entails less effort, to users, in interacting with the system, less time spent in learning how to use it, and lower risk of error. Therefore, it is reasonable to assume that the same principles apply when interaction is with an embodied animated character: Users cannot suspend their belief of interacting with a purely technical tool if this tool behaves inconsistently. Consistency and coherence in emotion elicitation in an individual would be ensured, again according to Ortony (2003), by a

> coherent and relatively stable value system in terms of which the environment is appraised.... Such a system is an amalgam of a goal hierarchy, in which at least some of the higher level goals are sufficiently enduring ..., a set of norms, standards and values that underlie judgements of appropriateness, fairness, morality and so on, and tastes and preferences.... (p. 194)

Although a major source of consistency derives from the fact that emotions induce associated response tendencies, personality traits establish individual variations in these tendencies. To assign a personality to an agent through a consistent (and psychologically plausible) trait combination, empirical evidence about the way traits cluster together may be employed—for instance, the famous five-factor model (McCrae & John, 1992).

The following part of this chapter focuses on a description of how an agent's behavior may be adapted to the cultural context in which it is applied by ensuring, at the same time, that consistency is not lost. This is not an easy undertaking. Hence, our suggestions should be taken as hints derived from our own experience in building a context-adapted embodied animated agent, rather than as general guidelines.

4. TRANSCULTURAL BELIEVABILITY   91

AN EXAMPLE: INFORMATION PROVISION SCENARIO

Let us consider the following situation: "A person (P1) is standing on the corner of a street, looking at an open map of the city." We may imagine the following interaction scenarios between P1 and a Person P2 coming in P1's direction:

Scenario 1:
P1: "Sorry, can you please tell me where is the Central Station?"
P2: "No, I don't know."

Scenario 2:
P1: "Sorry, can you please tell me where is the Central Station?"
P2: "No, sorry, I'm not living in this city."

Scenario 3:
P1: "Sorry, can you please tell me where is the Central Station?"
P2: "No, sorry, I'm not living in this city. But you may ask to that Bar, they probably know it."
Or
"Let me have a look at your map: Maybe we will find the place, together."
Or
"Yes, turn left and go straight along that road."

Scenario 4:
P2: approaches P1 and says: "May I help you?"

The four scenarios describe different attitudes for P2, with increasing levels of cooperativeness. In Scenario 1, P2 is not cooperative: Maybe he is not even sincere; he knows about the city, but does not want to lose his time in giving explanations to P1. Maybe he is sincere, but not very polite. The level of politeness of P2 is higher in Scenario 2 (he justifies why he cannot help P1), although the cooperation attitude is the same. In Scenario 3, P2 tries to go beyond the literal request of P1 to help her. P1's main goal is to come to know where the Central Station is irrespective of who will give her this information; P1 reasons about that and suggests to P1 an alternative way to achieve this goal. In Scenario 4, P2 anticipates P1's request: By seeing that she looks at the map, he infers that she needs help in locating herself in the place where she is. He therefore offers his help before P1 asks for it.

We might see the differences in the described scenarios in various terms: P2 might be seen as a person with a different personality in the four cases (cooperative vs. not cooperative and polite vs. not polite). Yet we might also see the scenarios as more or less plausible in different countries. The differences of attitudes would be due, in this case, to differences in the cultural context. For example, Scenario 3 or 4 might be more plausible in Naples (where people are, in general, very cooperative), whereas Scenario 1 or 2 might occur more frequently in Milan, which is characterized by a more typical Northern European attitude.

What is it that drives the difference of behavior in the four described scenarios? Which is the common core that may be described either in terms of personality or cultural context? In our opinion, this common core is the degree of P2's interest toward the interlocutor's needs and therefore the type of reasoning he makes about P1's mind—in particular, about her higher level goals.

When the cooperation attitude of P2 increases, he tries to reason, as far as he can, on P1's higher order goals and interprets her explicit request (or even her needing attitude) by looking at these goals. This may originate from several components of the scale of values in P2's mind: the desire to give a good image of himself, not show his incompetence or uncertainty, give to a foreign person the sensation of being welcome, or, alternatively, not invade the interlocutor's mental territory. These values are usually in contrast with each other, and a particular combination of them prevails in a given context. For instance, Scenarios 1 and 2 might be due to a high respect of the other's privacy, based on the desire to not invade the territory of others, which entails an attitude of minimizing interactions that are not explicitly required. On the contrary, Scenario 3 might be due to a great desire to establish contact with the other person for either individual reasons or because of cultural norms. The resulting attitude is a mixture of politeness and cooperation level, which is not universally considered optimal, but is the consequence of the scale of values established in the considered culture.

If we look at the described scenarios in terms of a culture-adapted agent, the consequence of a variation in the scale of values adopted by the agent is a different discourse or dialogue plan in the same situation. In our examples, cooperativeness leads to more complex plans that try to respond to what are perceived to be the specific needs of the interlocutor. However, it may happen as well that cooperativeness brings the interlocutor to understand the speaker's real goal faster, making the dialogue shorter.

Consistency and coherence in the agent's behavior are ensured by the relative stability of this scale of values. Therefore, by representing, in the agent's mind, its scale of values and belief system, we may be reasonably confident that, in front of similar situations, the agent will adopt similar cooperation attitudes and similar levels of politeness.

We did not say anything about the nonverbal behavior of P2 in the four described situations. How may we ensure that this behavior is consistent with the verbal part of P2's communication? What we may assume, in general, is that a highly cooperative attitude, which results from the attempt to understand the interlocutor's higher order goals and responding to them as far as possible, comes with the desire to enter (in some form) into contact with her; this desire may manifest itself in different nonverbal forms, again according to the cultural context. For instance, a highly cooperative person may tend to make larger gestures—to touch the interlocutor with his hands, look at her directly in the eyes, make large smiles, and bow or speak aloud. These nonverbal behaviors share a meaning of invading the interlocutor's territory, although each of them is typical of a specific culture. On the contrary, a low-cooperation attitude is accompanied by a kind of gesturing that conveys the meaning of separation from the territory of the interlocutor or, in the worst case, of refusing contact with her. In this case, gesturing and facial expressions of the speaker are more controlled and, at the extreme, express the meaning of trying to send the interlocutor far away from his own body.

REQUIREMENTS FOR BUILDING CULTURE-ADAPTED AGENTS

Agents may be built so as to be tailored to a particular culture or may be designed so as to adapt to different cultural contexts. In the first case, every time a new application context comes out, the agent has to be redesigned: Its way of thinking, appearance, and behavior have to be modified, and changes have to be introduced so as to ensure that consistency is not lost. In the second case, the long-lasting experience of user-adapted interfaces may be applied to design and build an agent whose mentioned characteristics change more or less naturally according to the context. In the most sophisticated case, changes may occur automatically: The agent observes the environment to understand the situation and progressively models it and decides how to behave. To have such a complex adaptation capability, the agent should hold an explicit knowledge of how to interpret situations and how to modify itself accordingly. For instance, if an agent takes the role of P2 in the information provision scenario, it should be able to recognize, from P1's speech or from the way she wears and gestures, that it is interacting with someone from southern Italy. If it wants to adapt to her culture, it will then try to behave as in Scenario 3, rather than as in Scenario 1 or 2, and will take a more verbose and warm attitude.

Interpretation and plan recognition are so far the most difficult components of user-adapted systems; one might then imagine systems that do not adapt automatically to the situation, but in which adaptation is introduced at the agent's design level. We discuss this situation in more detail later. The following are the principles on which our agent is based (de Rosis et al., 2003; Pelachaud et al., 2002):

- Although it is reasonable to assume that a person's body is a reflex of his or her mind and that this assumption should guide the generation of embodied animated agents as well, we claim that separation of the agent's mind from its body helps achieve a higher flexibility in adaptation of the agent's behavior, acting, and appearance to the cultural context. It offers the opportunity of varying its mental state and reasoning style according to the context and establishing the communication forms to be applied after considering the technical resources available;
- If the agent's mind is represented according to the BDI (belief, desire, intention) theory and architecture (Rao & Georgeff, 1991), the way that its mental state is related to the cultural context may be represented explicitly. This enables the agent to vary the decision taken (including the discourse that achieves a given communicative goal) and the emotions triggered and displayed according to its mental state's structure;
- Relations between the various components of a person's mind, and the way they control the affective state, cannot be established in a rigid and fixed way: Various forms of indeterminism govern this process from uncertainty attached to beliefs and their relations to weights associated with achieving terminal and instrumental goals. In addition, body expressions are not always specific: A gaze may have several meanings, and the same meaning may be conveyed through a combination of several signals. Again the way meanings and signals are related is not rigid and fixed, but governed by uncertainty. Therefore, some formalism is needed to appropriately represent these parameters in the agent's mental state and the way they affect its behavior.

THE ARCHITECTURE OF GRETA

We now describe in more detail what we propose as the architecture of a culture-adaptable embodied animated agent by extending the experience we gained in building the embodied agent GRETA in the scope of the EC Project "MagiCster."[2] GRETA is a realistic three-dimensional talking head who is able to converse with the user by harmonizing the verbal and nonverbal components of her dialogue move and adapting them to the context. The adaptation factors that may be considered include social relationship with the user and personality factors: GRETA's appearance and behavior are, therefore, once again, inspired by a specific cultural context. However, the architecture and the way her mind drives her body include the main ingredients that might enable us to adapt her to a different culture. In describing the main components of this agent, we therefore illustrate how her

---

[2]IST-1999-29078.

adaptation capability might be extended. Although GRETA is domain-independent, to make this description a bit clearer, we consider a simple example of medical dialogue, in which the agent plays the role of a doctor who explains to a patient (the user) the appropriate therapy to follow to treat his disease.

Mental State of the Interlocutors

The agent's mind and its (default and dynamically updateable) image of the interlocutor's mind are represented as structured sets of beliefs, desires, and intentions. A weight is associated with the goals, and a strength, measured in terms of uncertainty, is associated with the links among these elementary components. This representation of Greta as a BDI agent enables us to employ a unique knowledge repository to drive all the phases of the reasoning process—in particular, the more rational aspects, as well as the more reactive ones, which are related to the triggering of emotions and the decision of whether and how to display them (De Carolis et al., 2001; de Rosis et al., 2003; Ortony, 1988; Ortony, Clore, & Collins, 1988).

> The content of this knowledge component may be varied according to the most relevant features of the cultural context examined earlier. As we said, a culture may be described in terms of the weights assigned to instrumental goals and the prevailing beliefs. If beliefs and goals are not seen as separate mental components, but as items that influence each other with variable degrees of strength, the strength of these links is also a function of the cultural context.
>
> For instance: To employ one of Hofstede's dimensions of culture, in a collectivist society the importance given to friendship and the goal of achieving the good of others should be higher than in individualist societies. Therefore, in particular, triggering of an emotion of envy might be less likely in this type of society.

Main Input

The main input of the agent's behavior-generation process is the name of a task that Greta has to perform. In our example:

Task(S):= Persuade(S U Follow(U t))

"The system (S) has to persuade the user (U) to follow a therapy t."

Task Decomposition

By applying a strategy of planning by abstraction levels, the system's task is decomposed into a set of subtasks, each with its priority. In our example, one of the possible decomposition plans, with a weight (H/high, M/medium, L/low) attached to every component is the following:

Self-Introduction	H
Forall p (Has-got U p)	
Describe (p)	H
Inform (Treats t p)	H
Forall d | (Is-in d t)	
Request (Take U d)	H
Describe (How-to-take U d)	M
Describe (Side-Effects (d))	M

As a first step, GRETA (G) introduces herself to the user (U) to establish a first communication with him. In the main part of the plan, to convince U to follow the therapy (t), G describes his main health problems (p). She then informs him that a therapy (t) can help solve these problems. Then for each of the drugs (d) included in t, G asks U to take this drug and informs him on how to take it. Finally, G describes to U the side effects (s) that may be associated with the prescribed drugs. The example shows that the last two subgoals are given a lower priority (M) than the first ones (H).

This decomposition plan does not necessarily apply to all contexts. We already mentioned Brown and Nichols-English's suggestions of how to adapt the pharmacist's behavior to the culture of the patient with which they interact in the United States. In a previous study in which we compared drug explanation attitudes in Italy and the United Kingdom, we could also verify experimentally, at a small scale, that these considerations equally apply to cultural differences within the European community. We noticed that these explanations were not the same in the two countries, and that much less details were given, in general, by Italian doctors, who tended to minimize or hide the side effects of the prescribed drugs because they were convinced that this knowledge could affect patient compliance negatively (de Rosis et al., 1999). In another study, we could verify that, to increase the effectiveness of the explanation, the order of presentation of information in the text should not necessarily be the same, but should be varied according to the patient's characteristics and the main effect the speaker wanted to achieve on his mind (Berry et al., 1997, 1998). For instance, if the doctor wants to be sure that drug administration rules are applied correctly by the patient and the prescribed drugs have few side effects, he will first talk about these side effects to leave recommendations of how to take each drug for the end. So another possible decomposition plan is the following:

Self-Introduction	H
Forall p (Has-got U p)	
Describe (p)	H
Inform (Treats t p)	H
Forall d | (Is-in d t)	
Request (Take U d)	H
Describe (Side-Effects (d))	L
Describe (How-to-take U d)	L

As we saw in the two examples, task decomposition may be defined in a context-dependent way by varying the weight attached to every subtask. This enables adapting the overall discourse and dialogue strategy to the cultural context.

Discourse Planning

The main task decomposition is then transformed into a discourse plan by assembling precompiled recipes, each of which enables performing a specific subtask. For instance, the subtask:

Describe-Side-Effects (d)

may be translated into a plan that describes, for each side effect of a prescribed drug, how severe this side effect is, which is its frequency, in which conditions it is more likely to occur, to which category of patients it applies, how it can be relieved, and so on. This plan applies to the case in which the patient's goal of knowing about side effects has a high weight and when the doctor also rates the importance of the subject as high (as in the case of English doctors, according to our study). Less details should be provided when this goal is given a lower weight by one of the two participants in the dialogue (as in the case of Italy).

In the planning process, the following knowledge bases are employed:

- A *domain knowledge base*, which describes the main facts that are relevant in the dialogue generation. In the medical example, these facts concern diseases, drugs, and treatments with their properties and relations. For instance:

Is-a (ANGINA p)
Is-a (ASPIRINE d)
Is-a (HEART-PROTECTION-TREATMENT t)
Is-in (ASPIRINE HEART-PROTECTION-TREATMENT)
Treats (HEART-PROTECTION-TREATMENT ANGINA)

... and so on.

> The domain KB may depend on the context to represent the different therapeutic practices applied in different situations. For instance, Brown and Nichols-English (1999) reported that patients who believe in folk-based interpretations of diseases are more likely to treat their problems with home or folk remedies and less likely to be compliant with prescribed medications. They suggested that pharmacists faced with such situations should, whenever possible, incorporate home remedies or other alternative practices into the patient's therapy regimen.

- A *library of recipes*, which is of precompiled plans; each plan is characterized by the following features:

1. a task that the plan enables performing,
2. a set of conditions that define when the plan may be applied, and
3. a list of the performatives that build up the plan.

For instance, the following recipes formalize the different stereotypical attitudes toward explanations of side effects of drugs in British and Italian doctors:

```
name: Detailed Description of Side Effects
task:            Describe (Side-Effects (d))
conditions:      Goal U (KnowAbout U DRUG-SIDEFFECTS) H and
                 Goal U (KnowAbout U DRUG-SIDEFFECTS) H
decomposition:   Forall s|(SideEffect d s)
                 Inform (SideEffect d s)
                 Inform (Severity s x)
                 Inform (Frequency s y)
                 Inform (Occurs s z)
                 Inform (RiskPatients s v)
                 Inform (HowToRelieve s w)
```

>   name: Synthetic Description of Side Effects
>   task: Describe (Side-Effects (d))
>   conditions: Goal P (KnowAbout U DRUG-SIDEFFECTS) M and
>   Goal S (KnowAbout U DRUG-SIDEFFECTS) L
>   decomposition: Forall s|(SideEffect d s)and(Frequency s HIGH)
>   Inform (SideEffect d s)
>   Inform (HowToRelieve s w)

Notice that in this example the weight of the communicative goals in the speaker's mind (that are specified in the conditions part of the recipe) affect the way the task is performed. In the first case detailed information about all side effects is provided, whereas in the second case only the most frequent ones are mentioned, with an indication of how to relieve them.

A dialogue move of GRETA includes one or more performatives, possibly connected by rhetorical relations, that result from application of the two processes of task decomposition and discourse planning. In the two previous examples, all performatives were of the inform type, and their arguments were formulae. For instance, (SideEffect d s) stands for "s is a side effect of d," (Severity s x) for "the severity of s is x," and so on.

If we represent the agent's mind as a structured set of beliefs and goals, with a weight attached to goals, adaptation of every dialogue move to the context may be achieved rather naturally by selecting the set of recipes whose condition side matches the agent's mental state.

Move Tagging

The dialogue move is now ready to be translated into a discourse. To this aim, performatives and their rhetorical relations have to be translated into natural language texts whose formulation is a function of the context. At present, we include in the context not only language issues, but also factors that are related to the social relationship between the two interlocutors. For instance, in de Rosis et al. (1999) we described how doctors add empathy in a message when they want to establish a closer relationship with their patients by an appropriate use of terms, adjectives, and pronouns that emphasize or deemphasize positive or negative aspects of the message: "The *good news* is that we do have tablets that are *very effective* against TB," "*I'm sure* you will feel better once you've been put on treatment," "You will have to take a *little* tablet . . . ," "The *only* side effect this drug may have . . . ," and so on.

To pronounce the text with an appropriate intonation, the agent must know what it is pronouncing—that is, which is the meaning of the most significant parts of the discourse. For instance, to render the performatives:

Inform (Has-got U ANGINA) and Inform (Severity ANGINA MILD)

with the sentence:

"I'm sorry to inform you that you have been diagnosed as suffering from a mild form of what we call angina pectoris,"

she must know that these performatives belong to the category of *inform*—that the first one is accompanied by a feeling of an emotion of sorry for (again because the doctor wants to show an empathic attitude toward the patient) and that the concept it deals with (the patient's disease ANGINA) is in a mild form. This knowledge enables GRETA to pronounce the sentence with an appropriate intonation (emphasis on sorry for and mild) that will reinforce the achievement of the empathy included in the text. In addition, it allows her to decide how to employ nonverbal communication (facial expression, head movements, body posture, and/or gestures) consistently with the verbal part of the message. Meanings are attached as tags in a markup language to discourse spans—that is, to the whole move, to an individual clause or sentence, or even to a single word (De Carolis et al., 2001).

The meanings that may be attached to a message are elements of a finite set. In Poggi and Pelachaud (2000) and Poggi et al. (2000), we examine those that may be rendered through gaze expressions and described how they may be categorized into a limited number of classes: metacognitive, turn taking, affective, certainty, adjectival, deictic, and so on. Other authors described the meanings that may be conveyed through gestures. Whether an emotion should or should not be attached to a discourse span results from an emotion activation process. According to what we said earlier in our model of emotion elicitation, emotion intensities and the way they vary with time are a function of several factors: the characteristics of the event involved, the goal being threatened or facilitated by the event, and the social context in which the event occurs (for more details, see de Rosis et al., 2003).

Once an emotion has been elicited, the reflexive component of the agent decides whether to display or hide it. Here the consequences of displaying the emotion are considered in cost–benefit utility terms (De Carolis et al., 2001).

4. TRANSCULTURAL BELIEVABILITY

> Once again move tagging depends on the context. Although some of the meanings do not vary across the cultures (e.g., turn taking or deictic beliefs), others may be attached to the move only in specific situations: typically affective, certainty, or metacognitive beliefs.

Meaning Realization

A meaning attached to a discourse span may be translated into one or more nonverbal *signals*. For instance, at present a table links meanings to signals. So when deciding the face expressions and (in perspective) the gesture with which to express a given meaning, our agent reads the tag attached to the discourse span, matches its value with the value of an entry in the (meaning, signal) table, and sends an order to its body. A conflict-resolution strategy is employed when several meanings have to be manifested at the same time (Pelachaud & Poggi, 2003). As we said earlier, the (meaning, signal) relation strongly depends on the context to which the agent refers. At present this table is unique, but several tables might be defined to be employed in several contexts.

> (Meaning, signal) tables might be adapted to the cultural context so as to express both quantitative and qualitative differences among the cultures: gesturing more or less, making more or less expressive faces, using eye expressions rather than hand or body gesturing, and so on.

User Move Interpretation

After GRETA completed her move, control is passed to the user. The user move is interpreted by GRETA and may trigger a clarification subdialogue in which the previously established dialogue plan is revised or made temporarily dormant. It should be noticed that, in particular, the user's move may entail triggering of an emotion in GRETA, which produces in its turn a revision of her goal priorities and therefore also of the dialogue plan. For instance, if the patient says: "I definitely will not take this drug, I hate taking bitter pills," it is reasonable to assume that GRETA will feel a bit angry toward the patient or at least worried that treatment might not be followed appropriately. This will reinforce, in her subsequent move, the goal of convincing him to follow the treatment. The list of subtasks that GRETA tries to achieve is therefore considered an open list that is dynamically revised during the dialogue.

> Interpretation of the user reaction should once again be adapted to the culture. Ideally, not only the verbal part of the user move should be interpreted, but also his or her facial expression, body posture, and so on. In this interpretation, again, culture-dependent criteria should be applied as mentioned previously.

HOW TO ENSURE CONSISTENCY

In the previous section, we described in general terms how adaptation to the context is already introduced, in GRETA, during the dialogue-generation process. Several parameters guide the adaptation strategy in every step. To extend adaptation to the cultural context, many more parameters should be considered in the model.

The risk of building inconsistent agents is always high. However, this risk increases considerably when adaptation to the context is included as one of the requirements of the agent's behavior. As anticipated before, consistency in the agent's behavior is a core aspect of believability. It is therefore important to find an answer to the following question: How may designers ensure that inconsistency is not introduced in the different phases of adaptation and in the different aspects of the agent's behavior? For instance, how may we be sure that the agent's mental model is internally consistent? We need a psychologically grounded theory to set up the weights given to the various goals and the strengths of relations between beliefs and goals so that they correspond to a psychologically and socially plausible individual or category of individuals, like a cultural community. How may we ensure that the external behavior of the agent is consistent, coherent across similar situations in time, and corresponds to the agent's system of norms, values, beliefs, and goals, which shape its mental model? The idea of linguists is that every message is produced to achieve a communicative goal and brings a meaning in it. When the message includes verbal and nonverbal parts, the two parts may reinforce each other or only one of them may be employed to convey a given meaning. Thus, a consistent message is one that does not include contradictions either in the same component or among different components.

Let us make some (extreme) examples of inconsistency in the two examples described in this chapter. In the information provision scenario, it would be rather implausible to attach to the same agent a highly cooperative and highly dominant attitude (although this might happen in some individuals). This would correspond to the case of a person who proactively offers help to someone who is not requesting it explicitly, but at the same time demands that his or her suggestion is followed immediately and literally (Castelfranchi et al., 1998). For the same reason, it would be inconsistent to attach an extroverted behavior to a noncooperative person because gestures whose meaning is generically that of invading the territory of others are unlikely in people who do not pay attention to others' viewpoints or needs.

In the medical scenario, it would be inconsistent to show empathy in a part of the message (say the description of the patient's illness) and coldness in another part (say the description of possible side effects of the drug). It would also be inconsistent to say "I'm sorry to inform you . . ." and smile at the same time, say it with a neutral face (although the degree of inconsistency is much lower in this case), or to not synchronize a smile with the verbal part of the expression of empathy.

These are only examples that do not provide any general solution to the problem. As far as affective factor representation is concerned, Ortony suggested looking at theories like the Five Factor Model (McCrae & John, 1992) to drive the design process toward consistency. In this theory, personality traits tend to aggregate into a few factors: A consistent personality would therefore correspond to an aggregate of personality traits in the five-dimensional space. Hofstede's Five Dimensions of Culture might play the same role in building culturally consistent characters. It might help, for instance, in assigning consistent values to parameters that are associated with short-term versus long-term orientation, femininity versus masculinity, power-distance, collectivism versus individualism, and uncertainty avoidance. However, this is still an open problem that should be investigated more in depth.

REFERENCES

Argyle, M., & Cook, M. (1976). *Gaze and mutual gaze*. London: Cambridge University Press.

Bagozzi, R. P., Wong, N., & Yi, Y. (1999). The role of culture and gender in the relationship between positive and negative affect. *Cognition and Emotion, 13*(6), 641–672.

Bartneck, C. (2001). How convincing is Mr Data's smile: Affective expressions of machines. *User Modeling and User-Adapted Interaction* [Special Issue on "User Modeling and Adaptation in Affective Computing," F. de Rosis (Guest Editor)], *2*(4).

Bates, J. (1994). Realism and believable agents. *Lifelike Computer Characters '94*.

Berry, D. C., Michas, I., & de Rosis, F. (1998). Evaluating explanations about drug prescriptions. *Psychology and Health, 13*, 767–784.

Berry, D. C., Michas, I., Forster, M., & Gilli, T. (1997). What do patients want to know about their medicines and what do doctors want to tell them? A comparative study. *Psychology and Health, 12*, 467–480.

Bickmore, T., & Cassell, J. (2000, November 3–5). How about this weather? Social dialog with embodied conversational agents. In *Proceedings of the American Association for Artificial Intelligence (AAAI) Fall Symposium on "Narrative Intelligence,"* Cape Cod, MA.

Branham, S. (2001). Creating physical personalities for agents with faces: Modeling trait impressions of the face. In *Proceedings of the Workshop on "Attitudes, Personality and Emotions in User-*

Adapted Interaction," held in conjunction with the Conference on User Modeling 2001, http://aos2.di.uniba.it:8080/ws-um01.html (last visited August 26, 2003).

Brislin, R. (1993). *Understanding culture's influence on behavior.* New York: Harcourt Brace College Publishers.

Brown, C. M., & Nichols-English, G. (1999, September). Dealing with patient diversity in pharmacy practice. *Drug Topics,* http://www.drugtopics.com/be_core/d/index.jsp (last visited March 13, 2003).

Cassell, J. (2000). Nudge nudge wink wink: Elements of face-to-face conversation for embodied conversational agents. In J. Cassell, J. Sullivan, S. Prevost, & E. Churchill (Eds.), *Embodied conversational agents* (pp. 1–27). Cambridge, MA: The MIT Press.

Castelfranchi, C., de Rosis, F., Falcone, R., & Pizzutilo, S. (1998). Personality traits and social attitudes in multiagent cooperation. *Applied Artificial Intelligence, 12,* 7–8.

Condon, J. C., & Yousef, F. S. (1975). *Introduction to intercultural communication.* New York: Bobbs-Merrill Company.

De Carolis, B., Pelachaud, C., Poggi, I., & de Rosis, F. (2001). Behavior planning for a reflexive agent. *Proceedings of IJCAI.*

de Rosis, F., Grasso, F., & Berry, D. C. (1999). Refining instructional text generation after evaluation. *Artificial Intelligence in Medicine, 17.*

de Rosis, F., Pelachaud, C., Poggi, I., Carofiglio, V., & De Carolis, B. (2003). From Greta's mind to her face: Modeling the dynamics of affective states in a conversational embodied agent. *International Journal of Human–Computer Studies, 59,* 81–118.

Donath, J. (2001, August 6–9). Mediated faces. In M. Beynon, C. L. Nehaniv, & K. Dautenhahn (Eds.), *Cognitive technology: Instruments of mind. Proceedings of the 4th International Conference* (CI 2001), Warwick, United Kingdom.

Eibl-Eibesfeldt, I. (1974). Nonverbal communication. In S. Weitz (Ed.), *Similarities and differences between cultures in expressive movements.* Oxford, England: Oxford University Press.

Ekman, P. (1979). Human ethology: Claims and limits of a new discipline: Contributions to the Colloquium. In M. von Cranach, K. Foppa, W. Lepenies, & D. Ploog (Eds.), *About brows: Emotional and conversational signals* (pp. 169–248). Cambridge, New York: Cambridge University Press.

Ekman, P. (1999). Facial expressions. In T. Dalgleish & T. Power (Eds.), *The handbook of cognition and emotion* (pp. 301–320). New York: Wiley.

Ekman, P., & Friesen, W. (1969). The repertoire of nonverbal behaviors: Categories, origins, usage, and coding. *Semiotica, 1.*

Frijda, N., & Swagerman, J. (1987). Can computers feel? Theory and design of an emotional system. *Cognition & Emotion, 1*(3), 235–257.

Gudykunst, W. B., & Kim, Y. Y. (1992). *Communicating with strangers: An approach to intercultural communication.* New York: McGraw-Hill.

Hall, E. T., & Hall, M. R. (1990). *Understanding cultural differences.* Yarmouth, ME: Intercultural Press.

Hofstede, G. (1980). *Culture's consequences: International differences in work-related values.* Beverly Hills, CA: Sage.

Keltner, D. (1995). Signs of appeasement: Evidence for the distinct displays of embarrassment, amusement, and shame. *Journal of Personality and Social Psychology, 68,* 441–454.

Keltner, D., & Haidt, J. (1999). Social functions of emotions at four levels of analysis. *Cognition and Emotion, 13*(5), 505–521.

Kendon, A. (1995). Gestures as illocutionary and discourse structure markers in southern Italian conversation. *Journal of Pragmatics, 23,* 247–279.

Knapp, M. L., & Hall, J. A. (1997). *Nonverbal communication in human interaction* (4th ed.). New York: Harcourt Brace.

Lee, E.-J., & Nass, C. (1998). Does the ethnicity of a computer agent matter? An experimental comparison of human-computer interaction and computer-mediated communication. In S.

Prevost & E. Churchill (Eds.), Proceedings of the Workshop on "*Embodied Conversational Characters,*" Tahoe City.
Loyall, B., & Bates, J. (1997). Personality-rich believable agents that use language. *Proceedings of the First International Conference on Autonomous Agents (Agents'97)*, 106–113.
Magno Caldognetto, E., & Poggi, I. (1995). Creative iconic gestures. In R. Simone (Ed.), *Iconicity in language* (pp. 257–275). Amsterdam: John Benjamins.
McCrae, R., & John, D. P. (1992). An introduction to the Five Factor Model and its applications. *Journal of Personality, 60*, 175–215.
McNeill, D. (1992). *Hand and mind.* Chicago: University of Chicago Press.
Nass, C., Isbister, K., & Lee, E.-J. (2000). Truth is beauty: Researching embodied conversational agents. In S. Prevost, J. Cassell, J. Sullivan, & E. Churchill (Eds.), *Embodied conversational agents* (pp. 374–402). Cambridge, MA: MIT Press.
O'Neill-Brown, P. (1997). Setting the stage for the culturally adaptive agent. In K. Dautenhahn (Ed.), *Socially intelligent agents* (pp. 93–97; Technical Report FS-97-02). Menlo Park, CA: AAAI Press.
Ortony, A. (1988). Subjective importance and computational models of emotions. In V. Hamilton, G. H. Bower, & N. H. Frjida (Eds.), *Cognitive perspectives on emotion and motivation* (pp. 321–340). New York: Kluwer.
Ortony, A. (2003). On making believable emotional agents believable. In R. Trappl, P. Petta, & S. Payr (Eds.), *Emotions in humans and artifacts* (pp. 189–211). Cambridge, MA: MIT Press.
Ortony, A., Clore, G. L., & Collins, A. (1988). *The cognitive structure of emotions.* Cambridge: Cambridge University Press.
Pelachaud, C., Carofiglio, V., De Carolis, B., de Rosis, F., & Poggi, I. (2002). Embodied contextual agent in information delivering application. *Proceedings AAMAS'02*, Bologna.
Pelachaud, C., & Poggi, I. (2003). Subtleties of facial expressions in embodied agents. *Journal of Visualization and Computer Animation.*
Poggi, I. (2002). Symbolic gestures—The case of Italian gestronary. *Gesture, 2*(1), 71–98.
Poggi, I., & Pelachaud, C. (2000). Emotional meaning and expression in animated faces. In A. Paiva (Ed.), *Affective interactions* (pp. 182–195). Heidelberg: Springer.
Poggi, I., Pelachaud, C., & de Rosis, F. (2000). Eye communication in a conversational 3D synthetic agent. *AI Communications [Special Issue on Behavior Planning for Life-Like Characters and Avatars], 13*(3), 169–181.
Rao, A. S., & Georgeff, M. P. (1991). Modeling rational agents within a BDI-architecture. In J. Allen, R. Fikes, & R. Sandewall (Eds.), *Proceedings of the 2nd International Conference on Principles of Knowledge Representation and Reasoning.* San Francisco: Morgan Kaufmann.
Samovar, L. A., & Porter, R. E. (1972). *Intercultural communication: A reader.* Belmont, CA: Wadsworth.
Sengers, P. (1999). Designing comprehensible agents. In *Proceedings of the International Joint Conference on Artificial Intelligence.* Stockholm, Sweden.
Sheridan, E. F. (2001). *Cross-cultural web-site design. Multilingual computing and technology* (Vol. 12, Issue 7). Multilingual Computing.
Shneiderman, B. (1998). *Designing the user interface.* Boston: Addison-Wesley.

CHAPTER

5

Creating Embodied Agents With Cultural Context

Jan M. Allbeck
Norman I. Badler
University of Pennsylvania

It is said that cultural information is a minimum prerequisite for human interaction—in the absence of such information, communication becomes a trial-and-error process (Knapp & Hall, 1992). We have seen movies where previously isolated cultures are encountered by Westerners for the first time. Typically, there is a lot of pointing and gesturing and much miscommunication. Eventually some form of communication is established, but only after a somewhat lengthy interactive process. In modern times, many of us have been exposed to other cultures (at least through the mass media) and during interactions have some a priori information that shortens the amount of trial and error needed.

The same trial-and-error process takes place when people encounter virtual agents. They poke, prod, and try to find out what the agent can and cannot do and how it reacts to different stimuli. Although sometimes amusing, this often detracts from the purpose of the agent. If the agents encountered behaved as expected (including within a standard deviation from cultural norms), the observer of the agent may settle into the task at hand more quickly and be less distracted by inaccuracies. This might even lead to a level of trust in the agent.

The inclusion of culture in agents is needed in many different applications, including language learning, military training, games, and avatars. Too often language learning only involves learning the syntax and semantics of the formal language. Slang is neglected as are cultural norms. Would learning to speak flawless Japanese allow you to interact flawlessly with a person native to Japan? You would still need to understand the intricacies of bowing, manners, and respect of status. Creating a virtual character for language learning could enable more comprehensive and context-sensitive learning of more than just the verbal language.

Computer simulations for military training are currently popular endeavors (Bindiganavale et al., 2000; Jones et al., 1999; Silverman et al., 2002; Swartout et al., 2001). However, there is a great need to train military personnel about cultural differences and sensitize them to these differences.

This story was conveyed to us by a former member of the U.S. Army who is now involved in military training:

> An army medic happened upon a situation involving an injured young girl and an injured ox. Both were severely hurt and required immediate medical attention. The father of the young girl urged the medic to save the ox. The medic was appalled by the father's insistence to help the ox. Then the mother of the young girl came out of the house and also insisted that the ox be saved, saying, "I can have another daughter. I cannot have another ox."

We present this example not to determine what the medic should or should not have done or to discuss the ways of life and values of an impoverished, war-torn nation. It is presented to illustrate the need for cultural sensitivity in military personnel. Sometimes people rush to judgment particularly when a discussion or an idea is so clearly in conflict with their own values. In military applications, this can lead to inabilities to fully understand the situation and communicate effectively. There are too many cultures in the world to train soldiers effectively about everyone. Through the use of virtual environments and virtual agents, however, we may be able to sensitize them to cultures and enable them to understand and communicate more effectively by not prejudging and remaining open minded about situations. For a simpler example, how would you respond if a person walked up to you and spit at your feet? If the person were from some Eastern African tribes, this would be a sign of greeting not insult (Axtell, 1991). Such research and applications of culture in virtual training environments would also add realism and depth to computer games and their characters.

In many virtual worlds and games, the participant is represented in the world by their avatar. This avatar is essentially supposed to *be* the participant in the world. Although a person may want to be represented by an avatar completely unlike them, when realism is desired, the lack of cultural context in these avatars has two consequences. First, other participants in the world observing this avatar will be unable to get a realistic depiction of the person represented by the avatar. Most virtual worlds and games have a limited number of physical representations to choose from and a limited number of actions that can be performed, thereby forcing participants to assimilate to the culture depicted. In some environments and games this is a necessary part of the experience, whereas in others it just limits the expressivity of the participant and causes a misrepresentation. Second, many games rely on the participants forming a bond with their avatars. In adventure games, they are to keep their character safe and lead them through the mission of the game. In simulation games, like *The Sims*, they are to keep the avatar as happy and problem-free as possible. Ortony, Clore, and Collins (1988) described this bonding with another person or persons as forming a cognitive unit. The strength of your cognitive unit with a person or thing determines how much you care about them and how much you identify with them. Would it not be easier to identify with a virtual character that looked and behaved more like you? Would you not be more involved in the game or world if you identified more strongly with your representation there?

In this chapter, we discuss different aspects of culture and how they may be depicted in a virtual character. We base our discussion in the context of nonverbal communication. As related in the following passage from *Snow Crash* (Stephenson, 1992), facial expressions and body language are important forms of communication:

> They come here [The Black Sun] to talk turkey with suits from around the world, and they consider it just as good as face-to-face. They more or less ignore what is being said—a lot gets lost in the translation, after all. They pay attention to the facial expressions and body language of the people they are talking to. And that's how they know what's going on inside a person's head—by condensing fact from the vapor of nuance. (p. 64)

People grow up familiarizing themselves with the nuances of communication in their culture, particularly with people with whom they often interact. That sort of familiarity cannot be taught quickly in a language or culture learning session. However, we could begin the process and sensitize people to the differences.

MANIFESTATIONS OF NONVERBAL COMMUNICATION

Although verbal communication is the standard channel of communication used by people, nonverbal communication also contains valuable information. The culture of a person can manifest itself in all of the channels of nonverbal communication. According to Lewis (1998), the channels of nonverbal communication are:

- facial expressions (smiles, nods)
- gestures (especially hand and arm movements)
- body movements
- posture
- visual orientation (especially eye contact)
- physical contacts (handshakes, patting)
- spatial behavior (proximity, distance, positions)
- appearance (including clothes)
- nonverbal vocalizations

We briefly describe each of these channels along with some of the research done in the embodied agents community and illustrate how culture affects each of these channels.

Facial Expressions

There are two approaches to facial expressions and culture. Cultural universalists believe that facial expressions are innate, whereas cultural relativists believe that culture shapes facial expressions. Ekman and Friesen (1975) joined these two theories. They believe that, although people are predisposed to connect certain facial expressions with emotional states, rules for displaying these expressions are socially conditioned. Cultural display rules influence how, when, what, and with whom certain expressions should be displayed and suppressed.

There is general agreement that there is some universality in the decoding of basic facial emotions (Ekman et al., 1987). Pictures of six different facial emotional expressions have been consistently recognized around the world. Eibl-Eibesfeldt (1972) identified a universal eyebrow flash among many cultures including Europeans, Balinese, Papuans, Samoans, South American Indians, and Bushman. The eyebrow flash can be seen in greetings, flirting, thanking, and signs of approval.

Smiles, however, can carry different connotations in different cultures. In the United States, smiles are associated with good feelings and happiness. In Japan, they can also indicate the concealment of embarrassment, displeasure, or anger. Overall culture appears to play a powerful role in terms of the types of emotions that should be displayed or suppressed in different interactive situations (Ting-Toomey, 1999).

In Ting-Toomey (1999), cultures are characterized along different dimensions. One such dimension is *cultural identity*, which indicates how much the individual is valued over the group. Individualistic cultures emphasize individual identity, rights, and interests. Prototypical individualistic cultures are Australia, Belgium, Germany, Switzerland, Canada, and the United

States (Triandis, 1995). Collectivistic cultures emphasize group identity, group needs, and group interests. China, Japan, South Korea, Vietnam, Ghana, Saudi Arabia, and Mexico are considered collectivistic cultures (Triandis, 1995). Individualistic cultures tend to invite the display of a broad range of positive and negative emotions. Collectivistic cultures gravitate toward the display of modest positive emotions while suppressing the display of extreme negative emotions, such as anger and disgust. Collectivists also have a harder time decoding negative facial expressions.

Facial expressions are known to express emotion (Ekman & Friesen, 1977), but facial expression can indicate what a person is thinking as well as feeling. The face reflects interpersonal attitudes, provides nonverbal feedback on the comments of others, opens and closes channels of communication, complements or qualifies verbal responses, and augments or replaces speech (Knapp & Hall, 1992).

Both Brand (1999) and Poggi and Pelachaud (2000) researched ways of generating facial expressions for speech. Brand generated facial animation from information in an audio track. Poggi and Pelachaud concentrated on the visual display of intentions through facial animation based on semantic data. They modeled performatives, which are the type of action a sentence performs, such as requesting or informing. They also discussed how the degree of certainty, the power relationship, the type of social encounter, and the affective state affect the facial animation.

Cassell et al. (2000) presented a system that automatically generates and animates conversations among multiple agents. A dialogue planner creates the conversation and generates and synchronizes appropriate facial expressions, intonation, eye gaze, head motion, and arm gestures.

To link culture with facial expressions, one could begin with the universally accepted facial expressions. The frequency and intensity of the expressions could then be modulated by the type of culture (individualist vs. collectivist). Culture-specific facial gestures could be linked to semantic information and tied to context and speech as appropriate.

Gestures and Body Movements

Gestures are voluntary or involuntary movements intended to communicate. They may involve any part of the body and are used to emphasize, clarify, or amplify a verbal message. They can also regulate or control a human interaction or display affect (Lewis, 1998). Unlike gestures, body movements are not intended to convey information. Body movements include walking, reaching, turning, bending, and so on. The manner in which these actions are done can help convey the cognitive state of the performer. People can walk in dramatically different ways: fast, slow, straight, swerved, proudly, sadly, or joyfully (Rose et al., 1998).

The most well-documented cultural gestures are emblematic gestures (Axtell, 1991; Burgoon et al., 1989; Knapp & Hall, 1992). Some gestures exist in different cultures and have the same meaning. For example, head nodding means *yes* in many different cultures. The same gestures can, however, have different meanings in different cultures. Head nodding means *no* in parts of Greece, Turkey, Iran, and Bengal (Axtell, 1991). Another example is the *O.K.* gesture, which in the United States conveys agreement or well-being. In other parts of the world, this gesture is profane. Other types of gestures also differ from culture to culture. The pointing illustrator in the United States uses the index finger. In Germany, the little finger is used.

The frequency, manner, and number of gestures are also culturally dependent. Mediterranean peoples have far more gestures than North Americans do (Burgoon et al., 1989). Italians tend to use big gestures and gesture more frequently than the English or Japanese. Southern Europeans, Arabs, and Latin Americans are inclined to use animated hand illustrators, whereas many Asians and northern Europeans prefer quieter gestures (Ting-Toomey, 1999).

Body movements such as walking also illustrate cultural differences. Latin men who are friends may walk with arms linked. In the Far East, women traditionally walk a pace or two behind men. Western women walk with a longer gait and more upright posture, which is sometimes seen as aggressive in other cultures (Axtell, 1991).

Gestural communication and body language have been studied by several research groups (e.g., Kurlander et al., 1996; Morawetz & Calvert, 1990). Amaya et al. (1996) studied the expression of emotion on the body.

We have been building a system called EMOTE to parameterize and modulate action performance (Chi et al., 2000). It is based on a human movement observation system called *laban movement analysis*. EMOTE is not an action selector per se; it is used to modify the execution of a given behavior and thus change its movement qualities or character. The power of EMOTE arises from the relatively small number of parameters that control or affect a much larger set, and from new extensions to the original definitions that include the nonarticulated movements of the face (Byun & Badler, 2002).

Recent research on communicative agents has paid particular attention to speech, gestures, and their synchronization (Cassell et al., 2001; Hartmann et al., 2002). These systems have created gesture libraries that label gestures with semantic information. This semantic information is then used to automatically tie appropriate gestures with speech acts. Cultural information could be included with this semantic information so that gestures are chosen based on both the speech act and cultural context. As procedures for altering motion qualities (Badler et al., 2002; Rose et al., 1998; Zhao, 2001) become more advanced, it is important to link the qualities with cultural norms. By altering a distribution of EMOTE parameters, we should be able to take a baseline gesture and make it look more Italian or Japanese.

Postures

Posture is an indicator of the degree of involvement, the degree of status relative to the other participants, or the degree of liking for the other interactants. For example, a forward leaning posture can indicate higher involvement, more liking, and lower status in situations where the participants do not know each other very well. Posture is also a key indicator of the intensity of some emotional states. A drooping posture is associated with sadness, and a rigid, tense posture is associated with anger. The extent to which the communicators reflect each other's posture may also be an indication of rapport or an attempt to build rapport (Knapp & Hall, 1992).

Hewes (1957) documented over 1,000 culturally variant postures. Many of the postures were influenced by environmental factors such as furniture. Facial expressions and body posture have an impact on credibility or social influence power. In South Korea and Japan, restrained facial expressions and rigid postures indicate an influential person. In the United States, relaxed expressions and postures give the impression of credibility (Ting-Toomey, 1999). Cultural beliefs can also impact accepted postures. The Thai culture considers the feet to be the lowest, most revolting part of the body. A posture in which a foot is pointed toward them is considered insulting.

Becheiraz and Thalmann (1996) presented an animation model of nonverbal communication where agents react to one another in a virtual environment based on their postures. Relationships between the agents evolve based on the perceptions of postures. To include cultural variations in the simulation, the meanings of the postures would have to include cultural variation. Choosing postures for an agent could be done by altering a distribution of postures based on the culture of the agent and semantics of the postures in that culture.

Visual Orientation

What a person pays attention to and how much attention he or she pays is another channel of communication. A person's gaze and even the dilation and constriction or his or her pupils can be an indicator of interest, attention, or involvement (Knapp & Hall, 1992).

Eye contact often differs between contact and noncontact cultures. Contact cultures tend to face one another more directly, interact closer to one another, touch one another more, look one another in the eye more, and speak in a louder voice. Eye contact can differ both in frequency and duration of gaze. For example, Swedes tend to look less frequently, but for longer duration than the English (Knapp & Hall, 1992). Hispanic women tend to hold eye contact with others, whereas many Asians, Africans, and West Indians tend to avoid direct eye contact. Staring is considered impolite and intimidating in Asian countries such as Japan, Korea, and Thailand. In Saudi Arabia, strong eye contact is preferred, but the eyes appear sluggish and their appearance disinterested (Axtell, 1991).

Studies have also shown cultural differences of eye gaze during listening and speaking modes of conversation. African Americans tend to hold eye contact while speaking and break eye contact when listening. Euro-Americans show the opposite tendency (LaFrance & Mayo, 1978). It is a sign of respect in some cultures (e.g., Asian Americans and Native Americans) to avert eye contact when communicating with elders and persons of higher status.

Johnson and Rickel (1997) presented an animated pedagogical agent, which uses both gestures and attention to aid in the instruction of manual tasks. Cassell et al. (2000) created an interface for chat room avatars that allows the user to give conversational cues through attention control. If a user sees an agent that he is talking to begin to look away from his avatar more and more, this is probably an indication that the agent no longer wants to participate in the conversation. Basic contact versus noncontact cultural variations could be added to such models by default positions and angles for interaction for different cultures. Lee et al. (2002) constructed a stochastic eye movement model for conversation that includes both listening and speaking modes. The model is based on statistics gathered from eye tracking data. Their procedure could be used to capture data and create models for different cultural variations in eye gaze.

Physical Contacts and Spatial Behavior

Physical contacts may be self- or other-focused. Self-focused touching may reflect a person's cognitive state or a habit and include nervous mannerisms. There are many kinds of other-focused touching, including irritating, condescending, comforting, and electric. The meaning of a touch behavior is often derived more from its context and manner than from its configuration. Spatial behavior refers to social and personal space. Spatial behavior can vary based on many aspects of individuals, including age, gender, status, roles, culture, personality, and context. Studies show that conversational distance is related to general comfort level (Burgoon et al., 1989).

Hall (1966) distinguished cultures by the distances at which members interact and how frequently members touch. Latin Americans, the French, and Arabs live in contact cultures, whereas Germans and North Americans live in noncontact cultures. For example, friendly embraces are much more likely to occur between Latin men than American men. Such differences can also depend on gender. Shuter (1977) found that the Italian culture contains tactile men and nontactile women, whereas American cultures contain tactile women and nontactile men. North Americans tend to adopt large interaction distances, whereas studies show that Arabs, Venezuelans, Latin Americans, Italians, Mediterranean cultures, and Pakistanis all have smaller interaction distances (Burgoon et al., 1989).

Conversational distance also varies from culture to culture. The average conversational distance for Euro-Americans is 20 inches. The space for Latin American and Caribbean cultures is closer to 15 inches. The conversational distance for two Arabs is about 10 inches (Ferraro, 1990). Hence, Euro-Americans find Arabs rude for standing too close, whereas Arabs find Americans aloof for standing too far away.

Physical contacts and spatial behaviors are types of behaviors that animation artists do well, but on which embodied agents researchers do not focus. To create consistent cultural variations, embodied agent systems need to coordinate these two channels of communication with the other channels.

Appearance

Appearance can be the first indication of culture or ethnicity. Generally shorter stature individuals with straight black hair, narrow black eyes, and yellow skin tones are at first glance thought to be from an Asian culture. By observing other channels of communication, it might be determined which Asian culture they are a part of or even that they are not a part of an Asian culture at all. Although appearance is not the sole determiner of culture, it does influence our determination of culture.

The appearance of an embodied agent is normally created before any simulation or application containing the agent is launched. There are many companies and research laboratories, including Poser (Curious Labs) and MIRALab (MIRALab), working on modeling virtual human bodies, skin, hair, and clothing. Perhaps more attention needs to be paid to modeling ethnically diverse agents. Skin has different complexion. Hair has different texture. Clothing can have different patterns, textures, and styles. The modeling of these features for real-time simulation is still an active area of research in general.

Nonverbal Vocalizations

Nonverbal vocalizations are vocal sounds other than words. This includes tone of voice, which is known to convey emotional information (Knapp & Hall, 1992). Vocal volume varies among cultures. Arabs tend to be loud to show strength and sincerity. Europeans tend to use softer volumes (Burgoon et al., 1989). *Voice qualifiers* include accent, pitch range, inflection, articulation, resonance, and tempo. Americans often interpret the curt speech of the British as arrogant, whereas the British consider U.S. speech to be causal and lacking class (Ting-Toomey, 1999). Some vocalizations are accepted in some cultures and considered rude in others. For example, belching in public is accepted in some Asian cultures, but considered rude by many northern Europeans.

Southern European cultures tend to value an emotional, expressive tone of voice when important issues are discussed, whereas Asian cultures prefer a softer tone of voice. Arabs tend to believe that greater volume indicates greater sincerity. They often consider the more moderate tones of the German and U.S. cultures as distant and cold (Ting-Toomey, 1999).

The Sims is a good example of the use of nonverbal vocalizations in embodied agents. In this game, the characters live their daily lives, participating in polite conversations and angry discussions, but the characters have no discernible spoken language. The game's characters communicate through gestures, thought bubbles, and nonverbal vocalizations. In the game, it is easy to distinguish a polite conversation from a heated argument by the volume and frequency of the nonverbal vocalizations. The vocalizations found in this game are not well suited to all cultures. Different tones, pitches, and tempos could be recorded to be played with agents of different cultural backgrounds.

HUMAN AGENT HIERARCHY

The influence of culture in a person can be found in more than just the channels of nonverbal communication discussed earlier. Funge et al. (1999) depicted a hierarchy of computer graphics modeling (see Fig. 5.1). This hierarchy shows areas of computer graphics modeling for virtual characters. The bottom two layers were addressed early in computer graphics research with geometric models and inverse kinematics. Physical models generate realistic motion through dynamic simulation. Behavioral modeling involves characters that perceive environmental stimuli and react appropriately. Through cognitive modeling, autonomous characters can be given goals and react deliberately as well as reactively.

There is more research to be done, particularly on the cognitive modeling layer, but we reintroduce this hierarchy as a reference for modeling culture in agents. Culture is not found only in a person's appearance. It is also found in their movement quality, habits and behavior, and values and decision processes.

To better convey culture, each level of this hierarchy needs to be addressed. The channels of nonverbal communication correspond to the geometric, kinematic, and physical layers, leaving the behavioral and cognitive

5. EMBODIED AGENTS WITH CULTURAL CONTEXT

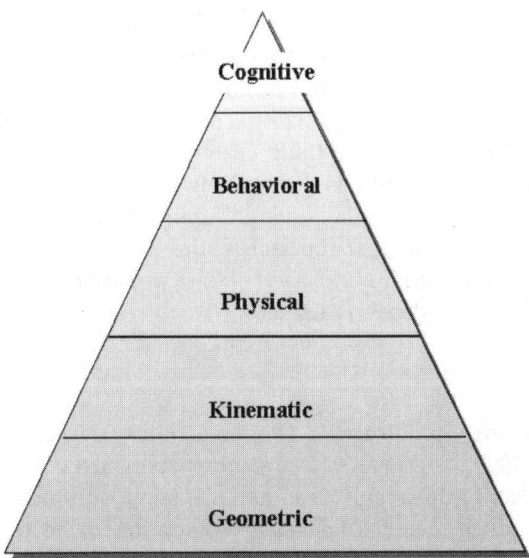

FIG. 5.1. Computer graphics modeling hierarchy from Funge et al. (1999).

layers yet to be addressed. The story of the injured girl and ox found at the beginning of this chapter did not include any information about the appearance, movement qualities, proxemics, or vocal qualities of the characters. It was a story of differing values in cultures. In addition to communicative difference, culture can also vary in values, priorities, preferences, and other cognitive aspects.

Status and Role

Hall (1976) claimed that human interaction, on a broad level, can be categorized into *low-context* and *high-context* communication systems. In low-context communication, people emphasize how intention or meaning is best expressed through explicit verbal messages. In high-context communication, people convey intention or meaning through the context (e.g., social roles or positions) and nonverbal channels (e.g., pauses, silence, tone of voice).

Similarly, communication style can be categorized as *person-oriented* or *status-oriented* (Ting-Toomey, 1999). The person-oriented style stresses individuality, informality, and role suspension. The status-oriented style stresses formality and the honoring of established power-based identities. Person-oriented communication is more symmetric, whereas status-oriented communication is asymmetric. Low-context cultures tend to use person-oriented communication, whereas high-context cultures tend to use status-oriented communication (Ting-Toomey, 1999).

English-speaking cultures tend to be person-oriented, whereas the Japanese culture tends to be status-oriented. Americans tend to treat others with informality and casualness, not honoring formal codes of conduct and ritualistic manners (Okabe, 1983). By contrast, the Japanese maintain proper roles and create a predictable interaction.

The value of status and role is particularly evident in the Korean culture (Yum, 1988). The Korean language has different vocabularies for each sex, and for different degrees of social status and familiarity. When a Korean communicates with a higher status person or a person to whom he or she is indebted, respectful language is used.

Gender

Another dimension of culture is the masculinity–femininity dimension (Ting-Toomey, 1999). Cultures where gender roles are clearly distinct are termed *masculine*. In these cultures, men are supposed to be tough, assertive, and focused on material success. Women are to be tender, modest, and concerned with the quality of life. In the feminine cultures, gender roles are less distinct. In these cultures, everyone is supposed to be modest, tender, and concerned with the quality of life. Japan, Austria, Venezuela, Italy, Switzerland, Mexico, and Ireland are considered masculine cultures. Sweden, Norway, Netherlands, Denmark, Costa Rica, and Finland are considered feminine cultures (Hofstede, 1991).

Emotions

Cultural display rules shape how, when, what, and with whom different emotional states are displayed. Ting-Toomey (1999) described how the cultural identity dimension can affect emotional expression. Individualists often feel freer to express themselves, whereas collectivists are more concerned with the opinions of others and guard their emotions more closely. Individualists and collectivists also internalize emotions differently. For example, a collectivist might feel more shame for the wrongdoings of a relative than an individualist would. Different actions also elicit different emotions in the cultures. For individualists, the successful achievement of goals that bring about personal pride and recognition makes them feel generally good. For collectivists, the achievement of goals that make members of the group feel good generates feelings of well-being.

Personality

One personality factor is whether someone is independent or interdependent (Ting-Toomey, 1999). This plays a role in how people regulate personal boundary issues. Independent concepts tend to dominate in individualistic cultures, whereas interdependent concepts tend to dominate in collectivistic cultures. Independent people view themselves as unique individuals separate from others; in contrast, interdependent people consider themselves part of a social network. Interdependent people define themselves by their relationships to others. Southeast Asians and Africans tend to be interdependent people (Ting-Toomey, 1999).

Behavioral and Cognitive Modeling

The OCC model is a model for generating emotions named after its authors (Ortony, Clore, & Collins, 1988). In this model, emotions are generated through the agent's construal of and reaction to the consequence of events, actions of agents, and aspects of objects. There are a limited number of other influences considered under each of these areas, including consequences for others and self, and prospects relevant and irrelevant (see Fig. 5.2). The model eventually grounds in 22 different emotions: love, hate, pride, admiration, shame, reproach, joy, distress, happy for, gloating, resentment, pity, satisfaction, relief, fears-confirmed, disappointment, gratification, gratitude, remorse, hope, fear, and anger. Although many researchers have based their work on this model (Bates, 1994; Elliot, 1992; Gratch & Marsella, 2001; Silverman et al., 2002), none has incorporated culture into the model.

The inputs for the OCC model are the standards and values of the agent, a set of goals, and the strength of its cognitive units. Standards, values, and goals should also be accompanied by some preference value so that priorities can be determined. Here values include the liking and disliking of objects and approving and disapproving of actions. The strength of the cognitive unit refers to the amount to which the agent identifies with or sympathizes with another person or persons. When joined with a planning algorithm, this model creates an interesting base for the construction of an embodied virtual agent (Gratch & Marsella, 2001). Because this model is widely known in the embodied agents research community, it provides a nice model to which to apply other aspects of culture. Essentially a different doctrine can be created for each culture and used as input to the OCC model.

In the doctrine, we could maintain an internal hierarchy or lattice of status relationships. Another part of the cultural doctrine is whether the culture of the agent is person- or status-oriented. This generates a value or standard for whether the honoring of higher status people is a priority for the agent.

Gender also shapes the values and standards included in the OCC cultural doctrine. In a feminine culture, there may be no distinction made between males and females, but in masculine cultures two sets of doctrine would have to be created. The male doctrine would value assertiveness and

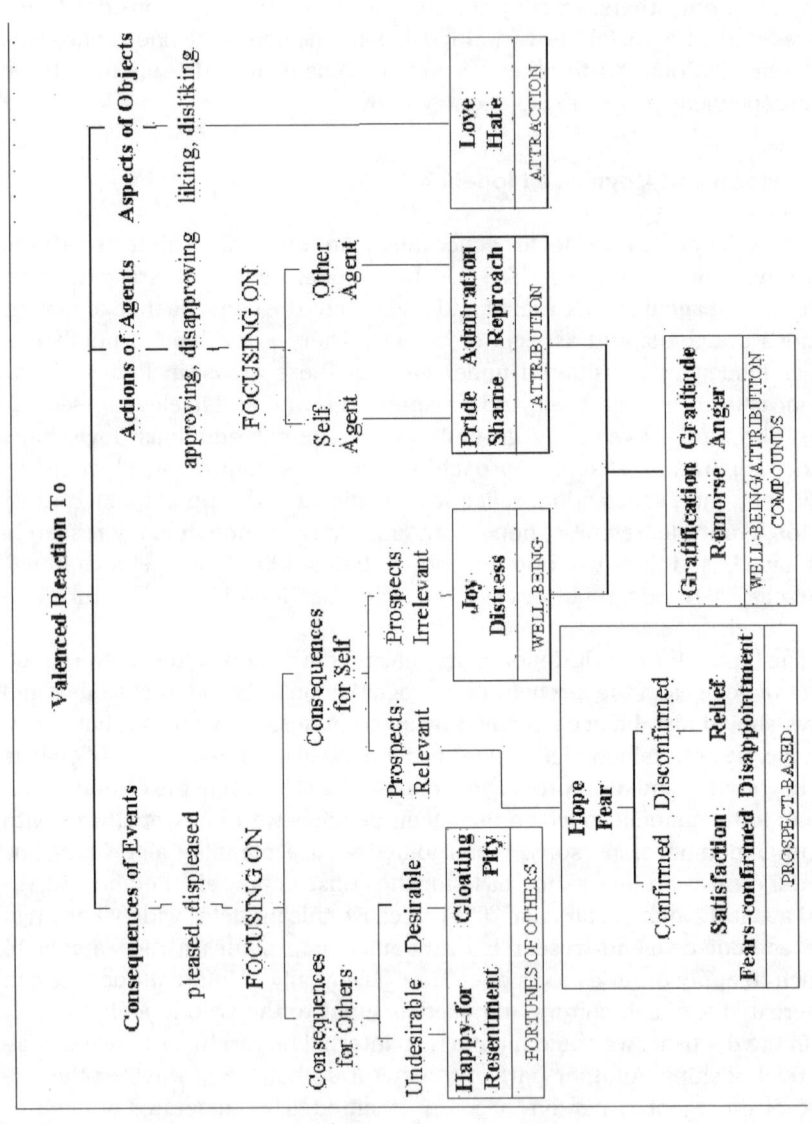

FIG. 5.2. The OCC model for generating emotions (Ortony, Clore, & Collins, 1988).

contain goals for material success, whereas the female doctrine would value modesty and contain goals for an overall good quality of life.

By grouping rules for standards, values, and goals into individualistic values and collectivist values, we could more easily customize the generation of emotions for cultural variations. For individualists, the successful achievement of goals that bring about personal pride and recognition makes them feel generally good. For collectivists, the achievement of goals that make members of the group feel good generates feelings of well-being. Personality factors can have an impact on cognitive unit strength. Although independent people tend to have fewer or weaker cognitive units, interdependent people tend to have stronger cognitive units.

CONSISTENCY

Consciously or not, people spend their entire lives observing other people. We have unconscious and cultural norms of human behavior and are more likely to notice the unexpected rather than the natural or expected. When we try to communicate with another culture, everything may appear unexpected or unnatural at first. Eventually, we learn aspects of the culture by noticing and remembering the consistencies in the behavior. This makes it particularly important to create embodied agents with consistent behavior in general (Allbeck & Badler, 2001; Badler et al., 2002; Isbister & Nass, 2000) and consistent cultural behavior in particular.

The agent's behavior must be consistent from moment to moment and from situation to situation. There should be no wild mood swings or complete loss of focus. Departures from consistency might be interpreted as dramatic effects or, more likely, as internal conflicts within the agent's own mind. Normally, we should expect the state of the agent to be consistent with every level of its behavior: expression on its face, affect of its movements, actions it performs, and goals it pursues. Also the state (and thus actions) must be consistent with the context or situation in which the agent finds itself.

DISTRIBUTIONS OF PARAMETERS

One problem that can result from parameterization is that rapid changes in the parameter values can cause inconsistent or unnatural looking movement. For example, an instantaneous change from a joint angle of 0 to 90 would appear quite unnatural. Treating the parameters as a distribution and altering this distribution (scaling, shifting, amplifying, etc.) based on culture parameters will lessen this inconsistency. A similar computational model has been used by Ball and Breese (2000) to model user mood based on user interface behaviors. For example, we could start with neutral EMOTE parameters and alter them according to culture types. Distributions of EMOTE parameters for different culture traits will be created (probably through a learning process based on many observations). During simulation, the agent begins interacting with its environment with actions modified by EMOTE values obtained from the parameter distributions.

For example, although individualistic cultures tend to display a broad range of positive and negative emotions, collectivistic cultures gravitate toward the display of modest positive emotions while suppressing the display of extreme negative emotions. Thus, we could shift and deamplify the distribution of EMOTE parameters for collectivistic cultures. Currently, we have such parameterization for gestures and facial expressions (Badler et al., 2002). As emotional responses arise, the EMOTE parameter distributions can be shifted or scaled to demonstrate the effect of culture on movement behavior.

CONCLUSION

The inclusion of culture in agents is needed in many different applications, including language learning, military training, games, and avatars. To be effective in these applications, culture needs to be included in many aspects of embodied agents from channels of nonverbal communication to verbal communication to decision making and values. In language learning applications, aspects of body language could be taught with verbal language. Body language could even be used as feedback during the learning process. Facial expressions and gestures could be used to indicate that the agent does or does not understand what is being communicated, thereby providing feedback similar to what would be experienced in that culture.

Military personal are deployed in many countries and cultures around the world with very short notice. It is not possible to train soldiers for every culture they will encounter. It may, however, be possible to sensitize them to possible cultural differences they may encounter and to be more objective when they encounter these differences. Differences include gestures such as the O.K. gesture, body actions such as spitting at someone's feet, facial expressions such as a smile to cover anger, and values such as valuing the life of an ox over the life of a child. Training for only one aspect of culture will not adequately prepare soldiers.

If done properly and consistently, culture can also add an interesting dimension to games. Consider once again *The Sims*. In the newer versions of *The Sims*, the characters can talk about their interests and increase or decrease their friendship level based on their combined interests. Similar increases and decreases in friendship could be based on cultural understanding. The more a *Sim* is exposed to a culture, the more it could learn about that culture and the better it could communicate with people of that culture. Here interests correspond to what the *Sims* value. The game would also be more visually interesting with different gestures, facial expressions, movement qualities, and proxemics for characters of different cultures.

Finally, the inclusion of culture in avatars, Internet sales agents, games, and other simulations may provide a sense of identity with and trust in these agents because their actions adhere to expectations. People have a tendency to trust what they know over what they are not familiar with. Would you trust a salesman whose behavior seemed unusual or foreign? To stay immersed in a game or virtual world, the behavior of the avatars and characters should be consistent and adhere to cultural expectations. We need to build agent models and motion generators with sufficient parameterization and flexibility to handle cultural variations.

ACKNOWLEDGMENTS

This research is partially supported by Office of Naval Research K-5-55043/3916-1552793, NSF IIS99-00297, and NASA 00-HEDS-01-052. Any opinions, findings, conclusions or recommendations expressed in this material are those of the author(s) and do not necessarily reflect the views of the National Science Foundation or any other sponsoring organization.

REFERENCES

Allbeck, J. M., & Badler, N. I. (2001). Consistent communication with control. *Workshop on Non-Verbal and Verbal Communicative Acts to Achieve Contextual Embodied Agents at Autonomous Agents'01*.

Amaya, K., Bruderlin, A., & Calvert, T. (1996). Emotion from motion. In *Proceedings of Graphics Interface '96* (pp. 222–229). Toronto, Ontario: Canadian Information Processing Society.

Axtell, R. E. (1991). *Gestures: The do's and taboos of body language around the world*. New York: Wiley.

Badler, N., Allbeck, J., Zhao, L., & Byun, M. (2002). Representing and parameterizing agent behaviors. In *Proceedings of Computer Animation* (pp. 133–143). Los Alamitos, CA: IEEE Press.

Ball, G., & Breese, J. (2000). Emotion and personality in a conversational agent. In Cassell, Sullivan, Prevost, & Churchill (Eds.), *Embodied conversational agents* (pp. 189–219). Cambridge, MA: MIT Press.

Bates, J. (1994). The role of emotion in believable agents. *Communications of the ACM, 37*(7), 122–124.

Becheiraz, P., & Thalmann, D. (1996). A model of nonverbal communication and interpersonal relationship between virtual actors. In *Proceedings of Computer Animation* (pp. 58–67). Los Alamitos, CA: IEEE Press.

Bindiganavale, R., Schuler, W., Allbeck, J. M., Badler, N. I., Joshi, A. K., & Palmer, M. (2000). Dynamically altering agent behaviors using natural language instructions. In *Proceedings of Autonomous Agents 2000* (pp. 293–300). New York: ACM Press.

Brand, M. (1999). Voice puppetry. In *Proceedings of SIGGRAPH '99* (pp. 21–28). New York: ACM Press.

Burgoon, J. K., Buller, D. B., & Woodall, W. G. (1989). *Nonverbal communication, the unspoken dialogue*. New York: Harper & Row.

Byun, M., & Badler, N. (2002). FacEMOTE: Qualitative parametric modifiers for facial animation. In *Symposium on Computer Graphics Animation* (pp. 65–71). New York: ACM Press.

Cassell, J., Bickmore, T., Campbell, L., Vilhjalmsson, H., & Yan, H. (2000). Human conversation as a system framework: Designing embodied conversational agents. In Cassell et al. (Eds.), *Embodied conversational agents* (pp. 29–63). Cambridge, MA: MIT Press.

Cassell, J., Vilhjalmsson, H. H., & Bickmore, T. (2001). BEAT: The Behavior Expression Animation Toolkit. In *Proceedings of SIGGRAPH'01* (pp. 477–486). New York: ACM Press.

Chi, D., Costa, M., Zhao, L., & Badler, N. (2000). The EMOTE model for effort and shape. In *Proceedings of SIGGRAPH '00* (pp. 173–182). New York: ACM Press.

Curious Labs. (http://www.curiouslabs.com. Last visited March 20, 2003.)

Eibl-Eibesfeldt, I. (1972). Similarities and differences between cultures in expressive movements. In R. Hinde (Ed.), *Non-verbal communication* (pp. 297–314). Cambridge: Cambridge University Press.

Ekman, P., & Friesen, W. (1975). *Unmasking the face*. Englewood Cliffs, NJ: Prentice-Hall.

Ekman, P., & Friesen, W. V. (1977). *Manual for the facial action coding system*. Palo Alto, CA: Consulting Psychologists Press.

Ekman, P., Friesen, W. F., O'Sullivan, M., Chan, A., Diacoyanni-Tarlatzis, I., Heider, K., Krause, R., Lcompte, W. A., Pitcairn, T., RicciBitti, P. E., Scherer, K., Tomita, M., & Tzavaras, A. (1987). Universals and cultural differences in the judgement of facial expressions of emotions. *Journal of Personality and Social Psychology, 53*, 712–717.

Elliot, C. (1992). *The affective reasoner: A process model of emotions in a multi-agent system*. Unpublished doctoral dissertation, Institute for the Learning Sciences, Northwestern University, Evanston, IL.

Ferraro, G. (1990). *The cultural dimension of international business*. Englewood Cliffs, NJ: Prentice-Hall.

Funge, J., Tu, X., & Terzopoulos, D. (1999). Cognitive modeling: Knowledge, reasoning and planning for intelligent characters. In *Proceedings of SIGGRAPH '99* (pp. 29–38). New York: ACM Press.

Gratch, J., & Marsella, S. (2001). Tears and fears: Modeling emotions and emotional behaviors in synthetic agents. In *Proceedings of Autonomous Agents '01* (pp. 278–285). New York: ACM Press.

Hall, E. T. (1966). *The hidden dimension* (2nd ed.). Garden City, NY: Anchor/Doubleday.

Hall, E. T. (1976). *Beyond culture*. New York: Doubleday.

Hartmann, B., Mancini, M., & Pelachaud, C. (2002). Formational parameters and adaptive prototype instantiation for MPEG-4 compliant gesture synthesis. In *Proceedings of Computer Animation 2002* (pp. 111–119). Los Alamitos, CA: IEEE Press.

Hewes, G. W. (1957). The anthropology of posture. *Scientific American, 196*, 123–132.

Hofstede, G. (1991). *Cultures and organizations: Software of the mind*. London: McGraw-Hill.

Isbister, K., & Nass, C. (2000). Consistency of personality in interactive characters: Verbal cues, non-verbal cues, and user characteristics. *International Journal of Human-Computer Studies, 53*, 251–267.

Johnson, W. L., & Rickel, J. (1997). Steve: An animated pedagogical agent for procedural training in virtual environments. *ACM SIGART Bulletin, 8*(1–4), 18–21.

Jones, R. M., Laird, J. E., Nielsen, P. E., Coulter, K. J., Kenny, P., & Koss, F. V. (1999, Spring). Automated intelligent pilots for combat flight simulation. *AI Magazine*, pp. 27–41.

Knapp, M. L., & Hall, J. A. (1992). *Nonverbal communication in human interaction*. Fort Worth, TX: Harcourt Brace Jovanovich College Publisher.
Kurlander, D., Skelly, T., & Salesin, D. (1996). Comic chat. In *Proceedings of SIGGRAPH '96* (pp. 225–236). New York: ACM Press.
LaFrance, M., & Mayo, C. (1978). Culture aspects of nonverbal behavior. *International Journal of Intercultural Relations, 2*, 71–89.
Lee, S., Badler, J., & Badler, N. (2002). Eyes alive. In *Proceedings of SIGGRAPH'02*. New York: ACM Press.
Lewis, H. (1998). *Body language, a guide for professionals*. New Delhi, India: Response Books.
MIRALab. (http://miralabwww.unige.ch/. Last visited March 20, 2003.)
Morawetz, C. L., & Calvert, T. W. (1990). Goal-directed human animation of multiple movements. In *Proceedings of Graphics Interface '90* (pp. 60–67).
Okabe, R. (1983). Cultural assumptions of East and West: Japan and the United States. In W. Gudykunst (Ed.), *Intercultural communication theory: Current perspectives* (pp. 21–44). Beverly Hills, CA: Sage.
Ortony, A., Clore, G. L., & Collins, A. (1988). *The cognitive structure of emotions*. Cambridge: Cambridge University Press.
Poggi, I., & Pelachaud, C. (2000). Performative facial expressions in animated faces. In Cassell et al. (Eds.), *Embodied conversational agents* (pp. 155–188). Cambridge, MA: MIT Press.
Rose, C., Cohen, M. F., & Bodenheimer, B. (1998). Verbs and adverbs: Multidimensional motion interpolation using radial basis functions. *IEEE Computer Graphics and Applications, 18*(5), 32–40.
Shuter, R. (1977). A field of non-verbal communication in Germany, Italy, & the United States. *Communication Monographs, 44*(4), 298–305.
Silverman, B. G., Johns, M., O'Brien, K., Weaver, R., & Cornwell, J. (2002). Constructing virtual asymmetric opponents from data and models in the literature: Case of crowd rioting. *11th Conference on Computer Generated Forces and Behavioral Representation*, SISO.
Stephenson, N. (1992). *Snow crash*. New York: Bantam.
Swartout, W., Hill, R., Gratch, J., Johnson, W. L., Kyriakakis, C., LaBore, C., Lindheim, R., Marsella, S., Miraglia, D., Moore, B., Morie, J., Rickel, J., Thiebaux, M., Tuch, L., Whitney, R., & Douglas, J. (2001). Toward the Holodeck: Integrating graphics, sound, character, and story. In *Proceedings of Autonomous Agents 2002* (pp. 409–416). New York: ACM Press.
The Sims. (http://thesims.ea.com/. Last visited March 20, 2003.)
Ting-Toomey, S. (1999). *Communicating across cultures*. New York: Guilford.
Triandis, H. (1995). *Individualism and collectivism*. Boulder, CO: Westview.
Yum, J. O. (1988). The impact of Confucianism in interpersonal relationships and communication in East Asia. *Communication Monographs, 55*, 374–388.
Zhao, L. (2001). *Synthesis and acquisition of Laban Movement Analysis qualitative parameters for communicative gestures*. Unpublished doctoral dissertation, Computer and Information Science, University of Pennsylvania, Philadelphia, PA.

CHAPTER

6

Enculturating Agents With Expressive Role Behavior

David R. Heise
Indiana University

Implementations of artificial intelligence (AI) often catalog a system's every state and event and use brute-force searches to find the event that is predefined as optimal for a given configuration of states. The search proceeds as follows: If this configuration of states is manifest, then select this event; if that configuration of states is manifest, then select that event; and so on. This approach—called *production-system modeling*—is remarkably successful with fast computers and well-bounded systems.

The production-system approach is less successful, however, when we move to virtual agents who have to become part of the human social world because social life is an open system of events rather than a bounded system. Humans generate such a large number of distinct events in social interaction that it is impractical to catalog every possible event. Consequently, a virtual agent based solely on production-system modeling would find itself impoverished in understanding and responding to human interaction partners.

An additional problem in employing the production-system approach to virtual agents is that if–then rules get bogged down in an explosion of special cases when states and events are related by nonlinear functions, as is common in social relations. Formulating nonlinear functions in terms of if–then rules requires partitioning each of multiple variables into minute states and setting outcomes for each combination of states on the different variables. The problem is not that modern computers cannot search the mammoth number of rules that results: they can. Rather, turning a nonlinear function into if–then rules raises a daunting engineering problem of accurately specifying rules for many subtle combinations of variables.

I argue that virtual agents must achieve the sophistication required for interaction with humans by incorporating quantitative models of generative processes in social life. The generative approach addresses the bounded-system problem by sectioning social events with an event frame that accepts various kinds of elements (e.g., agent, action, object, setting; Heise & Durig, 1997). An event is generated by filling the event frame with an element of each type. Although this approach still yields a bounded system, it enables consideration of a huge number of social events. For example, a somewhat limited model currently in use employs 713 kinds of agents or objects, 596 kinds of actions, and 344 kinds of settings, permitting the generation of more than a billion social events specified in terms of agent–action–object–setting combinations.

The quantitative modeling approach addresses the nonlinearity problem by retaining the nonlinear equations that describe social-psychological process. For example, a social event produces an impression of the event's agent in part via classic attitude-balance effects that can be represented by multiplying evaluations of the agent, action, and object involved in the event (Gollob, 1968; Heise, 1979). Equations incorporating these multiplicative terms can be applied directly in interpreting social actions, the equations can be manipulated mathematically to derive possible responses, and the equations can be combined with other equations to predict the agent's emotions (Smith-Lovin & Heise, 1988). Although the quantitative results have to be turned into qualitative elements—words or pictures—to be useful, retaining mathematical equations as long as possible provides great power and flexibility in analog modeling of the human mind.

In the remainder of this chapter, I try to illustrate the benefits of equipping virtual agents with a quantitative model of generative processes in social life. The next section provides an overview of the model to be used, discusses its application to creating an emoting face, and considers some problems in using the model as a foundation for virtual agents. The subsequent section offers a detailed illustration of using the model in a simple agent that might be operative at a World Wide Web site. The final section engages the distinctive issue raised in this volume: Do we need to make agents cross-culturally adaptive; if so, how do we do that?

ACT OVERVIEW

More than 25 years of research have established affect control theory (ACT) as a multifaceted basis for understanding microsociological processes. The theory's sociological insights, combined with its mathematical model, empirical databases from different cultures, and computer simulation program, give ACT considerable promise as one of the tools to be used in constructing multicultural virtual agents.

All aspects of ACT are documented in detail elsewhere (Heise, 1979, 1986, 2002; MacKinnon, 1994; Schneider & Heise, 1995; Smith-Lovin, 1990; Smith-Lovin & Heise, 1988). However, I begin with a précis to provide a shared framework for discussing how the theory might contribute to agent construction.

According to ACT, people construct and interpret social action so as to have important meanings affirmed by manifest experiences. ACT focuses especially on the affective component of meanings—on the sentiments attached to concepts. People cast themselves and others into specific identities during social encounters. They then engage in physical and mental work so that events create impressions that maintain the sentiments attached to their identities, as well as the sentiments attached to other categories of action like individual attributes, behaviors, and settings.

In ACT, sentiments are measured quantitatively on three culturally universal dimensions of affective meaning (Osgood et al., 1975): evaluation (the extent to which things seem good vs. bad), potency (impressions of powerfulness vs. powerlessness), and activity (impressions of activation vs. tranquillity). Measurements on each dimension vary from -4.3 (infinitely bad, powerless, or tranquil) to $+4.3$ (infinitely good, powerful, or active). Thus, a sentiment is represented by three numbers—evaluation measurement, potency measurement, and activity measurement—known collectively as an *EPA profile*. The same dimensions serve for measuring transient impressions of people, behaviors, and settings, and so transient impressions are also represented as EPA profiles.

ACT proposes specifically that people seek experiences in which transient impressions created by an event match preexisting sentiments about the event elements as much as possible. Impressions generated by an event are highly predictable transformations of feelings that exist before the event (Heise, 1979; Smith et al., 1994; Smith-Lovin, 1987). Therefore, people seek experiences that transform current feelings into new, sentiment-confirming feelings.

People protect their sentiments both through their own conduct and through their interpretations of past events. For example, a father could exonerate a daughter who disobeyed him, and this action by the father would begin to re-affirm fundamental sentiments about fathers and daughters. Alternatively, he could see the daughter who disobeyed him as greedy, manipulative, or mean—reconceptualizations that generate sentiments fitting the girl's behavior. Thus, in ACT, an event that deflects impressions away from sentiments might be repaired by implementing a new event or reinterpreting the interactants or other components of the problematic event.

In ACT, emotions are momentary personal states that reflect how events affect people. The emotion depends on the current impression of the person and on how that impression compares to the sentiment attached to the person's identity. For example, a person might feel anxious if made to seem bad and weak. However, an even more extreme response of feeling ashamed, desperate, or depressed is to be expected if the person has a particularly good and powerful identity in the situation.

ACT's mathematical model and empirical databases of sentiments are implemented in a computer program[1] called Interact, which allows prediction of normative actions, emotions, and interpretations (Schneider & Heise, 1995). For example, suppose we employ a database of U.S. sentiments and set up an analysis where one person is a father and the other a daughter. Interact indicates that the father might educate the daughter—a predicted behavior norm for father interacting with daughter in unexceptional circumstances. Interact also defines the emotion norms (Heise & Calhan, 1995; Heise & Weir, 1999) for the event: While educating his daughter, the father might have emotions like feeling generous, secure, or forgiving, whereas the daughter might feel humble, relaxed, or touched. Unexpected events can be examined with Interact to see what effects they have. For example, if a daughter disobeys her father, Interact predicts that the daughter must be feeling irate or angry and the father feels melancholy, apprehensive, or shocked. Interact additionally suggests that the disobeyed father might turn himself into a disciplinarian or see his daughter as greedy, manipulative, or mean.

The ACT simulation system currently allows analyses to be conducted for U.S., Canadian, Irish, German, Japanese, and Chinese (People's Republic) interactants so cross-cultural variations in interaction can be examined. For example, using the Japanese database of sentiments rather than American sentiments leads to subtly different predictions about a father and daughter. The Japanese father's normative action toward a daughter is less good and less active than the American father's, resulting in potential actions like counseling and reproving even in unexceptional circumstances. Were a Japanese daughter to disobey her father, her appropriate emotion would be feelings of irritation and impatience rather than anger, and the father might feel serene and peaceful in the face of her action, although ready to admonish her.

Graphic Emotion Display

Interact has another interesting feature with regard to virtual agents. It displays a facial expression for every emotion, and it constructs the facial ex-

---

[1]The Java program is available on the World Wide Web at the ACT Web site: www.indiana.edu/~socpsy/ACT/.

pression computationally from a three-number profile representing the pleasantness, vulnerability, and activation of the emotional state.

Paul Ekman's research (1982) indicates that emotional messages are constructed on the face mainly by the action of facial muscles shaping the mouth, eyes, and eyebrows. The brows may be in a neutral relaxed position or may be curved upward (as in surprise), flattened and raised (as in fear), flattened and lowered (as in sadness), or pulled down and inward (as in anger). The opened eyes may be neutral or wide open (as in surprise), have raised lower lids (as in disgust), have raised and tensed lower lids (as in fear), be squinting (as in anger), or have upper lids drooping and sloped (as in sadness). Aside from neutral, the mouth may be dropped open (as in surprise), corners pulled horizontal (as in fear), lips pressed tight (as in anger), squared outthrust lips baring teeth (as in anger), upper lip pulled up (as in disgust), corners down (as in sadness), or corners raised (as in happiness, with extra stretching for smiles, grins, or laughs). Blends can be formed by combining signs of two emotions; for example, arched eyebrows and a smile indicate surprised happiness. Other features of the face can be affected by these facial actions. The end of the nose may be raised by pressure from the upper lip, and the upper nose may get crinkled. Cheeks may get raised during laughter. The forehead may be wrinkled by pressures from the eyebrows.

Interact computes an EPA profile for the emotion of each interactant during an event from the sentiment attached to the interactant's identity and from the impression of the interactant being created by the event. That profile is converted into a facial expression by the following principles, which were surmised from Ekman's descriptions of facial expressions and his photographs of primary emotions:

- Curve the interactant's lips up when the interactant's emotion is positively evaluated, and curve the lips downward with negatively evaluated emotion.
- Increase the upward arching of the interactant's brows when the interactant's emotion is positive evaluated, and reduce the upward arching with negative evaluated emotion.
- Raise and separate the interactant's brows when the interactant's emotion has negative potency, and lower and close the brows with positive emotion potency.
- Move the interactant's lips—especially the upper lip—higher when the interactant's emotion registers positive potency, and move the lips downward with negative emotion potency.
- Widen the separation between the interactant's eyelids when the emotion is defined by positive activity, and reduce the separation with negative activity.

- Drop the interactant's lower lip and narrow the lips when the interactant's emotion involves positive activity, and raise the lower lip and draw the lips outward for negative emotion activity.

The adjustments of brows, eyelids, and lips are computed from formulas that correlate the amount of each movement to the size of the corresponding number in the EPA profile. For example, an emotion that has an evaluation of +1 causes upward curvature of the lips, and an emotion with an evaluation of +2 causes more pronounced upward curvature of the lips.

Each graphic head used in Interact is derived from a photograph of an actual person. Parameters in the formulas for computing motion of brows, eyelids, and lips are set artfully for different faces so that the computed expressions appear realistic. Some examples of Interact's virtual displays of emotions appear later in this chapter.

Design of an ACT Agent

ACT can contribute to the construction of a virtual agent in several ways. ACT defines behavior norms applying in a situation, with adjustments for the significance of recent happenings. Incorporating such understanding in a virtual agent would allow the agent to be socially intelligent in sequences of interaction with humans, even to the point of responding appropriately to a human's deviance. ACT also defines the emotions that individuals might have as they participate in events, and incorporating this facility into an agent allows the agent to seem emotionally responsive to ongoing events. Additionally, ACT indicates how people might change their minds about the identities of self and others as a result of happenings. Incorporating this facility into an agent would give the agent an apparent capacity for making judgments about human morality and character.

At the same time, ACT has limitations in serving as the infrastructure of an agent. ACT computes norms from interactants' definitions of the situation, but ACT does not furnish the definitions of situations. Thus, interactants (or a knowledgeable informant) must tell us what the situation is in terms of the identities being taken—for example, that the situation consists of an employer with an employee as opposed to a woman with a man. ACT predicts responses to events, but ACT does not recognize what events are occurring. Interactants (or an informant) must tell us the way in which each event is to be understood—for example, that one person is disciplining another as opposed to battering the other. ACT suggests what actions an individual might take to advance an interaction, but ACT does not specify how an action is elaborated into a hierarchy of subgoals and how those subgoals are unfolded instrumentally into specific verbal and nonverbal behaviors.

Despite its limitations, ACT can be used when building a virtual agent. The enabling principles for an ACT-based agent are as follows:

- Deal with a set of situations involving a limited number of reciprocal role identities. Have the human interactant select a starting identity at the beginning of the human–agent interaction, whereupon the agent's identity is set reciprocally.
- Deal with a set of potential events involving a limited number of behaviors. Have the human interactant behave by choosing among options that identify the human's intended action.
- Use production-system models to convert the human's actions or virtual agent's affectively generated responses to computer representations.

Having enabled an ACT agent in these ways, the steps in applying ACT are as follows:

- Create a database of empirically based sentiments—measured as EPA profiles—for the identities and behaviors.
- In the computer program that represents the virtual agent, implement ACT's impression formation equations to compute the impressions generated by events. Implement ACT's mathematically derived equations to compute normative behaviors, emotions, and reconceptualizations in emerging situations.
- Implement Interact's on-screen face to show the human interactant the virtual agent's emotional responses to events as they occur.

The Interact computer program for simulating social interaction can be used to build a rough mockup of an ACT agent. The following simple example is offered as a feasibility demonstration.

ILLUSTRATION OF AN ACT AGENT

Imagine a commercial Web site[2] with a virtual agent who has the identity of host. Humans arriving at the Web site are assigned the identity of visitor,

---

[2]The available dictionary of EPA profiles in the United States does not permit much subtlety in describing business relations—normal or deviant—so I have to use "gratify" rather than "buy from," "escape" rather than "leave the premises of," and "shoplifter" rather than a more appropriate label for deviants in Internet transactions. Work already is in progress that will improve this situation.

and a visitor can place an order—interpreted as gratifying the host—or the visitor can escape the host. The host can advise the visitor.

Additional role relations could be enabled by letting the visitor click buttons on the welcoming page—for example, a button to obtain information, invoking identities of novice for the human and expert for the agent, plus the behaviors of query and educate. I forgo additional normal relations in this illustration. I do want to include a deviant role relation, however, to show the agent's emotional capabilities. It allows the human to take the role of shoplifter and cheat the host, and the host can evict the shoplifter.

Immediately on the human's arrival, the host advises the visitor, say, by printing a table on the screen listing products and prices. The implementation of advising the visitor at different points of the interaction is a standard kind of programming problem that would be handled with flow-control structures and production-system models, and the procedures are not elaborated here.

After being advised, the visitor can escape or place an order. For this illustration, we suppose that the visitor places an order implemented via the familiar shopping cart paradigm. Now imagine that the shopping cart routine is so stupidly implemented that the visitor is able to change products' prices. The visitor notices this and does so, thereby cheating the host. Finally, imagine for the sake of illustration that, although the shopping cart is stupidly programmed, the programming of the agent is sophisticated enough to detect the felonious behavior. The host—realizing that the visitor is cheating—ejects the visitor from the Web site.

Interact Analysis

Simulating this interaction in Interact allows us to see how ACT contributes to implementation of an agent. First, we limit Interact to the required identities and behaviors, shown in Table 6.1 along with EPA profiles from males and females. (Interact contains hundreds of identities and behaviors beyond those in Table 6.1, but I pruned away all the others to simplify the analysis.[3])

---

[3]*Interact* setup:

1) Change "Basic Functions" to "Expert Functions".
2) Go to the "Find concepts" form. Enter 100.00 as maximum distance.
3) Go to the "Import entries" form. Check "Erase current entries". Leave the "Identities" radio button checked.
4) Enter the rows of text and numbers for each identity in Table 6.1, and then click the "Import entries below" button.
5) Check the "Behaviors" radio button, enter the rows of text and numbers for each behavior in Table 6.1, and then click the "Import entries below" button.

6. ENCULTURATING AGENTS

TABLE 6.1
EPA Profiles for Identities and Behaviors

	Males			Females		
	E	P	A	E	P	A
Host	1.20	1.10	0.28	1.51	1.18	0.45
Shoplifter	−2.02	−1.28	1.55	−1.93	−0.78	1.46
Visitor	1.39	0.44	−0.08	1.07	−0.22	0.01
Advise	1.12	1.32	−0.42	1.38	1.15	−0.66
Cheat	−2.13	0.02	0.77	−2.61	0.07	0.09
Evict	−1.64	1.41	0.83	−1.47	1.60	0.42
Gratify	1.49	1.56	0.17	1.25	0.94	−0.15
Escape	0.07	0.75	1.61	−0.48	0.46	1.33

Next we implement the sequence of events in the illustration.[4] Table 6.2 provides information printed by Interact about normative actions at each stage of the interaction. Figure 6.1 shows Interact's predictions about the agent's facial expressions of emotion as each event occurs.

When the human arrives at the Web site, the agent assigns the human the identity of visitor and assigns self the identity of host. Table 6.1 shows that these both are positively evaluated identities, but the host identity is substantially more potent and active than the visitor identity.

The agent initiates interaction by advising the visitor. The first numerical column in Table 6.2 shows that advising is a motivated behavior for the agent in the sense that it has a short distance (0.56) from the EPA profile for the agent's optimal behavior (1.55, 0.76, −0.18 as computed by Interact). Thus, advising the visitor is an action that confirms the meanings of the host and visitor identities, along with the meaning of the behavior, and the agent implements the event. Note that evicting the visitor—the other behav-

---

6) Go to the "Define situation" form. For viewer "Person 1" set: "Person 1" is "host" and "Person 2" is "visitor".

7) Go to the "Define events" form. Replace "No repeated behaviors" with "Behaviors retain meaning". Click the "Insert this event" button to generate an action of Person 1 toward Person 2 with behavior unspecified.

8) Go to the "Analyze events" form. Click the event to define the actor's optimal act. Click "advise" to implement the actor's optimal act.

9) Select behaviors and identities as needed to complete the transaction.

[4]Computations are based on male EPA profiles. Assume that male–female meaning differences are worth noting only if the mean ratings of males and females are greater than 0.5. Then Table 6.1 shows that males and females share essentially the same meanings for all of the identities and behaviors, with the following exceptions. The visitor identity is somewhat less potent for females than for males, cheating is somewhat more active for males, and escaping is somewhat more displeasing for females.

TABLE 6.2
Distances Between Actor's Ideal Behavior Toward Object and Available Behaviors

Prior Event	New Situation	After Host Advised Visitor	After Visitor Gratified Host		After Shoplifter Cheated Host
Actor–Object	Host–Visitor	Visitor–Host	Visitor–Host	Shoplifter–Host	Host–Shoplifter
Advise	0.56	1.01	3.02	25.54	3.20
Gratify	0.77	1.67	3.95	26.02	3.55
Escape	5.42	9.07	10.65	9.57	6.82
Evict	11.66	15.48	18.88	11.97	5.32
Cheat	15.03	19.20	20.46	6.00	12.59

Note. Implemented behavior underlined.

ior available to the agent—is far from the optimal behavior, with a distance of 11.66, and consequently the agent is unmotivated to perform this action.

Advising the newly arrived visitor generates a pleasurable, confident emotion in the host, although not one with much activation (EPA profile of 1.28, 1.24, 0.15 according to Interact calculations). Interact suggests some verbal labels, including feeling satisfied or appreciative. Interact additionally draws the facial expression for this emotional state as shown in the topmost image in Fig. 6.1. Such an emoting face presumably would be displayed on the Web page while advising the visitor, thereby personalizing the agent for the visitor.

The second numerical column of Table 6.2 gives information on the visitor's motivation for different acts following being advised by the host. The assumption is that the human also defines the first event as a visitor being advised by a host. The likelihood of this might be increased by a callout from the face saying something like, "Hi, visitor! I'm your host. Take a look at these special prices we're offering!"

The listed behavior that is closest to the visitor's ideal behavior at this point is advising, but that is a behavior in the host's repertoire, not the visitor's. The next smallest distance is for gratifying (a stand-in here for purchasing from). Gratifying the host is much closer to the visitor's ideal behavior than either of the other visitor's options—escaping or cheating—and so the visitor carries out computer activities for gratifying the host.

The visitor gratifying the host produces a new emotion for the host. Interact's descriptive words for the emotion include feeling at ease or contented. Interact's drawing of the facial expression for this emotion is the second from top in Fig. 6.1. This new expression would be displayed on screen as soon as the visitor begins placing an order. Note that this face can be viewed as politely attentive as you would expect from someone receiving your order.

6. ENCULTURATING AGENTS 137

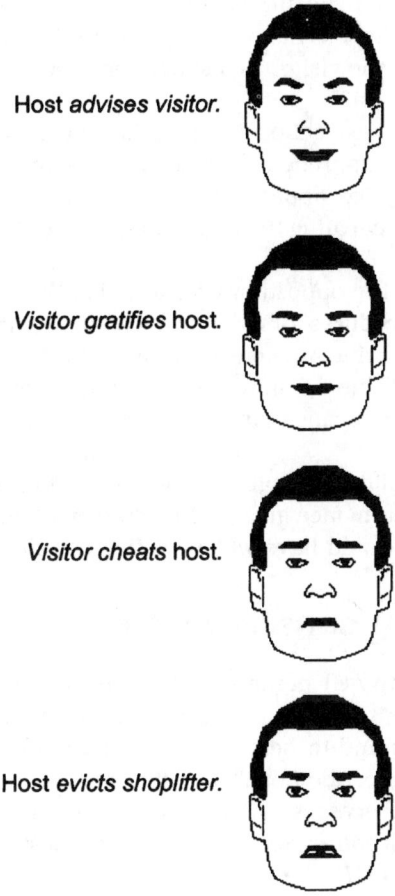

FIG. 6.1. Agent's facial expressions at each stage of the interaction.

According to Table 6.2, the visitor's next action should be gratifying the host again, and cheating the host would be the least likely of all available actions. However, seeing an opportunity for successful deviance, the visitor might switch to the shoplifter identity, whereupon cheating the host becomes the most likely possibility, as shown in the fourth numerical column of Table 6.2.

Assume that the visitor does change into a shoplifter and cheat the host. The host's emotion changes dramatically as soon as he realizes that the visitor is cheating him. Interact suggests that the host's new emotion might be called self-pity or perhaps embarrassment. The third face in Fig. 6.1 is the expression that Interact draws to represent this emotion, and this is the face that would appear on the screen when the visitor-shoplifter begins changing prices of items. This emotion continues almost unchanged until the host reidentifies the visitor as a shoplifter by asking, "What kind of person would cheat a host?"

The host's ideal next action is to advise the shoplifter, which would leave the host feeling apprehensive and uneasy. Advice perhaps would take the form of a dialog box appearing on the screen with information about the punishments for cheating the host. However, our illustration takes a different turn.

The host forgoes the opportunity for advising the shoplifter and instead implements the procedures for evicting the shoplifter from the Web site. Although the act of evicting previously was an unlikely behavior for the host, the last column of Table 6.2 shows that evicting now is fairly close to the host's ideal behavior because the host is dealing with a shoplifter who has just cheated him.

Evicting the shoplifter changes the host's emotion to feeling tense and aggravated. The bottom face in Fig. 6.1 is Interact's rendering of the expression that the agent would have while evicting the shoplifter.

CROSS-CULTURAL CONSIDERATIONS

Animating agents with ACT permits relatively easy adjustment of the agent's norms and emotions to fit local cultures. Sentiments attached to key roles like host and visitor and to behaviors like advise and cheat can be set to reflect the local culture. Additionally, coefficients in equations that model impression-formation processes can be changed to reflect local thought processes. Empirical procedures for both kinds of adjustments are well established (Smith-Lovin, 1987).

Heise (2001) analyzed sentiment measurements from six cultures— United States, Canada, Ireland, Germany, Japan, and People's Republic of China—to assess overall patterns of similarity. He found that sentiments about various kinds of people have surprisingly high correlations across cultures, implying that the cultures largely agree about what kind of people are relatively good or bad, relatively powerful or powerless, and relatively active or passive. At the same time, cultures diverge in absolute levels of sentiments. For example, mothers are evaluated higher than children in both Japan and China, reflecting the high correlation in sentiments. In China, however, mothers and children are both evaluated positively, whereas in Japan, mothers are rated as good, but children are rated as neither good nor bad. The nuclear family roles are more esteemed in China than in Japan.

Heise (2001) also found that sentiments associated with behaviors have high cross-cultural correlations on the evaluation dimension, implying that people around the world agree about what behaviors are relatively good and relatively bad. Notwithstanding the high correlations, cultures do vary in absolute evaluations of behaviors. For example, assisting someone is evaluated more highly than reprimanding someone in both Canada and Germany, but assisting is a much nicer act in Canada than in Germany, and reprimanding is a much nastier act in Germany than in Canada.

Heise (2001) found less cross-cultural agreement regarding the potency and activity of behaviors. Cross-cultural correlations of behavior potencies ranged from 0.74 down to 0.06. Cross-cultural correlations of behavior activities ranged from 0.81 down to −0.14. Thus, the major arena of cross-cultural diversity is not in the meanings of identities or the moralities of behaviors, but in perceptions of the strengths and activity levels of behaviors. This results in subtle cross-cultural differences in norms and emotions. For instance, behavioral norms attached to a particular identity might be different in different cultures because different behaviors are required to confirm the identity's potency and activity. Additionally, variations in behavior potency and activity could make an event with particular interactants and a specific behavior generate different emotions in different cultures—perhaps satisfaction and contentment in one place and excitement or even lustfulness in another place, or anger in one society and depression in another society.

Right from the beginning, researchers in the ACT tradition wondered whether processes involved in impression formation were the same across cultures. Smith-Lovin (1987) reported results of a meta-analysis based on studies of subjects from the United States, Ireland, and the Middle East. She found that basic processes in impression formation significantly influence evaluation of an actor in all of the subject groups she considered. Overall Smith-Lovin found considerable cross-cultural similarity in the equations predicting assessment of an actor's goodness and activity. The universality of core processes affecting evaluation of actors also has been confirmed in Canada by MacKinnon (1985/1988/1998) and in Japan by Smith, Matsuno, and Umino (1994).

However, Smith-Lovin (1987) reported interesting differences in how Arabic speakers assess an actor's potency as compared with English speakers. Subjects in the United States and Ireland (and also it turns out in Canada and Japan) feel that actors are especially powerful when they engage in potent actions, whereas the powerfulness of the object of action has little impact. For Arabs, however, resorting to strong actions implies that an actor is powerless. Meanwhile, object potency influences how potent an actor seems for Arabic speakers: Engaging powerful others makes an actor seem more potent, and acting on weak others costs an actor potency. Smith-Lovin forwent speculation about this difference because of methodological limitations in the studies she considered, but it is interesting to employ the technology of ACT to work out what such a difference could mean for international relations.

Here is an example.[5] Among people in the non-Arabic world, it is reasonable for a master nation like the United States to overwhelm a villain nation, like Iraq or Afghanistan. Doing so would make the winners feel satisfied and think of themselves as shrewd, whereas people in the overwhelmed villain nation should feel shocked and should start being more agreeable. However, this may not be how the same event plays for Arabic speakers even if they accept the same identifications of winner and villain nations. For Arabs, a winner overwhelming a villain is feeling anxiety, and such an actor could be labeled as cowardly, whereas the object of action is viewed as brave, although shocked by the action. Such an interpretation perhaps resonates with some rhetoric that comes out of the Middle East and might help explain why East–West disagreements are so intractable in the Middle East. At least the simulation result motivates research to confirm or disconfirm Arab differences in impression-formation processes.

Herman Smith has been exploring impression-formation processes in Japan and China. As I just mentioned, Smith, Matsuno, and Umino (1994) found that Japanese are actually similar to Americans in how they construct outcome evaluations about an event's actor, behavior, and object. However, Smith, Matsuno, and Ike (2000) found Japanese–American differences in interpreting an individual's state of being as designated by phrases like "angry admiral," "tactless doctor," or "rich professor." Overall, Americans seem to process states of being more simply than Japanese. First of all, Americans employ the same principles to assess how an identity combines with an emotion, as opposed to a trait, as opposed to a status characteristic. For example, Americans feel about the same whether encountering an angry admiral or a suspicious admiral. This is not so in Japan, where the admiral with a negative emotion would be evaluated more negatively than the admiral with a negative trait. Second, Americans average feelings about the modifier and identity to arrive at an evaluation of a modifier–identity combination and additionally employ a consistency principle in their assessment of the combination. Japanese also average and employ consistency. In addition, however, Japanese attend to intricate considerations, such as whether a potent or impotent modifier is describing a good, weak

---

[5]This paragraph reports results of Interact analyses (Schneider & Heise, 1995) using U.S. dictionaries with averaged male and female profiles. Simulation of the non-Arabic view was obtained with U.S. equations. Simulation of the Arabic view used the same materials except the equation for predicting actor potency was modified as follows. The coefficient for the effect of preevent behavior potency was changed from +0.47 to –0.47, and the coefficient for the effect of preevent object potency was changed from –0.04 to +0.20.

person as opposed to a bad, strong person. For example, this consideration leads Japanese to feel that a meek daughter is better than just a daughter and a meek gangster is more contemptible than just a gangster.

Smith's work in China is still in progress, so it is too early to review how Chinese impression-formation processes differ from processes in other cultures or how China's major regional cultures differ from one another. However, Smith's preliminary analyses suggest that Chinese impression-formation processes may differ from both Japanese and American processes.

CONCLUSION

Research by King (2001) suggests that an Internet subculture is evolving, such that only small differences exist among Internet users in sentiments regarding Internet concepts. This finding might encourage a hypothesis that no cultural adjustments are needed at all when virtual agents interact with humans on the Internet, the reasoning going like this. As an individual begins participating in Internet activities—e-mail, newsgroups, World Wide Web, and so on—that individual quickly acquires the standard sentiments attached to behaviors like spamming, cross-posting, and surfing; to identities like hacker, newbie, and Webmaster; and to settings like newsgroup, chatroom, or the World Wide Web. The individual gets socialized with essentially the same feelings about these entities as other people using the Internet. Consequently, a virtual agent need only implement the same subcultural sentiments to interpret the actions of experienced Internet users, act normatively with them, and display appropriate emotions.

At least one counterargument seems compelling.[6] The Internet subculture, like any other subculture, sets sentiments only for concepts that are directly related to the subculture, and Internet-related concepts may be too few to matter in building virtual agents who will serve commercial, medical, educational, political, and other functions. Virtual agents of these diverse types have to have the proper sentiments for non-Internet identities like salesperson, doctor, teacher, politician, and so on, and for non-Internet behaviors like guarantee, diagnose, flunk, vote for, and so on. Sentiments attached to these kinds of concepts are set by individuals' outside cultures. Thus, meaningful social interaction by an agent on the Internet depends on giving the agent sentiments that derive from the human interaction partner's locale and culture.

---

[6]Another counterargument is that individuals from different societies might process affective meanings differently, requiring that their mental processes be represented by different equations. However, it is not clear yet how important variations in mental processing are or whether socialization into a subculture homogenizes mental processes as well as sentiments.

REFERENCES

Ekman, P. (1982). *Emotion in the human face* (2nd ed.). New York: Cambridge University Press.
Gollob, H. F. (1968). Impression formation and word combination in sentences. *Journal of Personality and Social Psychology, 10,* 341–353.
Heise, D. R. (1979). *Understanding events: Affect and the construction of social action.* New York: Cambridge University Press.
Heise, D. R. (1986). Modeling symbolic interaction. In S. Lindenberg, J. S. Coleman, & S. Nowak (Eds.), *Approaches to social theory* (pp. 291–309). New York: Russell Sage Foundation.
Heise, D. R. (2001). Project Magellan: Collecting cross-cultural affective meanings via the Internet. *Electronic Journal of Sociology, 5*(3). (http://www.sociology.org/content/vol005.003/mag.html.)
Heise, D. R. (2002). Understanding social interaction with Affect Control Theory. In J. Berger & M. Zelditch (Eds.), *New directions in contemporary sociological theory* (pp. 17–40). Boulder, CO: Rowman & Littlefield.
Heise, D. R., & Calhan, C. (1995). Emotion norms in interpersonal events. *Social Psychology Quarterly, 58,* 223–240.
Heise, D. R., & Durig, A. (1997). A frame for organizational actions and macroactions. *Journal of Mathematical Sociology, 22,* 95–123.
Heise, D. R., & Weir, B. (1999). A test of symbolic interactionist predictions about emotions in imagined situations. *Symbolic Interaction, 22,* 129–161.
King, A. B. (2001). Affective dimensions of internet culture. *Social Science Computer Review, 19,* 414–430.
MacKinnon, N. J. (1985/1988/1998). *Final reports to Social Sciences and Humanities Research Council of Canada on Projects 410-81-0089, 410-86-0794, and 410-94-0087.* Guelph, Ontario: Department of Sociology and Anthropology, University of Guelph.
MacKinnon, N. J. (1994). *Symbolic interactionism as affect control.* Albany: State University of New York Press.
Osgood, C. E., May, W. H., & Miron, M. S. (1975). *Cross-cultural universals of affective meaning.* Urbana: University of Illinois Press.
Schneider, A., & Heise, D. R. (1995). Simulating symbolic interaction. *Journal of Mathematical Sociology, 20,* 271–287.
Smith, H. W., Matsuno, T., & Ike, S. (2000). The affective basis of attributional processes among Japanese and Americans. *Social Psychology Quarterly, 64,* 180–194.
Smith, H. W., Matsuno, T., & Umino, M. (1994). How similar are impression-formation processes among Japanese and Americans? *Social Psychology Quarterly, 57,* 124–139.
Smith-Lovin, L. (1987). Impressions from events. *Journal of Mathematical Sociology, 13,* 35–70. (Reprinted in Smith-Lovin & Heise, 1988.)
Smith-Lovin, L. (1990). Emotion as the confirmation and disconfirmation of identity: An affect control model. In T. D. Kemper (Ed.), *Research agendas in the sociology of emotions* (pp.). Albany: State University of New York Press.
Smith-Lovin, L., & Heise, D. R. (Eds.). (1988). *Analyzing social interaction: Advances in Affect Control Theory.* New York: Gordon & Breach.

CHAPTER

7

Toward Cross-Cultural Believability in Character Design

Heidy Maldonado
Stanford University

Barbara Hayes-Roth
Stanford University and Extempo Systems, Inc.

Most of the research within the field of interactive characters has concentrated on generating guidelines or models for believable individual variability among characters, often based on psychological personality models (such as Nass, Isbister, & Lee, 2000) or biological behaviors (see e.g., Blumberg, 1994). In contrast, we ascribe to Bates' (1994) proposition that "believability will not arise from copying reality," (p. 125) and suggest that recent research discoveries (such as those documented in Reeves & Nass, 1996) point toward a different fundamental orientation in our interactions with computerized systems, given the additional cognitive processing required to maintain awareness of the mediated and created nature of the interaction.

If—as this research suggests—what is primal of our interactions with the world is our suspension of disbelief, and ascription of human-like emotions to inanimate and barely animated objects, different research dimensions and design considerations become salient. Therefore, rather than concentrating solely on creating and fostering our users' "suspension of disbelief" through individual variability in synthetic characters, we suggest that simultaneous research efforts should explore the characteristics of breakdown moments when the deferment of questioning fails, when the "suspension of belief" is activated, so that character designers may avoid such pitfalls in the future.

Moreover, we advise that character design should expand its individual variability focus to include pursuits of cultural variability and summarize findings from existing research in cross-cultural communications that should inform the design of culturally specific characters. Characters as facilitators of cross-cultural relations and learning guides have been the focus of our research at Stanford University and Extempo Systems since 1996. Besides covering some of the applications developed, we also examine some of the success metrics for characters online that drive our work.

To that extent, we present a framework for the ten key characteristic qualities that animate characters—synthetic characters with lively autonomy and individual personas (Hayes-Roth & Doyle, 1998)—should possess, as perceived and ascribed by the interactors. These ten key qualities are: identity, backstory, appearance, content of speech, manner of speaking, manner of gesturing, emotional dynamics, social interaction patterns, role and role dynamics (Hayes-Roth, Maldonado, & Moraes, 2002). We briefly describe each before demonstrating their applicability to character design by localizing one of Extempo's animate learning guides to three different cultures: the United States, Brazil, and Venezuela.

Before proceeding, let us present an example of how the ten characteristics of animate characters become salient during a typical interaction with one of Extempo's Learning Guides: Kyra. Kyra is an animate character with whom visitors to the Extempo web-site (www.extempo.com) may interact by typing in textual utterances. Kyra was designed by students and researchers at Stanford University's School of Education and Computer Science Department and Extempo Systems Inc. The original team consisted of Oceana Blueskies, Heidy Maldonado, Jim Bequette, and Karen Amano. Since then Kyra has been localized to the to the Venezuelan culture by co-author Heidy Maldonado and to the Brazilian culture by Marcia Moraes, at the Computer Science Department of the Universidade Federal do Rio Grande Do Sul in Brazil.

Kyra's mysterious background and superpowers are designed to appeal to preteens, sometimes dubbed tweens in market research, because they are in between the childhood and adolescent market segments (see Hymowitz, 1998). Kyra seeks to motivate and educate this challenging demographic on artistic expression values and art history tendencies. She presents her enlightening Quest through an Internet browser's image displaying capabilities with her gestures, textual, and spoken utterances (please see Fig. 7.1 for a screenshot of Kyra's browser window). Moreover, Kyra has a complex natural language understanding engine, mood system, and learner's model that allow her to respond appropriately and adapt to even the most rude and stubborn students.

As she explains in her own words, Kyra will "never rest until she has conquered the invading forces from the realm of Negativity, Oppression and Ignorance who threaten to invade and contaminate our world of Art, Wisdom

7. CROSS-CULTURAL BELIEVABILITY

FIG. 7.1. Picture of the interface to interact with Kyra. Reproduced by permission from Extempo Systems, Inc.

and Creativity." Kyra is presented as a tween, slender but athletic, often executing martial arts maneuvers to illustrate her points, clothed in a futuristic uniform (see Fig. 7.2). However, having magical powers does not free Kyra from the typical preteen concerns and angst about which she empathizes with her audience. Kyra has her courses at the Academy, her best friends and her crush to worry about, and then there is that tell-tale colored streak in her hair that foretells her passion for the arts to even the most casual observer.

Table 7.1 features a sample record of Kyra's interactions—where a human visitor types in comments and the character reacts accordingly—to highlight each of the ten characteristics mentioned earlier, grouping qualities that are often expressed or presented simultaneously. We provide a textual description of Kyra's verbal responses—which appear in a text bubble above the character—and graphical actions.

FIG. 7.2. Kyra. Reproduced by permission from Extempo Systems, Inc.

TABLE 7.1
The Ten Key Perceived Qualities of Characters Expressed
in Kyra, the Extempo Art Learning Guide

Quality	Description	Expressed in Kyra through...
Identity	Who is Kyra?	Kyra is a trendy girl of about twelve, whose first statement to the visitors is: "Greetings Mortal! I am Kyra, Chief Defender of the Arts!"
Backstory	What shaped who Kyra is?	"Professor Yamamoto, alias Captain Y, found me on his doorstep as a baby and taught me all about defending creativity from ignorance's dark side!" Kyra's adoption by Captain Y and educational emphasis on artistic understanding drives her passion for the arts.
Appearance	How does Kyra's embodiment limit, expand, and communicate who she is?	Kyra is presented as an anime rendition of a trendy young teenager, with expressive eyes, a streak of painted hair, wearing a futuristic outfit.
Content of Speech	What does Kyra want to talk about, what does she avoid, and how does she say things?	Kyra enjoys discussing her present her friends, magic powers, favorite pastimes, and her job, often using trendy, teenage words such as "like," "as if," and "whatever."
Manner of Speaking and Manner of Gesturing	How does Kyra express herself verbally and non-verbally?	Kyra often teases and interrupts the interactor, accompanies some utterances with shrugs, winks, and even martial arts movements.
Emotional Dynamics	What angers or excites Kyra, how does she express it, and how long does this emotional charge last?	Kyra thoroughly resents derogatory comments about her intelligence, is embarrassed by any references to her secret crush at the Academy, and deeply saddened by the passing of Professor Yamamoto.
Social Interactions Patterns	How does Kyra address and react to those with whom she interacts? Does it change depending on gender, age, position, knowledge, or time she has known the interactor?	Kyra is a young teenager intent on educating those who interact with her on the importance of art and the accomplishments of many artists. She addresses interactors as peers. She chooses what to expound on based on her interactor's knowledge of the topic.
Role	What value does she add to the web-site? What is her "job"?	Kyra is employed at the Extempo site to be a learning guide introducing young teenagers to art history and appreciation. Currently, her most popular lesson is the life and times of Vincent Van Gogh. Her role affects every aspect of her performance as she constantly directs visitors to topics, activities, and web-pages related to it.
Role Dynamics	How does she relate to human interactors in accordance with her role?	Kyra is a tutor. How she reacts to users' errors and triumphs is dependent not only of her moods at the time, but also on her personality and cultural-specificity.

MOTIVATION

As some researchers in the field have recognized (see Perlin & Goldberg, 1996) an artistic emphasis in character design may be preferable to the psychological or biologically driven models of recognizable human behavior. Laurel (1991) observed that "we take pleasure when—and only when—even the surprises in the character's behavior are casually related to its traits," (p. 145) and the complexity in character that would result from an accurate model of human personality makes this enjoyable predictability harder to deploy in the formulation of the character's actions. Laurel therefore suggested that "somewhat ironically, dramatic characters are better suited to the roles of agents than full blown simulated personalities" (p. 145).

Biologically correct behavior may not only detract from our enjoyment, but additionally, as Hayes-Roth and Doyle (1998) remark, biologically correct behavior may be subtle and difficult for observers to interpret, and appear stilted and robotic when implemented through the present limitations of graphical animations. Psychological modeling may believably render individuals with significant deviations accurately, but generating everyday humans may require extensive interaction sessions to capture the uniqueness of each individual. In contrast, perhaps due to what Bates (1994) described as the cumulative experience of producing hundreds of thousands of individual, hand-drawn, flat-shaded line drawings, moved frame by frame, "forced animators to use extremely simple, nonrealistic imagery, and to seek and abstract a precisely that which was crucial . . . to express the essence of humanity in their constructions" (p. 122).

In the English language, the verb *to believe* has historically been linked to a person, object, or statements' credibility and trustworthiness, often associated with religious dogmas. Within the entertainment industry, enterprises associated with the production of fiction have gradually and etymologically disassociated *believability* from trustworthiness and linked it directly with the audience's engagement in the novel, story, film, or performance. This engagement is most often defined as empathy with the characters' emotions and predicaments, through the suspension of awareness of the narrative's created—or artificial—nature. The perception that believability is the result of a willing suspension of disbelief originates from British poet and literary critic Samuel T. Coleridge's (1772–1834) autobiography, *Biographia Literaria*.

Since then entire books have been written advising aspiring writers and filmmakers on how to achieve this suspension of disbelief—from character emotions to camera angles. Quickly adopted into the gaming industry by the 1995 Electronic Entertainment Exposition in Los Angeles, during which Sega CEO Tom Kalinske echoed the rallying cry of the entertainment industry for the next decade: "Consumers will demand immersive experiences that create suspension of disbelief." Sony spokespeople, not do be outdone, claimed that their new Playstation "no longer requires the video gamer to suspend disbelief."

Achieving believability continues to be the holy grail of character design, whether expressed through animation, acting, filmmaking, writing, or programming, and deservedly so. Widespread audience engagement leads not only to repeat visits and word-of-mouth recommendations, but also to commercial opportunities (as the Disney empire has demonstrated), brand loyalty, and even other less pecuniary returns. As Reeves (2001) reported, quantitative studies have shown that merely placing a character with natural language processing next to a form's text entry box increases dramatically the number of search words per query (from approximately 3.3 words when a character is not present next to the text-box to 8.5 words when a character appears next to the text box) and simultaneously increases the grammatical correctness of the statements. These invaluable aids to the search engine mechanisms were accompanied by appealing results for web-based commerce. The 15,000 visitors who interacted with a customer service representative character were three times more likely to reveal personal information than the 75,000 visitors to the dell.com web site who did not interact with the character. The visitors who interacted with a character were twice as likely to add items to the shopping cart and accepted twice as many suggestions or recommendations from the character. Moreover, visitors interacting with the character represented a 25% to 35% increase in up-selling merchandise and service agreements' rates.

Reeves' findings arise from his analysis of Finali Corporation's characters on the dell.com web site, characters supported by finite-state machine dialog models, such that visitors would choose from a small set of options to converse with the character. In contrast, Extempo's characters have natural language conversational abilities, and accordingly complex personalities and moods, to appropriately handle most of the visitors' freeform utterances. Unlike Reeves' Finali characters, often deployed in single-visit trouble shooting or purchasing pages, Extempo's characters tend to be deployed in sites where repeat visits are desirable with or without explicit commercial goals. A typical example of Extempo's characters is Mr. Clean, which Extempo designed for Procter and Gamble's forty-year-old brand icon and who generated 80% repeat visits from customers seeking "his" advice on cleaning challenges.

Perhaps a more telling example is found in another Extempo character, Jack, the a furry cream-colored wheaten terrier puppy who could converse with visitors through his free-form text entry box and text-to-speech engine at the unfortunately demised Petopia.com, a company specializing in home delivery of pet supplies. Jack's antics and barks persuaded approximately 75% of the site's visitors to reveal personal profile information. From e-mail addresses to hometown and gender, to information about their pets and hobbies, names, web experience, and experience with the company, visitors happily divulged personal details of their everyday life. Interacting with Jack led to a typical (90% of visitors) stay of twelve minutes, with each visitor typing in approximately fifteen sentences to Jack on topics as varied as discussing their own pets, Jack's haiku and gossip stories of Jack's animal friends, as well as Petopia's philanthropic "Bottomless Bowl Service" program and the "Million Pet Mercy Mission," and watching Jack perform animated tricks onscreen.

During the first four weeks of 2000, approximately 30% of the 1,000 unique visitors who interacted with the delightful Jack returned to interact with him within two weeks (often several times), and more than 60% accepted Jack's suggestions to visit specific target pages—from Petopia's philanthropy pages to more tailored content on products for the users' pets—compared with the 1% of visitors who clicked through to the philanthropy pages spontaneously. Jack has such a following that Extempo is hosting him at their own site, now that Jack's previous cyber home has disappeared. You can interact with him by following the "R&D" link from the "Demos" page at Extempo's website (www.extempo.com), and choosing the "Arena" link off the left navigation bar.

Despite the difference in interaction abilities, domain of deployment, character complexity, and goals, Reeves' Finali findings match those of Extempo and substantiate our claim that interacting with computer characters triggers certain behavior patterns in visitors. These patterns—such as the shift from transaction to conversation Reeves reported in his text-box example—signal an interaction that does not consider the created or artificial nature of the character. Rather than pausing and considering whether the computer character next to the text box is an entity, we interact in a primarily social context as if with another sentient being.

If—as the findings and research we have outlined thus far suggest—what is primal of our interactions with the world is our suspension of disbelief, rather than an awareness of the created nature of the objects we interact with in our daily activities, if this suspension of disbelief predates the stimulus and if, moreover, our primary mode of interaction with the world is to ascribe human like emotions to inanimate and barely animated objects, different research dimensions and design considerations become salient. Far from aiming at designing interactions with created media to elicit a willful *suspension of disbelief*, we should aim to design interactions where the deferment of questioning fails and *suspension of belief* is triggered. Moreover, we should consider that what may be willful and requires considerable cognitive effort is this *suspension of belief*. Rather than enumerating different possibilities for creating believable characters and experiences that create suspension of disbelief, if this suspension of disbelief is so primal and basic a human instinct, future believability studies should consider addressing the breakdowns that cause audiences to perceive the created nature of the characters and experiences.

Therefore we ground our exploration of the dimensions of character design on identifying situations where the *suspension of belief* may be disrupted, and join character animators in their quest for believability, rather than focusing on replicating or simulating realistic behavior. "That is what we were striving for . . . belief in the life of characters" (Jones, 1990, p. 13). Although there are identifiable advantages to mimicking and matching users' central psychological tendencies, in particular within the realm of persuasion (Nass & Reeves, 1996), we believe that for a character to be engaging and believable he or she need not match our personality. However, even though believability is not dependent on accurate realistic simulations, it is highly dependent on the viewers' ascription of emotion to the created characters, as these emotions are key to revealing how and when the characters "appear to think and make decisions and act of their own volition. It is what creates the illusion of life" (Thomas & Johnston, 1981, p. 9).

Although adept storytelling and narrative construction also play prominent roles in engaging the audience, as Flannery O'Connor observed, "it is the characters who make the story, and not the other way around." Believable characters drive every engaging story, yet the types of characters needed for an engaging story may occasionally have to be, of necessity, characters whom we may not wish to emulate, nor those who resemble us, just as engaging individuals we encounter every day need not match our personality or lead exemplary lives. Many of the greatest characters of literature have intentionally been crafted as scoundrels, including O'Connor's. Thomas and Johnston (1993), legendary Disney animators, even claimed that without their villains the beloved Disney heroes would lose their appeal because only a worthy nemesis forces a hero to rise in defense of what is right.

Even if we would prefer not to lead the life of Hamlet, Don Quixote, David Copperfield, Snow White's Queen, or Scarlett O'Hara, and would not make the same choices as Wile E. Coyote and Donald Duck, we nonetheless appreciate their ingenuity and endurance, enjoy their triumphs, and share their sorrows. Far from limiting the appeal of villains to static media, the success of recent interactive games—such as Rockstar Games' Grand Theft Auto and Electronic Arts' Dungeon Keeper—have shown that temporarily embodying evil roles can be as enthralling and enjoyable an experience for the audience as interacting in a fantasy world through a heroic character.

Therefore, rather than looking for a central tendency among all people and our characters, we seek instead a metaphorical match similar to that of characters in narratives, film, and television, giving our characters interesting variations, similar to the interesting variations we see among people, as this is an essential element of human—and characters'—appeal. Chuck Jones creator of many beloved Warner Brothers' cartoon characters, recommended to budding animators that "it is the individual, the oddity, the peculiarity that counts ... eschew the ordinary, disdain the commonplace. If you have a single-minded need for something, let it be the unusual, the esoteric, the bizarre, the unexpected, such as a cat hooked on grapefruit" (Jones, 1990, pp. 14, 20).

Interesting sources of variation need not just be individual, but may also be cultural despite the relative research preference in character design for exploring individual personality variations. The role of cultural sameness of interactors has been proved to be as important as psychological matching, particularly in terms of persuasion. In particular, the studies of Osbeck, Moghaddam, and Perreault (1997) across multiple cultures suggest that individuals and groups are more inclined to like those characters that they perceive as culturally similar to themselves.

Recent research by Nass et al. (2000) also shows that characters from the same ethnic background as the interactor are perceived to be more socially attractive and trustworthy than those from different backgrounds. Moreover, participants in these experiments also conformed more to the decisions of the ethnically matched character and perceived the character's arguments to be better than those of the ethnically divergent agent. Therefore, cultural localization is critical even for researchers and designers who craft their characters to actively match the user's ethnicity, and perhaps also central psychological tendency. Untapped roles for which characters of certain cultural backgrounds are desirable, independent of the cultural background of the interactor, are also plentiful even outside the realm of persuasion: teaching culturally specific subject matter—from language practice to cooking—castings for entertainment and narrative roles, among others.

Localization of characters must go beyond the obvious language translation. Just as adapting interfaces to different countries requires not only changing the language, but also redesigning the interface's appearance, content, and interactive behaviors that could be considered inappropriate for that country (Miller, Kozu, & Davis, 2001), adapting a character to a different culture involves careful reconsideration of each of the ten key characteristics identified earlier. These characteristics—*identity, backstory, appearance, content of speech, manner of speaking, manner of gesturing, emotional dynamics, social interaction patterns, role* and *role dynamics*—both define and are defined by each character's unique idiosyncratic behaviors and signature personality traits, as well as by the character's cultural grounding. Even as each human personality is unique, each culture tends to evoke specific modes of adjustment and reactions in different situations (Ewen, 1988).

These cultural variations among people are responsible for the considerable diversity of specific cultural norms even among populations located geographically near and for many embarrassing moments in multicultural exchanges. They give us cultural stereotypes that allow members of particular cultures to recognize each other and ascribe cultural backgrounds to individuals they encounter in everyday life. Cultural stereotypes need not be limiting, pejorative, negative definitions. Stereotypes can be positive constructs—ways of describing patters of diversity in human behavior that make our interactions with human beings rich, interesting, and delightful—without offending the represented culture. The ideal balance, where both the represented culture recognizes itself in the character and, simultaneously, members of other cultures also ascribe the desired background to the same character, without offending either group of potential users, hinges on extensive user testing of culturally- or group-specific characters with user populations of both the represented culture and other cultures.

A MULTICULTURAL PERSPECTIVE OF THE TEN KEY QUALITIES OF ANIMATE CHARACTERS

We find culturally specific tendencies in each of the ten key qualities of animate characters mentioned earlier, and to illustrate how these qualities can be instantiated toward providing a blueprint for the design of culturally specific characters, we have built three culturally specific instances of the character Kyra. American Kyra shares with Brazilian Kira and Venezuelan Kirita the same identity, embodiment, and teaching role, yet differs in the remaining seven key qualities of animate characters we enumerated: backstory, content of speech, manner of speaking, manner of gesturing, emotional dynamics, pattern of social interaction, and role relationships.

By maintaining the embodiment, identity, and teaching role constant, we aimed to highlight the significant degree of cultural specification possible within the typical project constraints and deadlines. That is, given an existing successful character, how can the key qualities we identify guide the localization of the character to other cultures? Ideally, the cultural specificity and focus of a character pervades every aspect of the character's development, yet consistency of character and role across the various cultures she or he operates in is critical for the purpose or branding of the character's sponsors. Moreover, often the underlying engine and animations prove too expensive and time-consuming to localize for every culture in which the animate character will operate.

Let us now describe and highlight how each of the qualities enumerated earlier and exemplified in Kyra are particularly salient within the framework of cultural specificity.

Ten Key Qualities of Animate Characters

1. Identity
2. Backstory
3. Appearance
4. Content of speech
5. Manner of speaking

6. Manner of gesturing
7. Emotional dynamics
8. Social interaction patterns
9. Role
10. Role dynamics

Identity

Who is the character? *Identity* as a category encompasses not only the character choice in terms of demographics and description, but also the personality traits and qualities of the character, including what she or he likes and dislikes, and the character's signature and idiosyncratic behaviors.

As successful implementations of text-based characters demonstrate, from Weizembaum's (1966) Eliza to the more recent Julia study by L. Foner's (1993), a character's *identity* is a distinct category from their *appearance*. Disembodied text characters amply convey a sense of personality and identity to the extent that participants often shared personal information—even private details—with both agents mentioned, and even asked the popular chatterbot Julia to go on off-line dates without realizing her digital nature. An animate character's appearance merely represents graphically certain physiological aspects—such as race, size, build, hair color and style, weight, age, and gender—and certain elements of the character's personal history and temperament—such as socioeconomic background and style. Theories of self abound, ascribing aspects of identity to genetic profile, personal history, and experience, even religiously determined. Yet the aim of this chapter is not to settle such a debate or to specify the artistic and theatrical requirements for engaging personalities, but rather to highlight the percolation of cultural norms and roles as they impact the character's identity in every facet of the interaction. For example, by choosing a child as a character, cultural conventions dictate the appropriateness and form of the character's activities expressed in the manner of addressing visitors of varying ages.

By choosing a female character to specialize culturally, in Kyra, we are tackling this challenge head on as the social roles of women vary across cultures with a greater spectrum and nuance than those of men. In some cultures, women have been excluded from explicit participation in the business world until recently, and in some fundamentalist societies, women are explicitly forbidden from interacting with members of the opposite sex outside the family unit. The most banal conversational exchange—such as appropriate responses to flirtation and banter between the genders—has the potential for breaking the engagement of the visitor. Even in the so-called developed countries, people still assign certain professions and roles to particular genders and, on average, express less confidence in a voice-over statement when the stereotypical gender assignment is reversed (Nass et al., 1997).

Backstory

We use the term *backstory* to refer not only to cultural variations in individual reactions, but to any self-recognized individual experience and history that had a direct influence on the character's personality, as well as current facts of the character's "life" outside the screen.

We can draw a parallel between the aspects that we group under a character's backstory and what a person would write in his or her diary or memoirs, as opposed to unrecognized and unconscious personality traits and quirks, which are classified under the *identity* category. Thus, a character's backstory encompasses family relations, friendships, favorite sports and colors, important celebrations, love interests, financial status, and political and religious affiliations, yet excludes the fact that the character unconsciously rubs his or her nose when embarrassed and tends to pedantically interrupt interactors, among other personality traits. Even as each human personality is unique, each culture tends to evoke specific modes of adjustment and reactions in different situations (Ewen, 1988). Therefore, every character inevitably highlights some cultural grounding from his or her backstory in their commonplace interactions.

When enumerating the ten key qualities of animate characters, we alluded to their intertwined nature, as in a well-crafted character every one of the eight remaining characteristics should convey aspects of the animate character's identity and backstory. Let us continue using our case study of Kyra's localization to three different cultures to illustrate the possibilities available to character designers in these eight remaining qualities: appearance, content of speech, manner of speaking, manner of gesturing, emotional dynamics, social interaction patterns, role, and role dynamics.

Appearance

Appearance refers to the encoding of each character's identifying physiological form—such as race, size, build, hair color and style, weight, age, gender—and certain elements of the character's personal history and temperament—

such as socioeconomic background and style—in the chosen embodiment of the character, as well as the representation of this embodiment.

Appearance affects the character's effectiveness and credibility at performing their assigned role and directs the patterns of interaction. Characters without graphical representations still have an appearance encoded by the word choices of their descriptions—for example, one can describe a character with the same identity as fat, chubby, round, pot-bellied, overweight, and even robust, with each word alluding to a different interpretation, and cultural implication.

Before the character utters a single word, before the web-page is completely loaded, the visitor has already processed the subliminal cues embedded in the characters' representation, such as the relative status and occupation of the interactors, and formed a model of what pattern the ensuing interaction will follow. Even a subtle distinction as status relations between peer-to-peer and superior-to-underling can be deduced from the social norms and cultural cues the character's appearance presents.

Character designers must take care to understand the gender roles, traditional attires, and cultural norms that impact every user's experience differently just by looking at—or reading a description of—the character. It is critical to choose a representation of the character that will appeal to the targeted population and an attire that will also be considered acceptable—or at least neutral—for the target population and role. For example, an American web-guide at an online wedding registry would probably wear a white dress, whereas in a site aimed at Chinese brides, the character would wear red clothing and avoid white, as white is considered a color of mourning in Chinese culture. Similarly, an edgy punk cartoon-rendered character may not be the best match for a role mediating executive conference calls.

The first design recommendation we can offer for a character's appearance is to design the character as attractive as possible for the target audience. In his famous book *Influence: The Psychology of Persuasion*, Cialdini (1993) explored the often untapped potential that physical attractiveness offers in social relations. Summarizing psychology research studies Cialdini showed how physical attractiveness often triggers a "halo effect," where one positive characteristic of a person dominates the way the person is perceived by others.

Cialdini evaluated how attractive people are seen as possessing better personality traits and intellectual capacities from early on; good-looking elementary school children are presumed to be more intelligent than their less attractive classmates by teachers, and their aggressive acts are perceived as less naughty. Attractive political candidates in Canadian elections received more than two and a half times as many votes as unattractive candidates. Attractive candidates for a private job offering were chosen over those with better qualifications; handsome men receive lighter penal sentences and reduced monetary penalties and are even twice as likely to avoid jail as unattractive defendants. Furthermore, attractive people are more likely to obtain help when in need, from members of either gender, and are more persuasive in changing the opinions of an audience even when the participants "denied in the strongest possible terms that their votes had been influenced by physical appearance" (Cialdini, 1993, p. 171).

Certain features and proportions believed to be universally appealing as evolution mechanisms and reproduction strategies include estrogenized female faces with small chins, androgenized male faces with large chins and cheek bones, and female bodies with optimal waist-to-hip ratios around 0.7. However, as the American adage states, "beauty is in the eye of the beholder," and we should keep in mind that attractive features and fashions are not only culturally specific, but also time specific. Standards of beauty are constantly changing, from Rubenesque curves to the rail-thin look sported by models Twiggy or Kate Moss; some symbols of status in African and Eastern culture may seem to the average Westerner as differing from the expected norm, as the prevalence, acceptability, range, and rate of hair color change in certain cultures to individuals from other cultures.

For example, in choosing to maintain Kyra'a appearance constant, we provoked stronger than anticipated reactions in our Brazilian testers because of the incongruity of her colored hair within their daily experiences and culture. This cultural response may be attributed to the greater homogeneity in appearance and behaviors of collectivist culture, which we discuss in greater detail later, and to the relative lack of exposure of the Japanese animation tradition of manga and anime in Brazil, which American Kyra was designed to emulate. Although not many Japanese animes may have been translated to Portuguese, American Kyra reflects this style and tradition to the extent that the character has been known to spout haiku poetry on occasion.

Gard, in a recent article for *Game Developer* magazine, added consistent costume and attitude poses to physical attractiveness in his recommendations for game character designers. He suggested exaggerating particular elements of the characters' personality to symbolize the essence of the character through simple clothing articles or patterns that make the character easier to remember and recognize, such as Hermes' winged feet and Spiderman's black on red web pattern. Gard suggests minimizing complexity of the costume and a limited palette of colors while increasing the dynamism of the character's trademark poses. We discuss identifiable poses and gestures in greater detail under the *Manner of gesturing* category, yet the importance of costume to convey cultural specificity may occlude a general character design guideline that Gard barely indicated.

Accoutrements and accessories of the character tend to convey significant non-verbal information about the character's *role* and *role dynamics*, yet these objects are imbued with varying cultural meaning. When designing a character, it is suggested that the gestures, clothing, and accessories extend the character's *role*, *role dynamics*, and *backstory* in such a way as to aid the visitor's understanding and identification of these facets of the character. To borrow an example from the movie industry where accessories convey the difference in role dynamics, Professor Henry Jones Sr. (Indiana Jones' father) and Lara Croft share the role of intrepid archaeologists embroiled in shady, dangerous research, in their signature films, yet the glasses and twill jacket of Professor Jones are a far contrast to the guns and tight, revealing clothing Lara wears.

Based on this research, the design recommendation we can offer is to choose representations whose appearance provokes no strong negative reactions in any culture, whenever the expectation of cultural localization arises, rather than representations that inspire positive reactions only in some cultures, but simultaneously inspire strong negative reactions in other cultures. Nonetheless, we must temper that recommendation with the warning that choosing too bland and adaptive a character might detract from the character's appeal in any culture. As the Microsoft paperclip character illustrates, designing a character within the zone of indifference of several cultures will not lead to its success as a localized character for any of those cultures (Reeves, 2001) unless other key aspects of his—or her—persona are localized to each culture.

Content of Speech

When considering how an agent should change between countries, language is the first porting issue that comes to mind, yet the available topic choices for the character may present even more salient cultural cues.

We analyze some of these topic knowledge requirements in detail shortly, but first let us address the importance of matching the language and dialect of the character with the culture it represents as closely as possible. Our Brazilian Kira's dialog was authored to reflect her Brazilian roots and upbringing, quite distinct from the continental Portuguese variations. The difference between these two forms of Portuguese is so noticeable to fluent speakers that it is common for Portugal-born travelers in the interior of Brazil to be mistakenly identified as Italian. Beyond spoken and written language, our characters should speak in a way that the intended audience understands, which encompasses idiomatic expressions, slang, and colloquialisms, as well as sensitively tailoring word choices to the semantic shades of meanings particular to each cultural group. Although they both speak English, an American character would utter sentences whose syntactic content would be very different than a British or Australian agent, even if the semantic content remains the same. For example, the French word "gosse" would mean "kids" to a Frenchman, but a very private male body part to someone from Quebec!

By concentrating on the syntactic content of the character's utterances, character designers run the risk of overlooking the cultural diversity present in the semantic content. Alluding to certain culturally diverse topics can enhance the character's perception as a representative of specific cultures—local historical incidents, geographical details known only to locals, sports, favorite pastimes, recent local events, holidays, and humor, among others (Axtell, 1985). Recent research has shown that certain topics previously considered as anathema to cross-cultural conversations—particularly money, politics, and religion—may have certain positive influences when raised by a helper agent. Isbister, Nakanishi, Ishida, and Nass (2000; see also chap. 11) studied the effects of a helper agent represented as a blue dog, which suggested either safe or unsafe topics at awkward pauses in the conversation between one Japanese and one American partner. In this study, American participants rated the interactions mediated by an agent who suggested unsafe topics more interesting, and Japanese found it more comfortable and engaging to continue than the interaction mediated by an agent who suggested safer conversational topics. Moreover, the agent's topic suggestions affected each participant's perception of self and of the other participant in the conversation, as well as the agent in surprising ways, underscoring the need for further research in cross cultural communication. For example, Japanese participants rated the agent who suggested unsafe topics as nicer and more competent than the agent who suggested safe topics of conversation, whereas American participants found the agent who suggested unsafe topics more blunt and domineering, less restrained, less friendly, and, not surprisingly, less appropriate than the agent who suggested safe topics of conversation.

The treatment of certain sensitive topics traditionally associated with high emotional content—topics dealing with embarrassing or dishonorable situations (e.g., losing your job), rites of passage (such as birth and death)—often vary across cultures. Each of our Kyra triplets is quite different from one another in terms of their respective cultural topic knowledge. For example, American Kyra composes her own haikus, loves Jackson Pollock's art and eats mint chocolate chip gelato. Venezuelan Kirita enjoys dancing Merengue, the art of Venezuelan kinetic sculptor Carlos Cruz Diez, and loves the traditional corn-flour arepa sandwich. Brazilian Kira prefers the martial arts inspired *capoeira* dance to soccer, although she shares her countrymates' pride in legendary soccer star Pele, and admires the art of Aleijadinho, a famous Brazilian Baroque artist from the 18th century.

A telling example of ways in which culture affects the characters' presentation and identity lies in how each of the Kyras approaches the issue of the character's abandonment and adoption as a toddler, which we maintained constant across the three implementations as a key feature of her backstory. For American Kyra, the issue is significantly less shameful than discussing her present crush. If asked about her family, American Kyra openly explains, "I'm an orphan, and no, I don't mind being one. There's an interesting story behind that. Would you like to hear it?" Meanwhile, Venezuelan Kirita hopefully explains, "Soy huérfana, pero no pierdo la esperanza de encontrar a mis padres algun día. Mi única familia es el Profesor Yamamoto, quien me crió." ("I am an orphan, but I haven't lost the hope of finding my parents one day. My only family has been Professor Yamamoto, who raised me.") In contrast, Brazilian Kira sadly categorizes her state: "Eu sou orfa. Nao eh bom ser orfa. Eu nao sei se meus pais verdadeiros estao vivos ou mortos." ("I am an orphan. It is not good to be an orphan. I don't know if my real parents are alive or dead.")

The phrasing of the character's utterances—beyond the use of colloquial expressions and culturally specific word choices—may also play an important role in establishing each character as a genuine representative of the culture they portray. Markus and Kitayama (1991), suggested that Americans may value autonomy and internal consistency with past actions, as well as personal expressions of uniqueness from others and from the environment. However, the researchers suggest that these cultural traits may not be present, or at least may be present to a much smaller degree in other cultures, in particular Asian cultures.

Individuals from individualistic cultures emphasize interpreting and explaining reasons for feeling, thinking, and acting in terms of attributes perceived to be internal to oneself and independent of external social forces. In contrast, in collectivist cultures, the individual's thoughts, feelings, and behaviors are organized, experienced, and explained in terms of social relationships, roles, and responsibilities (Miller, Kozu, & Davis, 2001). Characters can—and often do—conform to such norms of behavior. The earlier anecdote of Brazilian Kira's hair controversy illustrates the Brazilian rejection of emphasizing individuality and the greater degree of collectivism in the Latin American cultures. These observations lead us to advise designers of culturally specific characters to avoid questions and statements implying personal agency when localizing characters to collectivist cultures in favor of those utterances emphasizing relationships.

Manner of Speech

Not only is *what* is said important—the topics addressed—but also *how* and *when* it is said. The acoustic characteristics of speech, intonation, pronunciation, timbre, and range of vocal expressions, with the appropriate variation for the localization of the character, are constantly used in everyday activities to differentiate individuals from cultures where other cues do not convey the necessary degree of specificity. These conversational aids can be used to determine not only the geographical origin of a particular person or character, but even their cultural influences and places of residences, as phonetic expert Henry Higgins in George Bernard Shaw's play, *Pygmalion*, illustrates.

Even without the additional communicative channel that text-to-speech engines provide, there are several other characteristics of speech that transmit the essence of each individual character and their cultural specificity, such as timing, speed, and the frequency with which the character uses slang words; sentence length, choice of complex or simple words, frequency and choice usage of conversational crutches (such as "hmm," "uhm," and "like"), and stuttering. For example, a teenager may transmit the same semantic content than an adult with the same cultural background using similar culture-specific words and references, yet pepper the syntactic content of their speech with trendy age-appropriate words and conversational crutches.

For comparison purposes, in Table 7.2 we feature an exchange where each of the Kyras has waited for their human partner to answer their question for a full minute. Before continuing to wait, each Kyra reminds the interactor in her own individual voice not only of whose turn it is in the conversation, but also of how to interact with the character in case the interactor is confused as to how to communicate with her. As the interactor's inactivity period lengthens, the bored character decides to use the opportunity to recharge her energy, which decays during the interaction, and after warning the interactor accordingly, Kyra, Kira, and Kirita proceed to fall asleep. We should note that the inactivity period that enacts the sleep response varies in its duration for each of the three Kyras, although in the following transcript it has been edited to compare the utterances and actions of each character. When the interactor finally reacts to the sleeping character, she awakens and yawns with a characteristic utterance for each of their personalities.

Before we move onto nonlinguistic expressions of culturally specific animate characters, it is worthwhile to mention that particular care should be taken with greeting the interactor. The interactor's first impression of the character affects every exchange between them, and care should be taken to design this initial greeting according to the desired role and role dynamics. Greeting the visitors on their return visits as such is also key to preserve the characters' believability, particularly if the character is graphically represented with a face, and the programming ease of browser cookies has lowered the overhead of adding this social requirement. Although we may forgive memory lapses on conversations and discussion topics for both people and characters, the character's facial features give

7. CROSS-CULTURAL BELIEVABILITY

TABLE 7.2
Manner of Speaking: Kyra, Kirita, and Kira React to Inactivity

American Kyra	Venezuelan Kirita	Brazilian Kira
"I hope you haven't forgotten about me, superstar! Type something into my text box to get the ball rolling again."	"Espero que no me dejes aquí plantada esperando mucho tiempo. Escríbeme algo en la caja de texto de aquí al lado en cuanto puedas." ("I hope you don't leave me standing here too long. Write me something in the textbox next to me as soon as you can.")	"Nao se preocupe, leve o tempo que precisar. Quando quiser digite a sua resposta na minha caixa de texto." ("Don't worry, take the time you need. When you want to, type your answer in my text box.")
Interactor remains inactive. > The character yawns.	Interactor remains inactive. > Kirita bosteza.	Interactor remains inactive. > O personagem se expreguica.
Interactor remains inactive. "I'm going to get some inspirational thinking done while you do what you need to do, mortal. Just start typing when you want me again."	Interactor remains inactive. "Bueno, ya que tú tienes otras cosas que hacer, yo me voy a echar la siesta. Despiertame cuando quieras seguir hablando, tipeando en la caja de texto de aquí al lado." ("Well, as you have other things to do, I will take a nap. Wake me when you want to continue our conversation by typing in the text box next to me.")	Interactor remains inactive. "Vou tirar uma soneca para refrescar a cuca. Quando quiser falar comigo novamente, basta comecar a digitar na minha caixa de texto." ("I'm going to take a nap to refresh my head. When you want to speak with me again, just start typing in my text box.")
Interactor remains inactive. > The character goes to sleep. Interactor types "Hey!" "Well, that's enough of that!"	Interactor remains inactive. > Kirita se queda dormida. Interactor types "Epa!" "Gracias por despertarme—ya me sentía como la Bella Durmiente!" ("Thanks for waking me up—I was beginning to feel like Sleeping Beauty!")	Interactor remains inactive. > O personagem vai dormir. Interactor types "Oeh!" "Vamos voltar a nosssa atividade!" ("Let's get back to what we were doing!")

rise to the expectation that the character can somehow "see," recognize, and address visitors she or he has previously met accordingly.

These suitable greeting patterns include addressing the interactor appropriately from the first moment on: title usage and familiarity are key cultural constructs that designers of culturally specific characters must master. Some cultures prefer a first-name basis even in the most formal occasions (such as is the case in Thailand and Iceland) whereas others reject addressing someone by their first name unless several years of close communication have elapsed. Although Americans are notorious for avoiding most titles, many Latin American countries use "doctor" as a prefix for addressing anyone that has completed undergraduate studies or has a position of authority. Special care should be taken not to generalize across geographical boundaries, because, for example, within the same business context, Antonio Martinez Campos would respond to "Señor Martinez" in Venezuela and Bolivia, but "Sr. Campos" in neighboring Brazil, "Mr. Martinez" in Britain, and would be addressed as "Tony" in the United States.

Manner of Gesturing

Gestures are an integral transmitter of meaning in our everyday dialog; when deprived of them as semantic aids, we recur to auxiliary mechanisms such as describing emotions, actions, and reactions, or using typographical aids to transmit this information, as the proliferation of emoticons (emotional icons) in e-mail and text-messaging demonstrate.

Cassell and Stone (1999) suggested that we use our faces and hands as an integral part of our dialogue with others no matter what our language, cultural background, or age may be, but identified particular emblematic gestures that appear to constitute between 10% and 20% of the everyday gestures produced by speakers engaged in conversation as culturally defined and imbued with meaning. Certain cultures have been shown to exhibit a greater number of these emblematic gestures in their communicative repertoire—such as French and Italian—than others, with Italians speakers often substituting emblematic gestures for speech. Moreover, members of certain cultures exhibit a greater quantity of gestures per utterance than others, with British nationals often qualifying for the least number of gestures used in conversations. Cassell (2000) also recognized cultural variability among those gestures intended to represent a common metaphor. In particular, conduit metaphoric gestures, which depict abstract ideas as bounded containers that can be held and passed between conversation partners, have been shown to vary dramatically across cultures and are even absent in certain language communities' narrations such as Chinese and Swahili.

Although some gestures only acquire meaning within certain community-based conventions, identical gestures often have quite different semantic content among different societies. In particular, assent and dissent gestures should be carefully monitored. For example, what Americans understand as the symbols for "ok," with the thumb and forefinger forming a circle and the remaining fingers extended, is insulting for a wide range of cultures including Brazilians, Russians, and Germans. The same gesture is commonly used both to refer to money in Japan—alluding to a coin's shape—and to worthless items—zero shape—in France. Similarly, even the simplest head nod can be interpreted as formal assent, as a sign of attentive listening, as a turn-taking confirmation, and as a formal negative, depending on the cultural context in which it is used.

Although gestures are integral to and support spoken dialog, they are also imbued with culturally specific meaning when decoupled from the active speech acts. Feyereisen and De Lannoy (1991) argued for a culturally specific "technique of the body," believing that as each person learns the dialect and language of the group to which she or he belongs, she or he reproduces the gestures, face movements and corporal expression typical of that group. From the directness and length of eye contact, to gestures and postures asynchronous with speech, all are embedded with cultural cues and conventions. For example, Japanese and Koreans often interpret a direct, sustained gaze as insulting or even as an overtly sexual gesture, whereas in the United States averting the gaze is often interpreted as dishonest behavior, a weakness, or an expression of extreme shyness at best. Similarly, while addressing visitors from a standing position may express respect in certain settings, in other cultural contexts it can express undesirability toward the interactor, just as welcoming visitors from a sitting position can be negatively interpreted as a status sign.

Emotional Dynamics

Animate characters' emotional model should impact their behavior and in turn be affected by the interactor's comments and actions. A sentence with identical semantic and syntactic content is performed much differently—graphically or through a text-based description—depending on the character's emotional state at the time. Emotional dynamics affect what gets said, how it is said, and the reactions of the character in light of the interactor's utterances.

The appropriateness, frequency, degree of emotional outbursts (such as crying, yelling, seething), amount of stimuli required for the outburst to reach its performance threshold, and length of time an emotional state lasts, as well as the degree of comfort with direct confrontations, vary across cultures. However, beyond how they are expressed, whether all our emotions—barring the startle and innate affinity/disgust reactions—are socially constructed and learned or innate remains an open research question with important implications for the field of multi-cultural character design.

Recent research by Picard (1998) holds the promise of monitoring the interactor's physiological emotional state unobtrusively in the near future, thus allowing the characters to adapt their roles and behaviors in response or in preparation to some universal interactor's moods. Yet knowing the interactor's emotional states still leaves the question of how to appropriately interpret and then respond to these moods and emotions open for the character designers, because the character's response to the interactor's emotions are interpreted through the interactor's expectations of socially acceptable behavior. The muted emotional response of an Asian character may not be perceived as empathic by the Italian interactor; in fact, it may be completely misunderstood as indifference.

What dimensions of emotions each character should have, what emotional states should be specifically accounted for and populated with multiple behaviors within the emotional dimension coordinate space, as well as how these emotions should be represented and performed by the character depend largely on the emotional theoretical framework chosen by the character designer. As Brave and Nass (2002) summarized, evolutionary theorists argue for innate emotions evolved to address a specific environmental concern of our ancestors, whereas emotion theorists claim that the role of higher cortical processes in differentiating emotions points toward emotions' constructed origins, such that any cross-culture consistency would be the result of similarities across social structures, rather than biologically grounded.

Between these extremes is the often called basic emotions theory, which draws on primate and cross-cultural studies to argue that emotions such as fear, anger, sadness, joy, disgust, and perhaps even surprise and interest are shared and recognized by all humans (see Ewen, 1988). Although they may be perceived to be of differential intensity, the basic emotions theory allows us to craft certain believable behaviors in our characters that may be recognized across the world within those six basic emotions, in contrast to the culture-specific dimension space that the emotion theorists claim, and to the absence of the need to consider emotional dimensions' variability across cultures that evolutionists claim.

As for our case study, all our Kyras share two main mood dimensions: an emotional one ranging from happy to sad, and a physiological dimension ranging from peppy to tired. The Venezuelan Kirita and the Brazilian Kira, however, in keeping with their Latin American temperament, have a social dimension ranging from friendly to shy, which allows them to respond to the interactor's flirtatious or insulting comments quite dramatically, and to populate the space suggested by the basic emotions theorists. The three Kyras differ drastically in the period of time until every mood regresses to a neutral state, with the Venezuelan Kirita maintaining every emotional charge the longest among the three, as well as the extent to which they express these emotions, with the American Kyra showing the most dramatic displays of anger.

Social Interaction Patterns

Perhaps even more important than what is said, how it is pronounced, and how gestures support the utterance is the knowledge and timing (what colloquially is termed tact) of when and how to bring up certain topics, which varies dramatically across cultures. When to bring up business issues within an interaction—even if it is constrained to providing information to a web visitor—as well as when to knowingly breach the familiarity barrier within a multiple-visit relationship, are examples of culturally variable time periods that characters must respect.

Disclosure tends to correlate with increasing familiarity, yet the timing of the request for information can bring an otherwise enjoyable conversation to a dead halt. For example, personal questions and information about family are always off limits for many Arab and Asian cultures, in particular those regarding the females within the family, yet they are commonplace and often expected even among casual acquaintances in American and Western cultures. The Extempo character Jennifer Jones, a female spokesperson for a fictional car company, often chats with her visitors about her husband and child and asks similar information from the visitors. This interaction exchange pattern is quite common among American women, her target audience (for transcripts of Jennifer Jones' interactions, see Hayes-Roth & Doyle, 1998).

Our present technological limitations force us, as designers of interactive characters, to confront the reality that our characters may make mistakes more often than not, either understanding or responding adequately to their interactor's utterances. How the characters recuperate from these blunders, whether and how often they acknowledge a lack of understanding, and whether they apologize for it are highly dependent on each culture's perception of mistakes and appropriateness of continued apologies. Therefore, characters from individualistic cultures may believably use a defensive non apologetic response, such as: "Hey, I don't know everything yet, do you?" when the interactor's utterance cannot be matched to the character's expression library. In contrast, characters from collectivist cultures, would tend to acknowledge the blame and apologize, perhaps even using contextual cues to shift the focus away from individualistic performance, as this has been suggested to lead to greater empathy within collective culture members (Miller, Kozu, & Davis, 2001). For example, in the situation where the interactor's utterance stumps the character's backend, a character from a collectivist culture could say: "Sorry—what was that? If it is not too much trouble, let's try that again."

Other important culturally variable social interaction patterns that character designers should explicitly address are how frequently the character should take the initiative in the conversation, the appropriateness of interruptions, the pace of the turn-taking conversation, how the interactor's comments are acknowledged, and the frequency with which questions can be exchanged, among others. Our Kyras are particularly different in terms of the social interaction patterns they exhibit. American Kyra speaks with authority and urgency, turning every lesson into an exciting battle plan, and although she is quick to encourage, she is not afraid to reprimand visitors. Venezuelan Kirita is slightly more soft spoken, more flirtatious and curious than her American counterpart about the interactor's previous knowledge on her lessons and on current topics outside her role. Brazilian Kira is very polite, assuming a more relaxed pace for the lesson and evaluation, rarely asks questions, and refuses to speak teenage lingo—although she understands it quite well.

Role and Role Dynamics

What is the character's relation to the visitor? Each character is crafted with a role in mind, be it to advise, entertain, educate, guide, among others. In all of these applications, interacting with an animate character should provide a uniquely immersive and human experience, as much like reading a book or watching a film as it is like using a computer.

As character designers, we have a responsibility to ensure that the roles our characters are assigned match each of the character's eight other characteristics to the best of our abilities, within the usual technical constraints, because we run the risk of alienating users from interacting with any character after an unsatisfying exchange. For example, a talkative character with slow loading animations will not contribute to the user's experience in a efficiency-driven application, and may perhaps serve its advice purpose better through a text-based, emotionally muted response. Doyle (1999) explored the issue of where communicative characters may best showcase their abilities and pointed out that large-scale user testing with character tutors have shown that the mere presence of the character makes children more attentive and may lead to improved retention of the subject matter covered (Lester et al., 1997). Besides constraining the character's behaviors based on their role, we should also consider the constraints each role places onto the character and the communication patterns each role entails.

Janet Murray (1994) defined believability as directly linked to our familiarity with the interaction pattern portrayed and encouraged by the character while articulating character-design lessons from the success of Weizembaum's computer program Eliza. "Most people immediately know how to interact with [Eliza]. The psychiatric interview is a known pattern that people bring to the interaction." Therefore each role also carries with it a set of culturally defined patterns of interaction that direct the exchange between human interactor and character, and contribute to maintaining the deferment of questioning. Kyra's role is that of a teacher or tutor, with the explicit goal of interesting young teenagers in art history and art appreciation.

The role of teaching young children may be perceived as offering little cultural variation, because our species' survival depends on the success of this cross-generational, one-on-one, affectively high teaching in a pattern that has existed for longer than the psychological interview pattern Murray considers so well known. Yet as we keep the tutoring role constant, teaching practices across cultures vary dramatically from each other even within the highly isomorphic national educational systems. Our Kyras predictably differ significantly in their role dynamics. American Kyra treats all interactors as playmates, occasionally teasing and cajoling the right answers from them ("Don't worry mortal, I won't use my powers against you if you choose wrongly!"), whereas Brazilian Kyra encourages her students from a respectful distance with: "Nao se preocupe se voce nao tem certeza de qual eh a pintura certa. Voce esta aqui para aprender." ("Don't worry if you are not sure of which painting is the right one. You are here to learn.") Meanwhile, Venezuelan Kirita reassures her students by explaining what the goal of the assessment is: "Si no te la sabes, dímelo y asi yo se por donde tenemos que empezar hoy la lección." ("If you don't know the answer, tell me so that I know where we should be starting the lesson today.")

Of particular interest as the first study to document such variation is the recently released video comparison on the teaching practices for mathematics on 231 eighth-grade classrooms during the 1995 school year. These classrooms were chosen among schools in Japan, Germany, and the United States as part of the Third International Mathematics and Science Study published by the U.S. Department of Education (Stigler et al., 1999). Not only are each educational system's requirements, teaching level, and lesson plans distinct, but even without these constraints, given free reign in their classrooms, teachers differed dramatically across national boundaries in terms of their pedagogical practices and goals, with surprising consistency within each national boundary.

Teachers in Germany and the United States emphasized teaching math skills, whereas Japanese teachers emphasized thinking. Accordingly, Japanese teachers organized their lessons to begin with a problem-solving session before revealing the lesson's concepts. Teachers in Germany and the U.S., in contrast, began each lesson with a concept acquisition phase and ended with a practice—or application—phase. German and American students, as a result, spent the majority of their working time in class practicing routine procedures, whereas their Japanese peers spent the majority of their time problem solving. Moreover, American teachers asked more yes/no questions than their German and Japanese counterparts, who tended to ask mostly explanation or description questions. Delivery of mathematical content also differed across countries, with only one fifth of the topics presented in the U.S. lessons developed, in contrast to more than three fourths of the topics presented in German and Japanese classrooms.

RESEARCH DIRECTIONS

Some of our group's previous work at Stanford University explored the use of characters as facilitators of second-language learning and cross-cultural communication through the child's emotional engagement with the characters, given the additional cultural context that rich animated characters can provide. *Tigrito* and the *Funki Buniz Playground*, at Stanford, focused the development on children beginning their foreign language development, while Extempo's *Foreign Language Online Workshop*, currently in development, is a character-based foreign language learning and practice environment aimed at more advanced adult learners.

Tigrito (Maldonado, Picard, & Hayes-Roth, 1998a; Maldonado, Picard, Doyle, & Hayes-Roth, 1998b) allows children to direct the interactions and spoken utterances of their avatar to plush toys from different cultures, leveraging the critical sociolinguistic context for successful practice of second-language instruction in schools across the world (Pufahl, Rhodes, & Christian, 2001). The autonomous and user-controlled characters in *Tigrito* have transient moods influenced by the child's actions and unique, culturally specific personalities expressed through the distinctive ways the characters perform actions and interact with each other. The animated characters' moods are displayed as sliders underneath the character (please see Fig. 7.3) and influenced by the character's internal state, its perceptions of the other characters' actions, and may be directly controlled by the child through the sliders.

The characters' moods emulate those used in our group's earlier work, in particular, in the animated puppets of the *Virtual Theater* (Hayes-Roth & van Gent, 1996) varying along three continuous dimensions: an emotional one ranging from happy to sad, a physiological dimension ranging from peppy to tired, and a social dimension ranging from friendly to shy. We describe these independent moods to children as happiness, friendliness, and energy, and we use icons familiar to them such as smiley faces and batteries for happiness and energy, to label the sliders on the screen. Believability is maintained through avoidance of repetitive behaviors, dynamic mood changes with relation to the length of the interaction, physical exertion of the actions performed, and the interpretation of other character's actions. Each action performed or observed has an effect on the characters' moods,

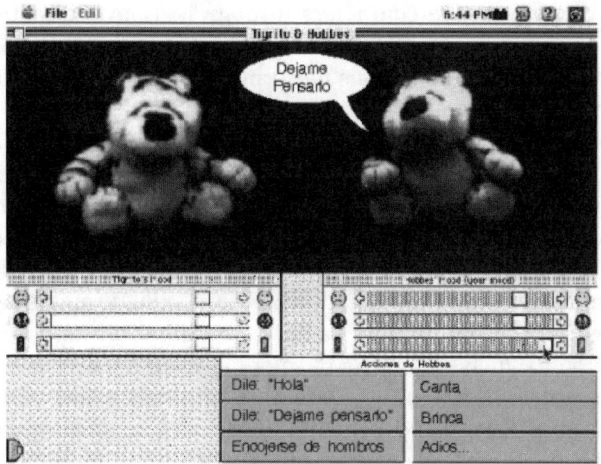

FIG. 7.3. Tigrito at play:
Interaction of a user-controlled avatar with an autonomous character.

increasing or decreasing their happiness, friendliness, and energy in accordance with the personality of each character involved. By having each character attribute a distinct meaning both to its own actions and those of the other characters, we seek to mirror human interactions where two individuals may recognize the same action but interpret it differently, as is often the case among individuals from different backgrounds.

Children can interact through their tiger cub avatar with cubs controlled by other children, or autonomously driven by the engine, because *Tigrito* allows for three different modes of interaction between the child and on-screen characters: First-Person Mode, Avatar Mode, and Movie Mode. In the First-Person mode, children interact directly with the autonomous animated character of their choice, choosing actions to perform without an embodied presence in the virtual environment. In Avatar Mode (illustrated in Fig. 7.3), the child directs the rightmost tiger in his interactions with the other character. This direction is delivered at a high level by moving the mood sliders below the child's avatar, and at a lower level by choosing actions for the tiger to perform, using the buttons in the lower right corner of the screen. These action buttons reflect the interaction by changing dynamically in accordance with the tiger's moods, and represent appropriate actions from *Tigrito*'s library. The character on the left in Fig. 7.3 is an autonomous character whom children can only influence through their avatar's actions. If the child decides to make the autonomous character happier he or she can do so, both by improving their tiger's moods through the sliders, and by choosing friendlier actions that their tiger can perform for the autonomous character. In the Movie Mode, both characters are on the stage, and the child controls the characters through high level directions, solely by altering the moods of both characters through the corresponding sliders. In this third mode, the actions performed by the characters are left entirely up to the autonomous agents to choose, in accordance with their moods.

Exploring the premise that expression of the six basic emotions—anger, fear, sadness, surprise, happiness/enjoyment, and disgust—are recognized across cultures, although cultural variations exist in the perceived intensity with which each emotion is felt (Ewen, 1988), another of our projects, the *Funki Buniz Playground* (Maldonado & Picard, 1999), encourages cross-cultural play between children of varied cultural backgrounds. By focusing the interaction exclusively on the developing affective relationships—both in the virtual world between the characters, and in the real world between the children—the *Playground* provides us with a unique opportunity to study cross-cultural affective responses and bicultural empathy.

As can be seen in Fig. 7.4, we opted to avoid a realistic representation of the characters within the *Playground*. Instead we designed a simple, cartoon like appearance for our characters' on-screen persona—as if the children's own drawings were coming to life. Our *Playground* was designed for the *interactive mural* at Stanford University's iRoom (for further details on the iRoom implementation, please see Johanson, Fox, & Winograd, 2002). This large high resolution display features alternative input methods to the traditional keyboard, allowing for direct interaction with a pen, laser, or ultrasound devices, on the mural, since the current touch-screen technology is limited to smaller displays. Because its particular physiognomy offered a natural mapping for the direct manipulation that the mural facilitates, we chose a rabbit for the lead role in our *Playground*. The traditional moods of the *Virtual Theater's* animated puppets, happiness, friendliness, and energy are represented through the rabbit's mouth, ears, and eyes, respectively. For example, friendliness is expressed by the degree to which the rabbit's

FIG. 7.4. The Funki Buniz Playground.

7. CROSS-CULTURAL BELIEVABILITY

long ears are open, and energy is expressed through the degree to which its eyes are open. Moreover, to model emotions believably, we linked these previously orthogonal mood dimensions in the underlying emotional model and displayed them accordingly on the character's facial features. Thus, when the Buni (as our young users referred to the rabbit characters) is angry, his ears are folded to denote a low friendliness level, his mouth set in a frown expressing unhappiness, and, additionally, his eyes slant in a suggestion of frowning.

The remaining direction control, action choosing, is similarly embedded; the top three most appropriate actions for the characters to execute are presented as three floating iridescent bubbles whose size reflects the appropriateness of the action contained. The bubbles reflect the constant dynamism of the character's moods, shrinking, growing, or popping as the appropriateness of the action they changes. As soon as a bubble pops, a new bubble is generated off screen and floats by, hovering above or to the side of the character.

Eliminating the linguistic content of the interactions proved key for the system's ability to explore cross-cultural affective responses, focusing the interaction exclusively on the developing affective relationships—both in the virtual world, between the characters, and in the real world between the children. The consequences of this elimination extended even to the virtual environment, stressing its role to be not only that of a key player in maintaining the child's suspension of disbelief and sense of engagement, as well as the role of facilitating the cultural exchange. The stage grew beyond the traditional black-curtained enclosure into a grass-covered meadow, complete with blue skies, cotton clouds, and butterflies (see Image 4). As the implementation progressed, butterflies and clouds gained a life of their own, with the former drifting in and out of the children's field of vision at their own volition, and the latter's emulation of natural randomness in their formation pattern, let children recognize shapes and animals in the water masses.

By rendering the affective responses in the avatar's facial expressions and eliminating the linguistic content in action choosing and execution, as well as providing a dynamic and content-filled environment for the avatars, we aimed to engage five- to seven-year-old children who shared limited dominion of a common language in open-ended collaborative play, in the hopes that through this experience their understanding their cultural differences of would increase—and maybe even be bridged.

In recent years, educational practice has shifted to emphasize constructivist learning in a discovery-based learning environment, and, simultaneously, recognizing the computer as an integral part of the schooling experience. Software characters allow us to carry this rich educational opportunity to a higher degree by adding a rich socio-cultural context and emotional motivation to the best pedagogical practices, with individualized, personally relevant content. As character designers, we may finally realize the potential of technology to offer a personalized learning experience, offering a range of representational models of concepts and events, as well as dynamic, adaptive assessment methods. Such a complete tutor (Gardner, 2000) would not displace the traditional teacher, but would complement the classroom interactions between students and teachers, as well as across the student body.

There is growing evidence to support such a role for characters: software agents have been shown to be successful tutors—in measures such as recall and problem-solving assessments, as well as likeability—when presenting curricular information through multiple modalities, such as text and voice, and when the characters' speech is personalized to their interactor as well as self-referential (Moreno & Mayer, 2000). Moreover, just as with human tutors, software agents have been shown to be more effective at teaching when dispersing principle-based advice than when dispersing more task-specific advice (Lester et al., 1997). These studies are part of a growing trend in research efforts at comparing the relative effectiveness of character coaching to that of traditional methods of instruction—such as books, workshops, and tutorials. Yet besides these traditional methods of assessment and measures of effectiveness, there may be an additional effect of lowering the learner's anxiety and social cost of mistakes when interacting with a forgiving synthetic character tutor.

The convenience of learning anywhere, anytime on the web is propelling several e-learning initiatives towards employing characters as tutors for curricular content matter where the social relationship with the character may provide additional depth of understanding and greater motivation for the learner. Extempo's *Foreign Language Online Workshop* (FLOW) is based on the premise that the unique expressive properties of animate characters can serve as catalysts of integrated language learning through verbal and affective interaction with learners (Hayes-Roth, Knodt, & Maldonado, 2002). Cultural nuances can be expressed through the key ten qualities identified earlier—not only to aid the learners in their foreign language acquisition and development, but also complement with a deeper understanding of the culture they are studying.

The gestures, beliefs, social roles, manners of speaking, and established norms of behavior are easily conveyed by characters designed to embody the best teaching principles, with the ability to repeat each interaction and explanation through different representations, until the learner masters the concepts and understands the culture. One avenue of such an exchange that we are presently exploring involves a community of animate tutors, referring learners to each other based on the specific needs, proficiencies, learning style preference, and interests of each learner. The learner's personal coach introduces new vocabulary and materials, guides the learner through related exercises and activities, and also selects a conversation partner for the learner depending on proficiency and interests. These conversational partners are complete characters with roles of their own who introduce the daily practices of each culture to the learner's understanding and perceptions. Preliminary evaluations lead us to expect Extempo's FLOW improvisational characters to effectively supplement any language acquisition course and function competently in medium to advanced conversational practice.

The work reported in this chapter was supported by a contract from the Advanced Technology Program (ATP) of the National Institute for Standards and Technology (NIST) in the U.S. Department of Commerce to Dr. Hayes-Roth at Extempo; a gift from Intel Corporation to Dr. Hayes-Roth at Stanford University; the Julian-Pillsbury and Ayacucho doctoral fellowship from the Stanford School of Education and Center for Latin American Studies to Ms. Maldonado, supporting her graduate research program.

REFERENCES

Axtell, R. E. (1985). *Do's and taboos around the world.* New York: Wiley.

Bates, J. (1994). The role of emotion in believable agents. *Communications of the Association for Computing Machinery, 37*(7), 122–125.

Blumberg, B. (1994). Action-selection in hamsterdam: lessons from ethology. In D. Cliff, P. Husbands, J.-A. Meyer, & S. W. Wilson (Eds.), *From Animals to Animats 3: Proceedings of the Third International Conference on Simulation of Adaptive Behavior* (pp. 108–117). Cambridge, MA: MIT Press/Bradford Books.

Brave, S., & Nass, C. (2002). Emotions in human computer interaction. In J. Jacko & A. Sears (Eds.), *The human–computer interaction handbook: Fundamentals, evolving technologies, and emerging applications* (pp. 81–96). Mahwah, NJ: Lawrence Erlbaum Associates.

Cassell, J. (2000). Nudge nudge wink wink: Elements of face-to-face conversation for embodied conversational agents. In J. Cassell, J. Sullivan, S. Prevost, & F. Churchill (Eds.), *Embodied conversational agents* (pp. 1–27). Cambridge, MA: MIT Press.

Cassell, J., & Stone, M. (1999). Living hand to mouth: Psychological theories about speech and gesture in interactive dialogue systems. In S. Brennan, A. Giboin, & D. Traum (Eds.), *Proceedings of the 1999 AAAI Fall Symposium on Psychological Models of Communication in Collaborative Systems* (pp. 34–42). North Falmouth, MA: AAAI Press.

Cialdini, R. (1993). *Influence: The psychology of persuasion.* New York: William Morrow.

Doyle, P. (1999, May). When is a communicative agent a good idea? *Workshop on Communicative Agents of the Third International Conference on Autonomous Agents.* Seattle, WA.

Ewen, R. (1988). *An introduction to theories of personality.* Mahwah, NJ: Lawrence Erlbaum Associates.

Feyereisen, P., & De Lannoy, J-D. (1991). *Gestures and speech: Psychological investigations.* New York: Cambridge University Press.

Foner, L. (1993, May). What's an agent, anyway? A sociological case study. *Technical Report Agents Memo 93-01.* Autonomous Agents Group, M.I.T. Media Lab. Cambridge, MA.

Gardner, H. (2000). The complete tutor. *TECHNOS. Agency for Instructional Technology, 9*(3), 10–13.
Hayes-Roth, B., & Doyle, P. (1998). Animate characters. *Autonomous Agents and Multi-Agent Systems, 1*(2), 195–230.
Hayes-Roth, B., Knodt, E., & Maldonado, H. (2002, March 28). *Language learning with an adaptive coach and lifelike conversation partners.* Presented at the 2002 Computer Assisted Language Instruction Consortium (CALICO) Courseware Showcase. University of California at Davis, CA.
Hayes-Roth, B., Maldonado, H., & Moraes, M. (2002). Designing for diversity: Multi-cultural characters for a multi-cultural world. In *Proceedings of IMAGINA 2002* (pp. 207–225). Monte Carlo, Monaco.
Hayes-Roth, B., & van Gent, R. (1996). Improvisational puppets, actors and avatars. In *Proceedings of the 1996 Computer Game Developers' Conference* (pp. 199–208). San Francisco, CA: Miller Freeman. Also available as KSL-96-09, Stanford Knowledge Systems Laboratory Report, Stanford University.
Hymowitz, K. (1998, October 28). Kids today are growing up way too fast. *The Wall Street Journal*, p. A22.
Isbister, K., Nakanishi, H., Ishida, T., & Nass, C. (2000). Helper agent: Designing an assistant for human-human interaction in a virtual meeting space. In *Proceedings of the 2000 Association for Computing Machinery's Conference on Human Computer Interaction, SIGCHI* (pp. 57–64). The Hague, The Netherlands: ACM Press.
Johanson, B., Fox, A., & Winograd, T. (2002, April–June). The interactive workspaces project: Experiences with ubiquitous computing rooms, *IEEE Pervasive Computing Magazine, 1*(2).
Jones, C. (1990). *Chuck Amuck.* New York: Avon.
Laurel, B. (1991). *Computers as theatre.* Reading, MA: Addison-Wesley.
Lester, S., Converse, S., Kahler, T., Barlow, B., Stone, & Bhogal, R. (1997). The persona effect: Affective impact of animated pedagogical agents. In *Proceedings of the Association for Computing Machinery's Conference on Human Computer Interaction, SIGCHI* (pp. 359–366). Atlanta, GA: ACM Press.
Maldonado, H., & Picard, A. (1999). The Funki Buniz Playground: Facilitating multi-cultural affective collaborative play. In *Extended Abstracts of the 1999 Association for Computing Machinery's Conference on Human Computer Interaction, SIGCHI* (pp. 328–329). Pittsburgh, PA: ACM Press.
Maldonado, H., Picard, A., Doyle, P., & Hayes-Roth, B. (1998a). Tigrito: A multi-mode interactive improvisational agent. In *Proceedings of the 1998 International Conference on Intelligent User Interfaces* (pp. 29–32). San Francisco, CA: ACM Press.
Maldonado, H., Picard, A., & Hayes-Roth, B. (1998b). Tigrito: A high affect virtual toy. In *Summary of the 1998 Association for Computing Machinery's Conference on Human Computer Interaction, SIGCHI* (pp. 367–368). Los Angeles, CA: ACM Press.
Markus, H. R., & Kitayama, S. (1991). Culture and the self: Implications for cognition, emotion, and motivation. *Psychological Review, 98*, 224–253.
Miller, P., Kozu, J., & Davis, A. (2001). Social influence, empathy, and prosocial behavior in cross-cultural perspective. In W. Wosinska, D. Barrett, R. Cialdini, & J. Reykowski (Eds.), *The practice of social influence in multiple cultures* (pp. 63–77). Mahwah, NJ: Lawrence Erlbaum Associates.
Moreno, R., & Mayer, R. E. (2000). *Pedagogical agents in constructivist multimedia environments: The role of image and language in the instructional communication.* Presentation, American Educational Research Association Annual Meeting. New Orleans, LA.
Murray, J. (1994). Flat and round believable agents. *Working Notes of the AAAI 1994 Spring Symposium Series* (pp. 65–68). Stanford University, CA. (March).
Nass, C., Isbister, K., & Lee, E. (2000). Truth is beauty: Researching embodied conversational agents. In J. Cassell, J. Sullivan, S. Prevost, & E. Churchill (Eds.), *Embodied conversational agents* (pp. 374–402). Cambridge, MA: MIT Press.

Nass, C. I., Moon, Y., Morkes, J., Kim, E., & Fogg, B. J. (1997). Computers are social actors: A review of current research. In B. Friedman (Ed.), *Human values and the design of computers technology* (pp. 137–162). New York: Cambridge University Press.

Osbeck, L., Perreault, S., & Moghaddam, F. (1997). Similarity and attraction among majority and minority groups in a multicultural context. *International Journal of Intercultural Relations, 21*(1), 113–123.

Perlin, K., & Goldberg, A. (1996). Improv: A system for scripting interactive actors in virtual worlds. In *Computer Graphics Proceedings (SIGGRAPH), 29*, 205–216.

Picard, R. (1998). Towards agents that recognize emotion. In *Proceedings of IMAGINA 1998* (pp. 153–165). Monte Carlo, Monaco.

Pufahl, I., Rhodes, N., & Christian, D. (2001). Foreign language teaching in 19 countries: Lessons to learn. *Techknowlogia, 3*(6), 39–42.

Reeves, B. (2001, November 15). *Conversational agents online: Automating human social intelligence in web transactions.* Invited talk, Symbolic Systems Forum, Stanford University, CA.

Reeves, B., & Nass, C. (1996). *The media equation: How people treat computers, televisions, and new media like real people and places.* New York: Cambridge University Press.

Stigler, J., Gonzalez, P., Kawanaka, T., Knoll, S., & Serrano, A. (1999). *The TIMSS videotape classroom study: Methods and findings from an exploratory research project on eighth-grade mathematics instruction in Germany, Japan, and the United States.* National Center for Educational Statistics, Office of Educational Research and Improvement, U.S. Dept. of Education.

Thomas, F., & Johnston, O. (1981). *The illusion of life: Disney animation.* New York: Hyperion Books.

Thomas, F., & Johnston, O. (1993). *The Disney villain.* New York: Hyperion Books.

Weizenbaum, J. (1966). ELIZA—a computer program for the study of natural language communication between man and machine. *Communications of the Association for Computing Machinery, 9*(1), 36–45.

CHAPTER

8

Recruiting a Virtual Employee: Adaptive and Personalized Agents in Corporate Communication

Benoît Morel
Cantoche, Paris

WHO ARE WE?

"You forgot to indicate your e-mail-address, your request cannot be processed," "You did not download the necessary plug-in," and so on. Who has never been disoriented by this type of message? These messages are often incomprehensible for the user and complicate or even obstruct navigation of Web sites. This fact has driven the company Cantoche to specialize, since 1996, in the humanization of all kinds of computer applications by proposing to its international clients interactive virtual characters, called *embodied agents*, as a new type of interface.

The varied media careers of Cantoche's two founders, Serge Vieillescaze and Benoît Morel, have contributed to this interest. Their previous professional experiences involved them in diverse domains such as image through a number of photographical and video portraits in different countries, sound through some music productions for radio and TV all over the world, animation through modeling claymation, and radio broadcast production as long-time producers for France's biggest radio news channel.

These different, but complementary, approaches made it possible for them to turn to video games. In the early 1990s, they discovered the first virtual worlds with their attraction strength and their interactive potential for players. They worked on games and edutainment applications for mainly European and American audiences, concentrating exclusively on the character and its behavior on the artistic, technical, and storyboard level.

This rich experience in virtual characters led them quickly to the field of agent technologies, with their potential to extend the application domains of interactive characters. New agent technologies have allowed them to bring the competences previously applied to games to the real communication needs of companies.

After having used Microsoft Agent Technology during the first few years, Cantoche created its own technology, Living Actor™, to respond more efficiently to current technical constraints and to the necessary customization of the agent to satisfy the needs of diverse audiences.

Today more than 100 characters created by Cantoche populate computer technology. For example, interactive characters such as Victor (for Hewlett-Packard) and Margarite (for Gateway) reside in your computer's operating system help software, whereas on Intranet/Internet sites or the Internet you will find Laure and Thomas at Gan Prévoyance, and Méthanie at Gaz of France. Cantoche has also created Bugs Bunny and Daffy Duck characters for Warner Bros CDROM applications, and the interactive Qmark, which resides directly on the Microsoft Windows XP operating system.

Cantoche characters impact the research sector as well. For example, James the butler (Fig. 8.1) is the most downloaded character in the world (http://www.msagentring.org) and is only one example of Cantoche embodied agents that are increasingly used in a research setting (Witkowski et al., 2003). Cantoche has been identified as one of the world leaders in the ani-

FIG. 8.1. James the butler.

mation of interactive characters. The company is firmly grounded in its two unique fields of expertise: one in the artistic domain of the creation and animation of agents, and the other in technology through the development of its Living Actor™ software solution.

MAKING INTERFACES MORE LIVELY AND SOCIABLE WITH VIRTUAL ACTORS

In our daily lives, we must constantly communicate and strive to improve our exchanges, be they emotional, cultural, or commercial. I speak about exchange intentionally because any relationship limited to the notion of sender/receiver is short lived if it is not accompanied by the familiarization that makes bonding possible. The more sociable our relationships become, the more these exchanges will be fruitful and efficient.

Take the example of a commercial relationship where we distinguish two different goals—that of buyer and vendor. The buyer enters a shop with the purpose to satisfy her needs or simply her curiosity. If no vendor is present or available, the buyer may waste precious time looking for the appropriate product on the shelves without being sure to find it. She does not know the shop's organization as well as a person working there and thus risks overlooking what she really needs. A vendor plays the role of companion: He will analyze and interpret her needs and orient the client in the shop, reassuring and consulting her as well as satisfying her needs.

The goal of the shop owner is to sell and establish a relationship of trust between the clients and the brand. To achieve this goal, the company works in different stages: (a) a presale stage, where it attracts clients, making them aware of the existence of its products and services; (b) a sale stage, where it motivates the buyers to acquire the product and directs them toward new products; and (c) an after-sale stage, where it tries to ensure the clients' loyalty.

These roles are played by different persons who need different competencies to address the wide variety of clients and audiences. A real exchange is established during all the stages to come close to the clients' expectations. Good expertise and adaptation to the client increase satisfaction of both vendor and buyer.

Today communication and exchange have to be delegated to virtual support. The example of the commercial relationship that we have given is not the only type of exchange concerned. Companies, universities, but also individuals use new media increasingly as they enable access to more information and numerous exchanges despite geographical constraints. The man–machine relationship is pushed toward more interactivity and increased personalization. With the arrival of the Internet (in particular) and new technologies (in general), we are experiencing a real invasion of the virtual into our living habits.

Companies that want to reach out to a larger audience use ICT networks for communication by implementing complex and costly solutions on their Web sites, but are confronted with their audiences' reticence. A study conducted by Creative Good, the leading company in Customer Experience, shows that 43% of attempts to buy online are unsuccessful. According to Jakob Nielsen, one of the leading experts in interface usability, 50% of sales potential is lost because of deficient information (http://www.creativegood.com/creativegood-whitepaper.pdf).

The exchanges made possible by the new information media certainly benefit from a multitude of innovative and highly performative solutions and from globalized networks. However, the challenge now is to simply give the right information and make these new solutions acceptable to the user.

A frequently heard critique is the austere and cold appearance of the computer and the lack of natural communication offered by a direct communication partner. On the existing solutions, perception and relational needs of the user have been ignored. In our everyday lives, seduction plays an important role in furthering our relationships, both professional and personal. The taste of a meal, the melody of a music, and the scent of a perfume are all elements that create an atmosphere that promotes better exchanges, if they are well balanced and adapted to the situation. These aesthetics, while taking into account the origins and cultures of the individual, facilitate communication by creating a climate of trust and favoring more natural interaction.

Mass reception of technological development increasingly shows that we are aiming at more and more convivial solutions. The most natural solution, then, would be to reproduce what exists in real life by providing applications with a personalized and convivial interface to create a humanized relationship just as in everyday life. The final goal is to make the access to information, hard and tiresome in classical search and retrieval, a highly pleasurable and playful experience while transporting a lively image that is stronger than a long discourse. This access then might develop into a real exchange, thus radically transforming the user's experiences. The embodied agent offers its services to its interlocutors by welcoming and accompanying them according to their requests and expectations.

The embodied agent is an animated interactive character that connects the user and application. It is able to interpret the desires of the user as well as the communication goals of the software developer. To adapt to the user and function adequately in its environment, this character has to be a real actor that is able to play different roles depending on its audience and task.

IMAGINARY CHARACTERS IN CORPORATE COMMUNICATION

The use of imaginary characters as part of a corporate communication strategy is not new. Numerous companies use mascots to visualize certain values. The mascot evolves with the enterprise, its localization, and its products in ways that are sensitive to and intelligible by its audience. The character is not simply a playful response, but has to be considered a personification of the company that ensures its continuity and coherence.

Some big brands are lucky to have a well-known character they can use in their marketing strategy (e.g., Bébé Cadum, the American green giant, the Marlboro cowboy, or Michelin's Bibendum). Let us take the latter case: The Michelin mascot attracts attention by its presence alone. As an efficient representative of the Michelin brand, the Bibendum is now used to consult the brand's clients. It has seen numerous transformations since its creation in 1898, losing some tires over the years as others lose weight, thus reflecting cultural tendencies of each epoch. At the occasion of its 100th anniversary, Bibendum moved from two to three dimensions. It is now ready to conquer all media—not only print, but also TV, movies, and computers. Bibendum is highly popular, has a friendly aspect, and has conquered the whole planet.

Other companies that do not own an emblematic mascot license characters to seduce their audience, such as Mickey, Bugs Bunny, Tintin, or Pokemon, or they use historical characters known all over the world.

Coca-Cola thus used Santa Claus, formerly known as Saint Nicolas, for advertising. In 1931, the Atlanta-based company wanted to conquer the children's market; to do so, it remodeled Santa Claus to look more human by giving him a normal size, a big belly, a friendly face, and a jovial air. Needless to say that the colors red and white are identical to the brand's colors. Thus, the image of Santa Claus in today's form was born. The attraction of children to this new Santa Claus was immediate, and Coca-Cola continued to use the character in its advertising campaigns for a long time. This same Santa Claus has known numerous modernizations and transformations not only with regard to the audience, but also due to the companies that use this character. Excom feminized the character, creating a female Santa Claus by shortening the coat to mid-thigh. He traded his sled and reindeer for a vespa to deliver spectacles for Gody, he wore ReWatch watches. Santa Claus has thus been transformed into a communication tool that makes it possible to address a vast audience.

Today imaginary characters are widely used and prove to be one of the best means to establish satisfactory and persistent relationships with the audience. The imaginary character brings the company close to the people by suggesting a relationship that takes into account emotions and expectations. These characters transport values attached to a brand, company, or community and are seductive for all ages.

FROM THE REAL TO THE VIRTUAL: THE IMPORTANCE OF THE ACTOR

The digitization of the media and the possibilities offered by the new technologies lead to new trends. Static elements are replaced by a new type of actor following the laws of movie making.

The success of a movie depends on its graphical and acoustic representation, the quality of its storyboard, and the direction of its elements, where all these components are embedded in a certain context and fulfill certain expectations. The first few minutes of a movie are decisive for seducing the observer. The image plays an essential role at this stage of seduction because it creates an atmosphere that lets the observer enter into the story (Noake, 1989). The actors use this atmosphere to play their roles and pass on their messages. They express themselves not only by the lines of the script they have to recite, but also in their interpretation, drawing on behaviors and appropriate expressions and emotions. The better the interpretation, the better the observer will like the film—up to the point where he identifies himself with the actor.

Some years ago, virtuality made its appearance especially through the success of video games. Thanks to the rapid technical development, its characters have become more lively and believable through better graphics and fluid motion. Hardly 10 years ago, virtuality was only used to add slightly surrealistic effects to movies. Nowadays producers often use virtual characters to re-create a reality that would be too difficult or costly to realize without the artifact. In the movie "Titanic," for example, the majority of passengers are completely three-dimensionally generated. It would have been impossible to assemble so many persons on a ship sinking into the Atlantic Ocean with such striking realism.

Another example is "Final Fantasy," where all main characters, including doctor Aki Ross, are in three dimensions. The producers of this movie even want to re-use the actors, especially the main actress, and give her other roles in other movies, just as they would with Julia Roberts.

Even if such movies require huge budgets today, it is obvious that the producers see in them an enormous gain in productivity for the future. One can easily imagine veritable virtual stars replacing today's real stars. These virtual characters could play more and more different roles without conflicting agendas, availability or aging problems, or any other trouble of real actors. Depending of the audience, they can be easily adapted because their variations are simple and low cost.

The success of three-dimensional characters with the audience leads to applications not only in movies, but also in advertising. Lara Croft, the famous heroine of the "Tomb Raider" game, has already been used to promote a French car in a commercial. Video games, as forerunners and responsible for the rise of new actors, also bring in the notion of interactivity. The player can identify quickly with his hero and participate actively in his adventure. Interactivity is of primary importance to immerse in a location or an environment. The player acts in this game and evolves among the scenes.

The idea of taking these concepts to other applications seems logical. One could use virtual actors in other types of software that need to be more personal and better adapted to their users. Until now it has been difficult to mix three-dimensional animation with other applications on ordinary computers. Machines are now powerful enough, and new three-dimensional technologies are better adapted to the technical constraints. Simultaneously, it is interesting to follow the evolution of the characters' appearance that have been created with a view to technical limits. This appearance has strongly influenced the taste of a young audience.

Several three-dimensional solutions such as Cantoche's Living Actor™ make it possible to combine a real actor with any type of application. As an actor, the character has a physical appearance and costumes that vary with its role and audience. It talks the language of its interlocutor, and it answers with precise gestures and emotions adapted to the situation. It is directed easily by the Web master, the application developer, or even a broad public on the desktop of their PC. One simply takes on the role of director to tell the actor what to do, what to say, where to go, how to behave, and to stage it on one's interface.

THE FUNCTION OF AN AGENT

The agents will thus make our relationship with the computer or Web easier and more comfortable. They play roles in our everyday lives as active assistants, service providers, or simply companions that we adopt. Virtually, we establish relationships with emissaries of commercial sites, vendors in online shops, professors in e-Learning environments, nurses and doctors on health sites, or just off-line office companions, speakers for institutions or communities that want to get into contact with us.

We distinguish four main types of usage for internal and external corporate communication:

1. *Receptionist, guide*: The agent becomes the ambassador of the corporation by transporting its image and values. The agent guides its interlocutors, presents the company's activities, accompanies them during their visit to a site or through an application, and takes them to information that they

would not have accessed in a simple visit. The users are taken by the hand at the first contact, be it a visitor or an employee. The agent motivates the employee to communicate and share his knowledge.

2. *Animation*: In commercial relations as well as in communities, the agent is an asset for the company by attracting the users' attention. The agent can be a messenger sent through a newsletter or an e-mail. The sales agent helps the clients make their purchase and simplifies the process; the publicity agent passes the message in a more friendly and convivial way.

3. *Training*: The teacher agent follows the learners in their reflections and reacts in real time to give them answers and solve their problems. The cyberteacher makes online support for self-study more attractive and adapts to different students.

4. *Assistance*: The agent supports the users in their research and presents them with the results. Once it is trained for a subject area, the agent answers intelligently and naturally to the users who are lost in cyberspace or who want to know more about a product.

First of all, it is necessary to define the role that the agent will take to determine its graphical characteristics, behavior, voice, and knowledge. After defining the role, the exact identification of the audience the agent will have to deal with is essential. Numerous criteria have to be considered and substantially change the representation sketched at the beginning.

One of the first questions to ask is whether the company addresses an internal or external audience. Internal and external communication strategies differ greatly because the goals are not the same. The agent has primarily the role of mediator, and its representation cannot be perceived in the same way by the staff and the broad audience. One could also start by raising this question: What is the agent supposed to represent? Should it represent a member of the company or rather a client? Certain companies have a strong culture and image, such as Gateway, which, in their requirement analysis, specified their desire for the representation of a real employee called Margarite (Fig. 8.2). They sent us photos of their candidate, which we interpreted according to her role and other criteria of the target audience. Other companies might prefer to represent their client. Other characteristics of the target audience like age, social group, gender, and obviously country strongly influence the graphical image of the agent.

One of our characters has been used in different countries—France, Argentina, and Italy. The target audiences everywhere were both men and women ages 18 to 50 from any social background. Our client asked us to modify Gena's clothing for each country (Fig. 8.3). These criteria are essential for the representation of the agent, but are not exhaustive. They define

FIG. 8.2. Margarite.

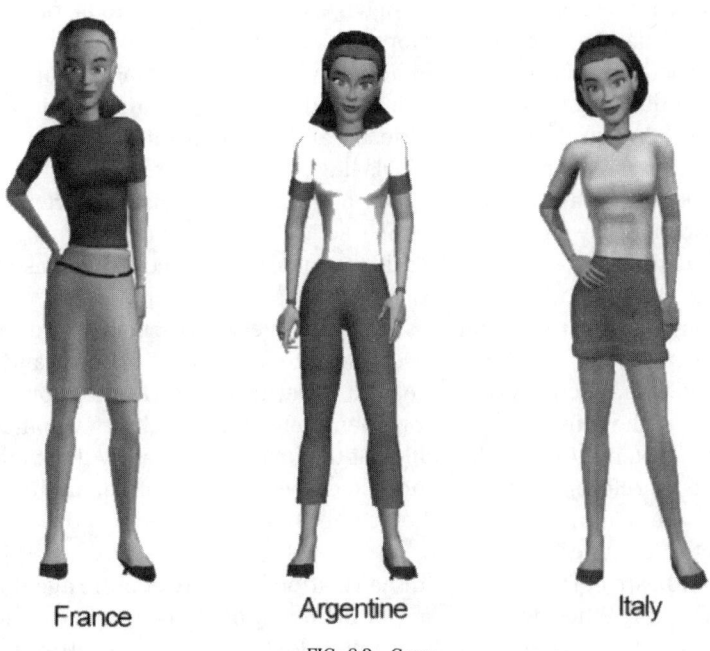

France　　　　　　Argentine　　　　　　Italy

FIG. 8.3. Gena.

the identity of the agent and influence every one of its aspects—its look, emotions, attitudes, motion, way of interaction, tone of voice, and speed of talking.

THE GRAPHICAL REPRESENTATION OF THE AGENT

Although at the beginning of this chapter we said that we try to re-create a real lifelike relationship by humanizing the interface, we remain in the virtual world. Our first task was to study which graphical representations were best suited and most efficient for a relationship between a machine and its user. It is still difficult to create a completely realistic character in looks, behavior, and intelligence. We are still in an artificial domain, and the smallest imperfection throws us back into reality. The character is and remains virtual, which, as a consequence, breaks the magic link between the user and a purely imaginary character.

For example, look at the work done in artificial intelligence (AI). To respond to user questions in natural language, the software has to be prepared by creating a database for one or more subjects. These solutions start to perform satisfactorily, but they still do not allow for an intelligence comparable to that of individuals who have been trained and uncultured by their social group, education, professional experience, and environment. The same holds for the character's behavior. Individuals have multiple expressions, attitudes, and reactions that are consistent with a situation, but also coherent with their personalities, which influence their action without being externally controlled (see de Rosis et al., chap. 4, this volume). Every individual has emotional states and moods at any moment that cannot be suppressed.

This fact leads us to a semirealist or even a very cartoonish approach in the graphical style of our agents. Users need to immediately understand the identity, role, and actions of the character because, contrary to movies or video games, the interaction is short. Generally, it lasts only a few minutes, and these few minutes must be sufficient to convince the user. All the movements are therefore slightly, but not excessively, exaggerated, and the expressions are reinforced so that the action can be interpreted and the state of the actor identified rapidly.

A second aspect is important in the relationship between the agent and its interlocutor: What does the fact of facing an obviously virtual character cause in the user? Users will certainly feel freer to call on this agent and be less preoccupied about what the virtual actor thinks of them. They will fear less being judged by a veritable person because the agent they face is only imaginationary. AI developers who use graphical representations (man, woman) log a vast amount of user questions that have nothing

8. RECRUITING A VIRTUAL EMPLOYEE

at all to do with the context. Accordingly, we plan for out of normal reactions to humor the user.

Our main approach is to work on cartoon characters that facilitate the establishment of a playful relationship with the user. The playful aspect is important even in so-called *serious* applications. An important portion of our customers are in the financial sector (banking, insurance), and a number of them have chosen cartoonish characters, such as Republic Mortgage Insurance Company (RMIC) in the United States, which selected a frog to help its users fill in a form, or the bank Crédit Mutuel, which required us to use their logo to create an agent that assists users to manage their bank accounts (Fig. 8.4).

The recipes of cartoon animation facilitate the transmission of a message especially for a highly diverse audience. This is illustrated by the case of Packard Bell, which, in February 2002, launched its new line of computers—namely, Internet Dre@m M@chine—together with Netissimo and Intel. For sales promotion, we wrote an interactive script played by Capt'n Surf, a sort of superhero who praises the power of the machine (Fig. 8.5).

Cartoon representations allow us to play with the form and proportions of the characters. According to the target audience, a character like Mickey can be made an accomplice with soft rounded forms, or it can appear more aggressive like Pokemon (Montigneaux, 2002). The size of the head in relation to the rest of the body provides valuable indications of the character, but also makes the perception of attitudes and expressions, and thus understanding the message, easier. Eyes, mouth, and eyebrows are indispensable elements for understanding the state of an agent. The majority of cartoon characters have big eyes like Bugs Bunny or the Manga type characters. These tricks are useful for the creation of an embodied agent.

FIG. 8.4. Agent for Crédit Mutuel.

FIG. 8.5. Capt'n Surf.

An overly playful aspect of cartoonish characters could be unacceptable in certain cases, so semirealist characters are sometimes more adequate. The more realistic appearance avoids any possible ambiguity and allows for professional discourse. In these cases, we create a character on the basis of a human character, but slightly exaggerate the important facial elements. In these cases, we limit ourselves to a talking head to avoid having to represent the behavior of a realistic character in full length, which does not integrate well with an application. Thus was the case of the character Angela created for the company AsAnAngel (Fig. 8.6).

FIG. 8.6. Angela.

8. RECRUITING A VIRTUAL EMPLOYEE

The choice of an agent for a diverse audience regularly leads to substantial reflections and discussions. Thus was the case of the agent integrated into Microsoft's Windows XP operating system. This system is used by all, young and old, managers and clerks, women and men, and obviously all over the world with all its different cultural identities. Which representation should be given to this agent, whose task is to assist with the installation of Windows XP? The research for a design took many months, and possible character solutions included men, women, animals, and objects until we finally arrived at a white question mark on blue background. Each proposal raised arguments because there was always a group in the audience who rejected it mainly because of a symbolic value that was too strong. Finally Qmark was chosen as the assistant agent because it *speaks* to all without hurting or offending anyone. However, a few people objected to this character even though it seems to be innocent enough: In some countries, the question mark is represented the other way round. Every representation is possible. The agent can appear as an animated logo, an object, a man, a women, a child, or an animal.

BEHAVIOR AND EMOTIONS

Besides the graphical appearance, the character's behavior gives it a strong personality. Always remember that character is all that matters in the making of great comedians, in animation, and in live action (Jones, 1990). Its behavior first depends on its morphology, but can vary according to the image we want it to have. Contrary to real life, the agent is always in the desired physical form. Its behavior develops with the dialog, but must also adapt to its audience. The concept of *actor* is relevant here because in an interactive scenario, the agent will react exactly the way its director desires.

Like actors, animators communicate through the language of the movements (Lord, 1998). All persons have standard postures and expressions like *pointing to something in 4 directions, moving, smiling, being sad*, and so on. Each agent, therefore, has a set of standard animations and behaviors as well as a broad range of expressions according to the contents of the dialog with the user. This set generally consists of about 50 animations and about 10 basic expressions. The set is limited, but mixing the elements and varying their intensity multiplies the potential behavior considerably. Of course the agent can be equipped with additional behaviors in accordance with the application.

The agent behaves more or less like a real actor. It has idling states (but not static), appropriate movements, listening states, and so on that give it some primary behavioral intelligence. If it is to move from A to B, the agent is sufficiently intelligent to know whether it should jump down or turn, walk, and turn again to face the user. The distance between the two points can also be interpreted so that the agent walks slowly or fast, runs or teleports. An agent's reaction to user requests has to be fast and must not in any way obstruct navigation.

Users have individual experiences and, even if the majority of agent behaviors are understood by all in the same way, some of them can be misinterpreted by some target groups. A sign of the head meaning "no" in one country may mean the opposite in another. Additionally, a young audience can interpret slow movement negatively.

If our customers want to reach a cosmopolitan audience, we suggest a casting of behaviors for them to test the different user reactions. Recently, this procedure was accomplished with the agent Lola (Fig. 8.7), who was chosen by a big food distributor represented in 12 countries. Lola is the eTrainer for the new employees (about 14,000 every year); she trains them in their new tasks while also explaining the corporate culture. Whether the trainee is in France, the USA, China, or Brazil, the interpretations of each behavior have to be the same. We interpret the differences resulting from the test to readapt Lola's behaviors.

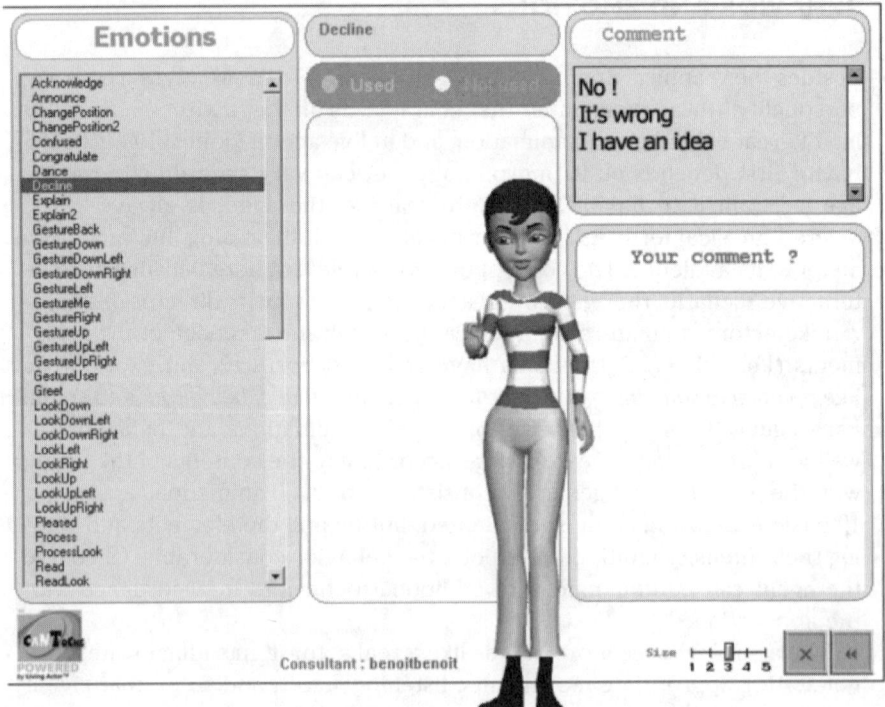

FIG. 8.7. Lola.

THE PROCESS OF CREATING AN AGENT

We advise our customers to "recruit" a new employee, and thus present the agent-creation process as a traditional recruitment with the following stages: head hunting (search for a candidate), training followed by integration, and a contract (short- or long-term contract) for the time of the mission the customer wants the agent to fulfill.

Head hunting: We define the profile of the candidate. After studying the corporate culture of the company the future employee will work for, we present to the customer an extensive questionnaire to understand the objective and subjective elements that define the character the customer wants to recruit, to determine its function, the experience needed, and its workplace. On the basis of the results, we work on the design of an existing character or create one from scratch, which is then translated to three dimensional.

Training: Once the candidate has been definitively selected, we enter into the training stage, which brings it to life. We teach the agent to behave according to its audience and role, to talk and express itself in any language. We animate the three-dimensional character by giving it a store of animations, expressions, and mouth movements. We also program the character so that it reacts intelligently to user requests.

Contract: The candidate is now ready to take over its functions on the chosen platform. The longer it remains in place, the more knowledge it will acquire, the better it will understand its interlocutors, and the more it will behave according to these knowledge. In this stage, we help the customer design the interventions of the character, and to write (storyboard) and integrate them (scripts) into the application.

CASE STUDIES

Next we highlight six Living Actor™ characters that we created for IFIC in the United States and several well-known European companies such as EDF, Gaz of France, Arcelor, Natoora, and Saint-Gobain Weber in 2002. The characters we have chosen to highlight illustrate the expectations our clients have depending on the audience target and the roles that can be entrusted to Cantoche interactive actors.

The International Food Information Council (IFIC) foundation, in partnerships with Coca-Cola, McDonalds, Kellogs, Kraft Foods, Pepsi Co, and Procter&Gamble, selected Busy Buddy as the new companion of American children to help them discover the advantage of a healthy life. In just 3 months, 127,000 children adopted their own Busy Buddy and have customized him

FIG. 8.8. Busy Buddy, learning companion of the Kidnetic Web site.

on the Kidnetic Web site. Day after day, and directly on the desktop, Busy Buddy teaches each child some new tips about nutrition and invites every boy and girl to connect to the Web site to access new games, services, quizzes, and so on. Busy Buddy exploits the push software of Living Actor™ Manager, offered exclusively by Cantoche. By embedding Cantoche's Living Actor™ Manager in their clients' Web site, the Leo Burnett advertising company, Chemistri, brings Busy Buddy to life for millions of children.

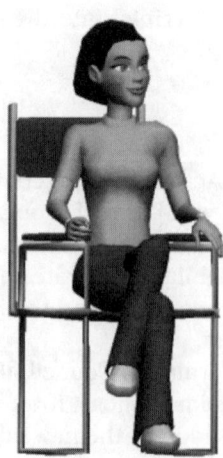

FIG. 8.9. Laura, adviser on the general public site of EDF.

8. RECRUITING A VIRTUAL EMPLOYEE

Laura was hired by EDF for a long-term contract in December 2002 to give advice and provide suggestions for maximizing home comfort though effective use of electricity on the site www.mamaison.edf.fr. The EDF team trained Laura to accompany and assist Internet users in their visit to a virtual home. With each click on one of the elements of the environment (water heater, hi-fi system, radiator, TV, etc.), Laura gives advice on how to consume electricity more intelligently. She also participates in an interactive quiz to make it possible for the Internet user to better include/understand intelligent use of electricity in daily living. Laura makes learning fun for people of all ages.

Ugine Savoy Imphy launched a large restoration project of its deployment and information systems, SAP, called the *Symbiose project*. Nathy Nox was specifically created to join the team of organizational change agents that had been assembled for the restoration project. In fact Cantoche conducted a true head-hunting design search to select a candidate (create the character) who precisely met the goals of the image and expectations of the customer. Today Nathy is the mascot of the project and its communications—she intervenes by mail, on the Intranet, and also in a newsletter called, "NathyNews," which was launched in November 2002.

Not only is Natoo the interactive character who can be trusted to know all of Natoora's products, but also Natoo is the human interface behind a revolutionary one-to-one communication tool called B-to-C (Business-to-

FIG. 8.10. Nathy Nox, in charge of communication for the "Symbiose" project at ARCELOR.

FIG. 8.11. Natoo, in charge of sales follow-up at NATOORA.

Customer). When installed on his customers' PC, Natoo takes care and checks if new messages are available, if your order is ready, or if temporary offers can be proposed to you (for the holidays, such as St. Valentine's day, or even your anniversary or birthday). Natoo exploits the push software of Living Actor™ Manager offered exclusively by Cantoche.

Hired with a temporary contract in October 2002, then in a long-term contract since January 2003, Bob is integrated on the Saint-Gobain Weber

FIG. 8.12. Bob, adviser of the global Intranet of Saint Gobain Weber.

Intranet, where he serves in the roles of assistant (tour guide portal, contextual help at the user request, etc.) and messenger (broadcasts messages and announcements). Bob is also used in various internal communication supports, such as in PowerPoint presentations, video, and so on. Although Bob currently speaks French as his first language, he will quickly be able to assist the collaborators of Saint-Gobain Weber who are present in more than 20 countries and make up six different language communities.

THE FUTURE OF EMBODIED AGENTS

Customer feedback shows that the presence of animated characters on a Web site or in any other application makes the relationship between the user and site more emotional and affective. Some of our customers have done tests where different user reactions to embodied agents were filmed. The main user reactions show first that nobody remains indifferent and all the characteristics and reactions of the character are interpreted immediately: "... he should be satisfied ... but doesn't smile ...," "you've seen how he looks at me ...," or "... it is me who puts the questions, not he who guides me...." Certain users even compare themselves to the agents in the same way as individuals identify with their heroes; others reject the agents because they do not correspond to their criteria. Work on the character—its warmth, physical appearance with regard to the role it should play, and the atmosphere it should create with its audience—remains extremely important. Users want to attribute meaning to everything that has to do with the character: why it does or does not give an answer, what change of place means, why it makes follow-up in some cases but not in others, and why it is positioned at a certain place. Lack of response or movement equally incite interpretations. Humorous traits may be taken for irony or mockery in certain contexts, but might be appreciated in others.

Obviously, it is impossible to have an all-purpose character that can play any role and reach any audience while being sure that the message is interpreted identically every time. Yet is that not the same as in real life? It is always possible to limit misunderstandings or misinterpretations to make the agent more credible by giving it one role only or adapting each agent to its audience, but it remains difficult to create one agent for all even with the means to adapt it to every audience and every culture.

Technically, there exist possibilities to take into account data provided by the application using cookies, profiling, and AI, and to identify the user to adapt the embodied agent. Recognizing the user is hard because of the large number of criteria to consider and not everything can be foreseen, but solutions are evolving. Currently Cantoche is working on a system of behavioral intelligence that allows the agent to adapt its behavior automatically according to information that it gathers and the more or less dynamic dialogs that it interprets. This revolutionary system will provide the actor with a high degree of autonomy. It will generate automatically and in realtime behaviors that are adapted to its audience. The agent will be sufficiently intelligent to translate the identity of its interlocutor into attitudes and expressions. It will have a sad expression if disappointed with the user's attitude, for example. This expression of sadness will be different, more or less pronounced depending on the recognized audience. It is possible to parametrize a maximum of elements in real time to respond effectively to cultural differences. What remains to be done is to plan for the multitude of possible combinations.

The technological revolution guided by the Internet tends to spread from the simple desktop computer to PDA, mobile computing, game consoles, palmtops, and even household appliances. The agent will have an important role in this revolution. For example, everyone will be able to check the contents of their refrigerator through an agent and directly command missing or deteriorated food once household appliances are connected to the Internet. It will be necessary to humanize these new modes of interaction by adapting to new support designs and cultural changes.

ACKNOWLEDGMENTS

I thank Sabine Payr and Elaine Raybourn for their comments on previous drafts of this chapter.

REFERENCES

Jones, C. (1990). *Chuck amuck*. New York: Avon.
Lord, P., & Sibley, B. (1998). *Cracking animation*. London: Thames & Hudson.
Montigneaux, N. (2002). *Les Marques parlent aux enfants* [The major brands talk to the young customers]. Paris: Editions d'Organisation.
Noake, R. (1989). *Animation*. Grenoble: Editions Glénat.
Witkowski, M., Neville, B., & Pitt, J. (2003). Agent mediated retailing in the connected local community. *Interacting with Computers, 15*(1), 5–32.

CHAPTER

9

Lifelike Agents for the Internet: A Cross-Cultural Case Study

Brigitte Krenn
Austrian Research Institute for Artificial Intelligence (ÖFAI),
Vienna (Austria)

Barbara Neumayr
Erich Gstrein
sysis interactive simulations ag, Vienna (Austria)

Martine Grice
Saarland University, Saarbrücken (Germany)

The following describes Flirtboat, a commercial multiuser application for the Internet where users design their representatives and send them off to a virtual community (a *net environment*) and we present user data we have collected so far from launches of this application in three different countries: Austria, the United Kingdom, and Croatia.

A net environment in our definition is a virtual space inhabited by embodied characters that have been created and are subsequently visited and instructed by users via the Internet. We refer to these user-defined characters both as avatars and as agents—avatars because they correspond to templates filled by the user, agents because each avatar, after creation, has its autonomous existence in the net environment. The terms *avatar* and *agent* are differentiated from the term *embodied conversational agent* (ECA; Cassell et al., 2000) because the latter focuses on the verbal and nonverbal communicative potential of the agent, whereas in the current implementation of the agents verbal communication is restricted to written text and animation is rudimentary (i.e., some smiling and eye blinking of the avatars). Despite that verbal and nonverbal channels of communication are not yet integrated, the net environments discussed here are a useful means to study user behavior.

We have collected basic usage data from cross-cultural launches of the Flirtboat application, such as the number of agents in an application, frequency of user logins to their avatars, distribution of female and male characters, assignment of a certain age, and personality and particular looks to one's virtual representative. We have gathered data from the same type of application, but from launches in different countries; this helps us lay a foundation for a better understanding of the potential of applications featuring animated characters.

The roles that can be filled by ECAs in Internet applications are already manifold. They can support customers in e-warehouses and e-showrooms; they can offer guidance in decision-making processes (e.g., acting as an interface to job or partner-matching algorithms); they can simply present information as do virtual newscasters; or they can enact the role of a character in a game as do avatars in computer games. Although users can be entertained, supported, informed without the use of embodied agents, such agents introduce a social dimension to the applications. The success of such an application, however, depends on its appropriateness or believability (i.e., agents must be identifiable as belonging to a particular sociocultural group), and they must be recognizable as individual characters requiring a notion of personality and affective agent behavior.

The main reason for bringing lifelike characters onto the Internet is commercial. They may attract users to a particular page and encourage them to log on more often and/or stay longer than otherwise in the application. This is of course advantageous for providers selling advertising.

If users are able to interact with a sales ECA in an intuitive and natural way, they may be more comfortable buying goods and services from this source. Moreover, because users tend to build relationships with such characters, they may prefer to buy from an ECA they *know* rather than from any other source, and they tend to trust ECAs they perceive to be similar to themselves in terms of age, personality, culture, and standing. Furthermore, because interaction with ECAs can be entertaining, they may also prefer to return to a site they know is fun.

Although the Flirtboat data have been collected with regard to business customer interests, such as how large is the number of visitors to the application, how persistent are users, and what are the user characteristics, the data are also helpful with respect to questions concerning the sociocultural grounding of animated character technology and the individuation and design of animated characters. This is explored in the remainder of this chapter. First, we summarize the main factors that contribute to making agents lifelike. Second, we give some examples of existing ECAs. In the following section, we introduce our notion of net environments and describe Flirtboat. The user statistics we have gathered based on the cross-cultural launches of Flirtboat are presented and analyzed later.

LIFELIKE AGENTS

What Makes an Agent Lifelike

The degree to which animated characters appear lifelike depends on how they function in face-to-face interaction with either the user or other agents, or both, depending on the application. They need to be able to simulate establishing contact and forming social relationships; to do so, they need to exhibit personality and express emotion. Furthermore, their communicative behavior needs to fit into a specific (application-driven) sociocultural setting.

In full scale, this requires analysis of human communicative input, including high-quality speech, gesture and facial expression recognition, and generation of animated conversation, including high-quality speech synthesis and animation of facial expressions, posture, and gesture. In general, the more lifelike a character appears, the more important it is to provide it with a sociocultural identity.

Although verbal aspects of communication, including discourse-related ones, have been studied in great detail in computational linguistics, nonverbal aspects of communication have been accounted for only recently. In the following, we discuss a number of aspects relevant for the display of appropriate communicative behavior.

Audio Channel—Speech. The naturalness of spoken communication is strongly determined by the quality of the speech produced. The currently favored option is to use canned speech because the quality is naturally high. However, this is not an option for sophisticated applications because it denies flexibility and makes further development of a system dependent on the availability of the person whose voice is used. Forward-looking systems rely instead on the state of the art in speech synthesis technology. Apart from ensuring the intelligibility of the verbal message or actual words spoken, speech synthesis has to be able to express the illocutionary force (or dialogue act) of an utterance. Speech synthesis must also be able to provide appropriate cues for turn management and discourse structure.

The interpretation of spoken utterances is also influenced by the sociocultural expectations of the spectators/listeners. There are many differences even across languages that are closely related. Take Standard Northern German and RP English, for example. Superimposing the typically higher pitch range of English females onto a German female voice makes the voice sound "aggressive and over-excited" (Gibbon, 1998). As for voice quality, it is well known that both male and female Northern German speakers have a tenser voice than their English counterparts. Such a voice quality does not imply the same personality, attitude, or emotion in both languages.

Animation—Facial Expression, Gestures, Posture, and so on. It is crucial that animation is closely timed with speech. This synchronization ranges from rudimentary synchronizations of lip movement and speech sounds (rounded lips on and around [u] sounds; spread lips on and around [i] sounds; cf. visemes in MPEG-4 [e.g., Ostermann et al., 1998]) to timing of intonational events with facial actions (such as raised eyebrows, eye flashes, or head nods on accented syllables to highlight words; Pelachaud et al., 1996). The use of facial expressions, gesture, and posture (and even eye-gaze [Colburn et al., 2000]) are crucial for realistic dialogue simulation (especially for turn taking) and for the expression of emotion and affect. Although Ekman's (1999) research on facial expression of emotion has suggested that there are universally recognized expressions, the specific connotations and implications of such expressions are largely culturally determined and are greatly affected by social conventions. (Also largely variable across cultures are how the distance between interlocutors and the extent of eye gaze are interpreted.) In addition, each culture has restrictions on the degree to which specific emotions are allowed to be expressed (cf. display rules; Ekman, 1979). Therefore, ECAs have to be situated in a particular cultural environment to function successfully. This is in part influenced by their graphical representation.

The way speakers manage turn taking is to some extent culture specific depending on a given culture's interactional norms (e.g., how much overlap across speakers is allowed). There is a close interaction between signals from the different channels (facial expressions, gaze, nods, hand gestures, posture on the visual channel, and intonation and voice quality on the auditory channel).

Graphical Representation of the Virtual Character. It is not clear where to best position the depiction of characters on a continuum between being completely lifelike and full-fledged cartoons and what the relevance of three-dimensional versus two-dimensional representation is. For Internet applications, a major aspect to be considered is bandwidth restrictions. With respect to sociocultural aspects, situation-specific dress codes need to be considered in the graphical representation especially as the depiction of the characters gets closer to the lifelike end of the continuum (cf. chap. 8).

Emotion. Modeling of emotion in ECAs is closely tied to the notion of believability of ECAs. A definition of *believability* that refers to the emotion aspect of ECAs is given in Badler et al. (2002). It says, "A character is believable if we can infer emotional or mental state by observing its behavior (even if is not portrayed as a human form)." This, however, does not necessarily mean that the agent actually has an internal representation of emotion; for instance, a smiling ECA is likely to be interpreted by the user as being happy (cf. Stronks, 2002). This observation closely relates to the findings of the Computers Are Social Actors (CASA) studies, where it is shown that (experienced) users interact with computers as if they were human (see Nass et al., 1994, for a brief overview of studies).

For consistency of behavior, however, explicit modeling of conditions that evoke emotional state in the agent is necessary. A widely used approach to the computation of emotional state is the OCC model (Ortony, Clore, & Collins, 1988), where three aspects of the world are distinguished as underlying 22 emotion types: desirability of events, praiseworthiness of actions, and appealingness of objects. The OCC model has been further extended by two other emotions—love and hate—and has been used in a system that is capable of reasoning about the emotional states of agents and emotion-induced actions in a multiagent world (cf. Elliott, 1992).

Another approach to emotion modeling is based on emotion dimensions (see e.g., Cowie et al., 1999) and is particularly useful in speech synthesis, where shades of emotions need to be expressed in the voice quality and changes of emotional tone over time need to be conveyed. An analysis of a database of emotional speech (Schröder et al., 2001) has found correlations between emotion dimensions and acoustic parameters, in particular the activation dimension, which correlates positively with higher F0 mean and range, longer phrases, shorter pauses, larger and faster F0 rises and falls, increased intensity, and a flatter spectral slope.

In addition to manipulation of speech parameters such as pitch, intensity, articulation rate, and voice quality, emotion/emotional state is also expressed by means of animation mainly in the facial expression, but also via posture.

For instance, MPEG-4 specifies high-level animation parameters for the visual representation of emotion allowing affective states such as anger, joy, disgust, sadness, fear, and surprise to be specified. Facial animation in MPEG-4 is based on work by Ekman and Friesen (1978).

The animation markup language MPML provides functions mapping the 22 emotion types defined in the OCC model to certain actions and particular voice characteristics. These mappings, however, are specified "by common sense (intuition) rather than according to empirical investigation" (Zong et al., 2000).

Personality. The incorporation of personality is indispensable for modeling lifelike agents because it functions "as a generative engine that contributes to coherence, consistency, and predictability in emotional reactions and responses" of agents (Ortony, 2003).

A study with North American probands presented in Cassell and Bickmore (2002) provided evidence that human assessment of trust, familiarity, and naturalness of interaction with an ECA is correlated with the personality (extroversion, introversion) of the human assessor. In particular, small talk is an important factor for extroverts to establish a feeling of trust and familiarity, and to consider communication successful. That is, it appears that an agent simulating similar degrees of extroversion as the user is more easily accepted, where degree of extroversion is modeled in terms of amount of small talk. These findings can also be made use of in the design of avatar–avatar communication.

However, there are also cross-cultural factors influencing the correlation of extro- and introversion with the appreciation or dislike of social communication, which need to be taken into account. For instance, Järvenpää and Immonen (2002) reported that small talk has a different importance for Americans than it has for Finns in cross-cultural business environments. Small talk is an important factor in conversation for Americans, whereas this is not the case for Finns. In particular, "Americans find silence and long breaks during discussion negative" and take this as an indicator "that the other party does not know the issue or has not considered it enough."

Other evidence for the relation between user/spectator/listener expectation/perception and personality traits comes from natural language studies. For instance, Moon and Nass (1996) found that users prefer to work with computers that produce natural language messages adapted to the personality type of the user (i.e., dominant-type users prefer computers presenting dominant-type messages). Oberlander and Brew (2000) used these findings to promote natural language generation, which controls the output language such that a personality is projected which matches the personality of the user. Evidence presented in Nass et al. (2000) also supports the assumption that users generally trust ECAs that are similar to them in age, culture, and standing because they can identify with them more easily.

Situational and Sociocultural Grounding. One vastly important area in communication that is heavily affected by cultural differences is nonverbal behavior, such as the distance between communication partners, body orientation, touching behavior, eye gaze patterns, length of pauses, and intonation of turn taking in speech (see e.g., chap. 11, this volume; Allbeck & Badler, 2001).

Another important aspect of situational and sociocultural grounding is the role being played by an agent. Allbeck and Badler (2001) assigned a role to every character in a virtual environment because roles involve expectations from (a) the character playing the role, and (b) the others interacting with that character, thus constraining the actions to be taken. Role assignment to ECAs is particularly useful to constrain the users' actions as is pointed out in Isbister and Hayes-Roth (1998). To enable successful communication, the roles must be agreed on by the communication partners. In terms of Ruttkay et al. (2002), "The ECA is believable if it acts according to the expectations of the user."

The expression of emotional states is governed by sociocultural norms—so-called *display rules*—that have a significant impact on the intensity of emotion expression (Ekman, 1972). Display rules are cultural conventions about withholding, disguising, or exaggerating expressions. According to several theorists, spontaneous emotional expressions do not convey accurate emotional information because people have been socialized to cover up their natural expressiveness in many circumstances. The kind and extent of this socialization is thought to vary from culture to culture. There is some empirical evidence that different types of events make different groups of people appear happy, sad, joyful, and so on.

The White Anglo-Saxon culture, for example, attaches great importance to the notion of independence (satisfying one's own needs creates positive feelings), whereas in the Asian culture interdependency appears to be more important (group conformity feels good; see chap. 2, this volume).

Some Examples of Current ECAs

As far as research systems are concerned, one of the most elaborate, full-body ECAs to date is REA (Cassell, 2000). REA is a virtual real estate agent that is able to engage in sales talk with one human client at a time. REA is a life-size, animated, lifelike character that is able to analyze human verbal and nonverbal communication and generate humanlike conversational output. On the generation side, it is capable of presenting facial displays, eye gaze, head moves, and hand gestures. REA has a speech synthesis component and conversational skills such as the ability to produce nonverbal cues for discourse structure (Cassell et al., 2001) and establish social relationships through engaging in small talk (Cassell & Bickmore, 2002).

An example for the realization of an ECA by means of a talking head is the medical agent developed in the Magicster[1] project. In this project, a system has been developed that focuses on the expression of communicative functions via facial animation and speech. For speech synthesis, the Festival[2] system is used. Greatest attention is paid to the generation of subtle facial expression (see De Carolis et al., 2001; Pelachaud, 2001; Pasquariello & Pelachaud, 2001).

Agneta and Frida are two animated female characters, mother and daughter, which sit on the desktop and watch the browser like one would

---
[1]Magicster (IST-1999-29078) is an EU-funded project.
[2]http://www.cstr.ed.ac.uk/projects/festival.

watch TV. They comment on browser functions, make remarks on computer technology in general, and provide the user with their own background story. The approach works with a library of prerecorded short films with the texts spoken by two actors, which enables the communication to be fairly critical and ironic in tone (cf. Höök et al., 2000; Persson, 1999).

Examples of commercial ECAs are the newscasters Ananova and Chase Walker. Ananova (Hopper, 2000) is a female talking head with speech synthesis and facial expression that allows for the display of emotion. Emotions are triggered by the emotionality of the news text. Chase Walker (McCarthy, 2000) is a male character with speech, facial expression, and hand gesture—an example of further integration of communicative channels.

Another type of commercial ECA are chatterbots. Chatterbots simulate humanlike conversation with a machine. Typically chatterbots sit on company's Web pages and engage the user in communication. Usually the user is presented with a cartoonlike character with a rudimentarily animated head and chest. Communication with chatterbots is heavily text centered. The user may type arbitrary text into a window/line to which the chatterbot reacts by producing text intended to create the illusion of a reply. The communicative behavior of a chatterbot is comparable to ELIZA (Weizenbaum, 1966). A representative example for a current chatterbot is Cybelle (http://www.agentland.com/).

Two fundamental observations can be made concerning the design of these ECAs:

1. Cultural aspects are in the first place indirect (i.e., the cultural background of the designers, programmers, and researchers influences the design and modeling of the respective agents).

2. Research in the field of ECAs is at its beginnings, and there are also still shortcomings concerning the state of the art in related areas of research and technology such as: natural language and speech processing; graphical representation, especially surface realization of personality and emotion; theories concerning the integration of verbal and nonverbal behavior; as well as (computational) models for sociocultural aspects of ECAs.

NET ENVIRONMENTS

The General Concept and Its Realization in *sysis* NetLife

In the context of the *sysis* NetLife platform, the basis of the applications described next, we define a *net environment* as a multiuser application for the Internet, where

9. LIFELIKE AGENTS FOR THE INTERNET

- the users are represented by avatars situated in a virtual location, engage in social relations, and fulfill specific tasks depending on what is required of them in a given application;
- the user is able to design her or his avatar with respect to its graphical representation, personality traits, and emotional disposition, as well as its interests. The amount of freedom the user has for defining her or his avatar is application-specific;
- the agents are autonomous after creation by the user; the user may influence the agents by giving them advice, which the agents may or may not take into account, depending on the agent's personality and mood or on parameters set within the application.

The *sysis* NetLife platform is the basis for implementations of such a net environment. The motivation of any action performed by the SAFE agents is the wish/necessity to satisfy needs and pursue specific goals. Needs arise from a multidimensional need system inside the agent, covering aspects like hunger, thirst, arousal, and curiosity. The environment or habitat in which the agents are situated offers all resources to satisfy the needs of the agents. The simulation of a NetLife application is 24 hours a day, 7 days a week, although there are reserved time slots where agents normally rest (i.e., from 02:00 to 07:00). Goals are determined by the particular application scenario and relate to the gamelike character of NetLife applications. For illustration, see the following examples.[3]

Austropolis is a virtual, democratic state and a playing field for politically minded users. The agents have their individual ideology and use democratic methods. Their overall goal is to run for president.

Cool School deals with teenage life on and around the topic of school. The avatars are schoolboys and schoolgirls who have a cool time trying to gain popularity among their peers.

Flirtboat is an application where the agents are on a cruise and try to find interesting partners for a flirt. In the following, we discuss Flirtboat in more detail because this application already exists in three versions—for the Austrian, the United Kingdom, and (more recently) the Croatian markets. As already mentioned, in the current chapter, we present user data from two Austrian, one United Kingdom, and one Croatian launch.

From the user's point of view, an important aspect of *sysis* NetLife applications is virtual storytelling. The stories created by the system can be influenced by the user interactions with their avatars. These stories are the core of the individual applications and create the application-specific feeling.

---

[3]More information can be found on the *sysis* home page http://www.sysis.at.

Sysis NetLife applications are good community-building tools because they are designed for particular interest and user groups. At the same time, they are flexible enough to allow the user to actively take part in the environment. They are a promising interface for e-commerce because they can offer information and customer support in a discursive and jocular manner. At a further stage of sophistication, once character animation has improved, natural language processing and expressive speech have been incorporated, and verbal and nonverbal aspects of communication have been integrated, *sysis* NetLife applications will also be a valuable test bed for sociocultural modeling and learning. At the current stage, they are already a useful tool for various kinds of data collection because questions can be seamlessly integrated into the applications, thus increasing the chances of collecting responses.

The Flirtboat Application

In the following section, we give a short overview of the most important characteristics of a Flirtboat application from a user perspective (i.e., we describe the steps a user typically takes through the application).

As a first step, the user creates her or his own avatar—a virtual representation of the user's self by answering a number of questions about the avatar's personality. Based on the user's answers, the model generates a personal profile for the avatar. This profile can be refined throughout the game as the user answers additional questions about the avatar's preferences and way of thinking.

A question might be, for example:

Got up late, missed the bus, left my shoes at home. Do I go berserk or do I keep calm?

The user is also asked to add personal interests and a motto, and to choose a graphical appearance for the avatar. Personality and looks can be changed at any point during the game. The profile is visible for other users during the game and makes it possible for users to see at a glance what they have in common.

One of the most interesting features of Flirtboat is the possibility of getting into contact with other users while remaining anonymous. The avatar will meet other user avatars, will make friends, and maybe will even find the partner of her or his dreams, but also dislike others. All this is reported by the avatar to her or his user, who is asked to decide on the next steps (e.g., arrange another meeting, send e-mail to another avatar, and select a certain action her or his avatar shall take during the next meeting). It is also possible to cancel previously arranged dates.

After every meeting, the user is informed of what happened by means of a message in a setting as shown in Fig. 9.1. During the cruise, an avatar/user will repeatedly be approached by other virtual characters that will try to talk to her or him. These avatars are part of the game logic and provide a particular contextual setting (see e.g., Pick-up Pete). They turn up regularly, tell funny stories, give advice, or ask questions. The users are awarded points for answering these questions.

Pick-up Pete is an expert on everything concerning flirting and dating. His task is to give advice to the users. He asks one question per day, and the user is awarded points for answering these questions. A question from Pick-up Pete might be:

Your date could talk the hind leg off a donkey. By the end of the evening, you know every detail of their life, even all about their Uncle Arthur's hip operation. Can you handle it?

The user gets two possible answers:

I let the jabbering go on for as long as it takes and just switch off.

I listen attentively and comment on several details.

Depending on which one is chosen, the user is given feedback on how clever the particular course of action is:

Be careful! Your partner will see through you straight away. This behavior suggests you're interested in a one night stand rather than a relationship.

Even if it seems tiring sometimes, that´s the right option. And it´s also a way to get to know your date.

This basic functionality is common to all Flirtboat applications and represents the core of this particular metaphor. In the following sections, we look into those aspects that are prone to cultural differences—the design of the user interface on the one hand, and usage and user behavior on the other.

Cultural Diversity in the User Interface Design of Flirtboat

The Flirtboat was originally conceived in early 2000. At that time, Web sites that covered aspects such as flirting and finding a partner were among the most sought after ones. This fact supported the choice of a fun metaphor, where communication between users was put into the center.

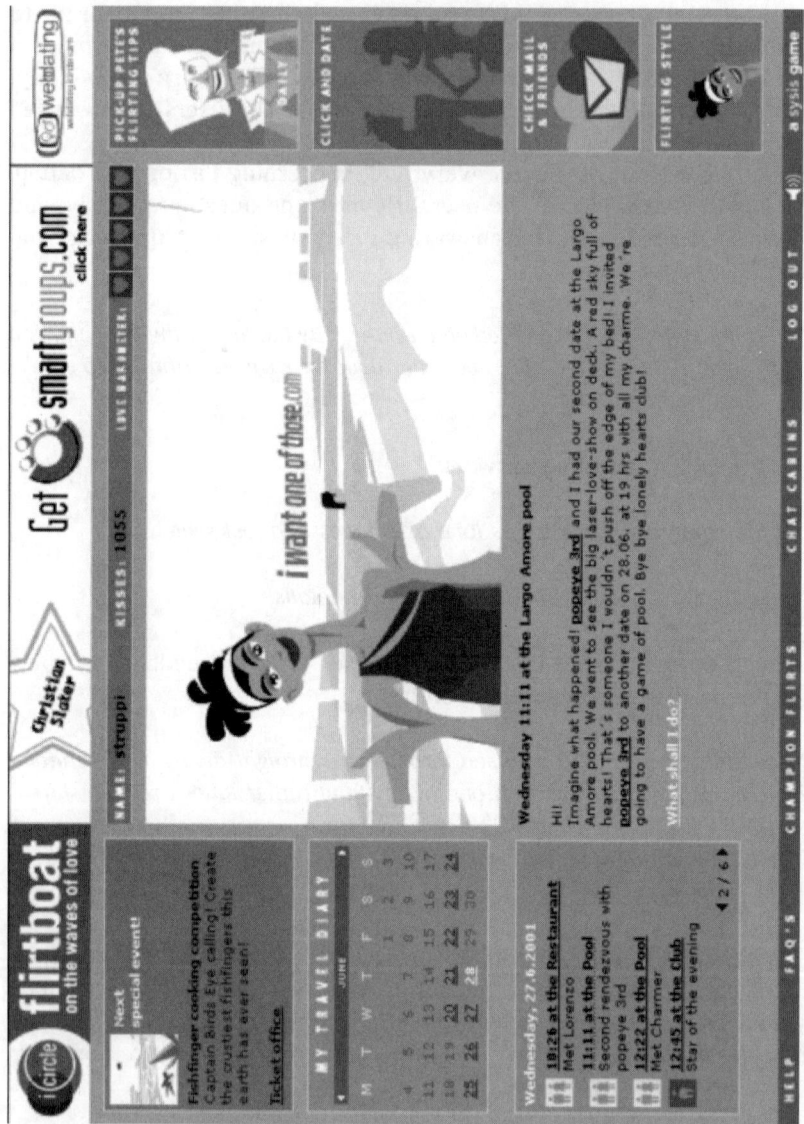

FIG. 9.1. Interface to the user visiting her or his avatar.

The next step was to define the target group. Given the basic idea of autonomous agents representing the user within a virtual environment and given the strategic goal of creating an application that would be accessible for a large number of users, above all attention was paid to common aspects rather than dividing ones. However, strategically, localization was taken into consideration from the beginning.

As regards the socioeconomic characteristics of the user group, the design principles were defined as follows. The application should be:

- Easy to use and intuitive: This aspect was particularly important because the simulation of a parallel world with autonomous creatures can be rather complex and difficult to grasp.
- Individual: The user is given the opportunity to equip a virtual representative with some of her or his characteristics and make them accessible to other users without being personally exposed.
- Young, progressive, and socializing: The application mainly aims at addressing a younger, progressive audience who use the Internet as a tool to communicate and establish social relations.
- Playful: The focus is on the fun aspect in dating. Dating is treated as a game.
- Encouraging: Special emphasis is placed on addressing women who are underrepresented in many online communities.

Because the application is conceived to be largely culture-independent, the graphical design—especially the characters—are also designed to be culture-independent at least to a certain degree. Figure 9.2 illustrates the importance of design depending on the definition of the target group. The characters on the left were created for Austropolis, a specifically Austrian application. Their appearance has a lot of associations for Austrian users (i.e., users can easily glean information as to social background and character traits because of their stereotypical representation). The characters to the right, taken from the Flirtboat application, bear much less culture-specific but more target group-specific associations. It is common for the Flirtboat characters to depict young, fashionable adults with a variety of styles to dress, different skin and hair colors, and hair styles. The intention here is not to communicate stereotypes via graphical representation.

A central issue in user interface design is text, especially in a system like Flirtboat, where feeling and model output are mainly communicated to the user via text. This is also the aspect where cultural considerations become most important. To ensure the cultural validity of the application, the translators were invited to create new stories that were more appropriate to convey the intended meaning in the target country (the United Kingdom and Croatia, respectively). The following is an example of comparing text

FIG. 9.2. Character design in Austropolis and Flirtboat.

created to convey one and the same story/information in UK Flirtboat and in AUT Flirtboat.

UK Flirtboat: *"I went to sculpture classes at the South Pole. The most beautiful snow sculpture won a prize. Pete had got the wrong end of the stick and was looking for powdered snow, while the captain collected icicles to stick in his drinks. Brrr! Babe Brighton got stuck to her ice dressing table. I showed what a megalomaniac I really am and made a penguin the size of the Taj Mahal. Had a photo done with all the stiff-upper-lip guys. It was like a royal visit."*

AUT Flirtboat: *"War beim Bildhauer-Wettbewerb auf dem Südpol. Gekürt wurde die schönste Schnee-Figur. Sascha Hahn gleich am falschen Dampfer. Sucht den 'Pulver-Schnee'. Käpt'n Flinn warf sich erst mal Eiszapfen in seinen Drink. Brrr! Bibi Bibione klebte an ihrem aus Eis gehauenen Schminktisch fest. Ich, im Größenwahn: zwei Meter hohen Pinguin gebaut. Siegerfoto mit hundert Frackträgern. Wie Lugner-Stargast auf dem Opernball!"*

9. LIFELIKE AGENTS FOR THE INTERNET

(Literal translation: Was at a sculpture competition at the South Pole. Elected was the most beautiful snow sculpture. Sascha Hahn immediately gets it wrong. Looks for the powdered snow. Captain Flinn first of all threw icicles in his drink. Brr! Bibi Bibione got stuck to her ice dressing table. Myself in megalomania mood: built a two meter high penguin. Winner photo with hundred tail-coat-wearers. Like Lugner's celebrity guest at the Opernball!)

Localization in the example story is achieved via a number of well-known Austria- and United Kingdom-specific references: To characterize the photo shooting with the winner as a rather posh event, in AUT Flirtboat a reference is made to the Opernball, an annual ball at the Viennese opera house that is *the* Austrian high society event. To provide the right connotation, the winner feels like a celebrity guest of Richard Lugner (Lugner-Stargast). Lugner is a well-known local Viennese figure—an owner of a shopping center—who for publicity reasons each year invites a female celebrity, preferably a movie star, to the Opernball. These guests are always discussed controversially in the Austrian media. In UK Flirtboat, the photo shooting is characterized as being "like a royal visit." The posh connotation is further achieved via the expressions "stiff upper-lip guys" (UK) versus tail coat wearers (Frackträger). To characterize the winning sculpture (i.e., an oversized penguin), in the UK version it is described as "the size of the Taj Mahal," whereas in the Austrian version, it simply says that the penguin is 2 meters high.

Another example of localization are the names of the virtual characters who approach the user time and again, two of which appear in the example story—namely, the charming steward Sascha Hahn versus Pete (alias Pickup Pete; Fig. 9.3) and the Flirtboat beauty Bibi Bibione versus Babe Brighton (Fig. 9.4). See Table 9.1 for a more detailed discussion of the meaning conveyed by the individual names. In the Croatian version, Sascha Hahn is Frane Galeb and Bibi Bibione is Marina Neverina.

As can be seen from the prior examples, although Flirtboat was conceived to be culture-independent, a lot of culture-dependent issues have arisen during conception and implementation of the applications. The next section looks into data collected during run time of three versions of Flirtboat—one launched in Austria (AUT Flirtboat), the second run in Austria (unless otherwise mentioned), one in the United Kingdom (UK Flirtboat, the first run in the UK), and one launched in Croatia (CRO Flirtboat, the first run in Croatia). It should be noted that data collection was not performed to identify inter-cultural differences. Nevertheless such differences can be observed from the available data. The current data allow for impressionistic interpretation, but are also a valuable resource for further experimentation.

FIG. 9.3. Sascha Hahn/Pick-up Pete/Frane Galeb.

FIG. 9.4. Bibi Bibione/Babe Brighton/Marina Neverina.

9. LIFELIKE AGENTS FOR THE INTERNET

TABLE 9.1
Naming of Avatars Built Into the System

Sascha Hahn	in the Austrian version	Sascha Hahn refers to a steward character who was featured in an Austrian and German TV series called "Traumschiff"[4] played by an actor called Sascha Hehn. *Hahn* means "rooster" in German—the name implies that he rules the roost.
Pick-up Pete	in the United Kingdom version	Pick-up Pete represents a completely different approach. His name implies that this character is master of the art of "picking up" a flirt.
Frane Galeb	in the Croatian version	*Galeb* means the same as Hahn and implies the same in Croatian as in German.
Bibi Bibione	in the Austrian version	Bibione is one of the northern Italian beach resorts mainly frequented by Austrians. The name *Bibi* implies a beach babe.
Babe Brighton	in the United Kingdom version	The English equivalent of Bibione is Brighton.
Marina Neverina	in the Croatian version	The literal meaning of *neverina* is "bad weather" or "stormy weather," but it also means trouble. Used together with the name *Marina*, the meaning "party girl" or "girl that is full of action" is conveyed. In the Flirtboat connotation, "girl that is experienced with respect to the sea and to love."

CULTURAL DIVERSITY IN USAGE

Data Collected

With the Flirtboat data, we have for the first time a resource based on a large number of users and launches in three different countries. Apart from the launches mentioned earlier, results from the first and third launch in Austria are also given where appropriate. Thus, cross-linguistic and cross-cultural data are available as well as data from subsequent launches of the application in one country.

Generally, the data available can be divided into the following groups:

- user attraction and persistence (i.e., the number of registered agents [users] over time, the number of user visits [logins] to their avatars); and
- user definition/selection of avatar characteristics such as gender, age, personality, and the look of the avatar.

---

[4]"Loveboat."

It has to be understood that the data on age, gender, and personality used in the following analyses were attributed by users to their avatar. However, informal feedback as well as the survey on user satisfaction based on AUT Flirtboat 1 indicate that a large number of users understand the avatar to be their virtual representative. Thus, there is some justification for the assumption that avatar design, at least to some extent, reflects user characteristics. However, it is still an open question as to how we can reliably assess how far this is the case. Thus, the data on age, gender, and personality discussed next are taken primarily as avatar and not as user characteristics.

User Attraction and Persistence

As previously mentioned, user commitment and persistence is measured according to the following parameters: number of registered agents (users) over time, and user visits (logins) to their avatars.

Number of Agents in the Game. AUT Flirtboat had 11,053 agents registered, 4,233 (38.3%) of which were inactive (i.e., the user did not visit her or his avatar [log into the application] for more than 1 day). For comparison, UK Flirtboat had 22,681 agents registered, 12,421 (54.8%) of which were inactive. For CRO Flirtboat, 6,718 avatars were registered, with 2,126 (31.65%) inactive. The figures show a clear difference in user commitment among the individual launches of the Flirtboat application in the three countries.

A potential explanation for this user behavior is that initial launches of Web applications tend to attract large numbers of users simply for curiosity reasons. This assumption is supported by the data on user commitment from the first launch of Flirtboat in Austria (AUT Flirtboat 1), where 16,165 avatars were registered of which 7,190 avatars (44.48%) were inactive. However, the Croatian numbers imply that we are in fact looking at a cultural difference in this case, which is most likely determined by the state of development of the online market.

An interesting feature of the Flirtboat application is that the number of registered users seems to rise constantly over time. Figure 9.5 shows the development of the number of registrations during the first 90 days run time of AUT Flirtboat, CRO Flirtboat, and UK Flirtboat. According to the data, the effect holds for an initial launch as well as a relaunch. From experience with the Austrian Flirtboat, the development of user numbers over 1 year can be shown (the data refer to AUT Flirtboat 3 and the time period from November 7, 2001 to November 7, 2002). The development of these basic usage figures shows that marketing measures are much more important than the question of how long the application is kept running. As Figs. 9.5 and 9.6 show, there is a small increase in total registrations. (Note that AUT

9. LIFELIKE AGENTS FOR THE INTERNET

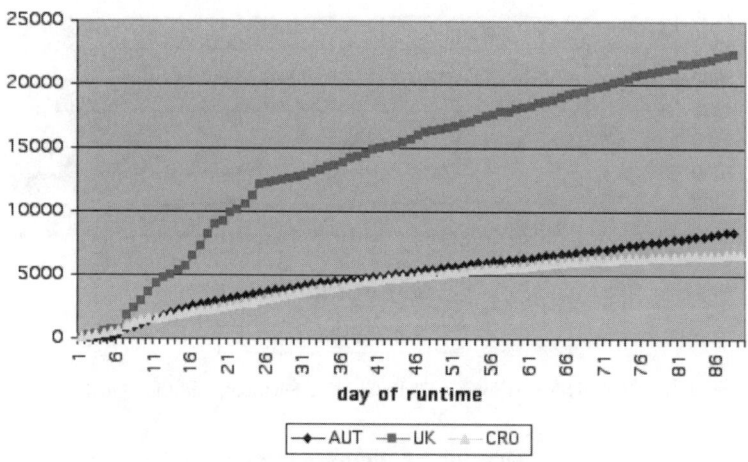

FIG. 9.5. Number of registered users for 90 days of runtime.

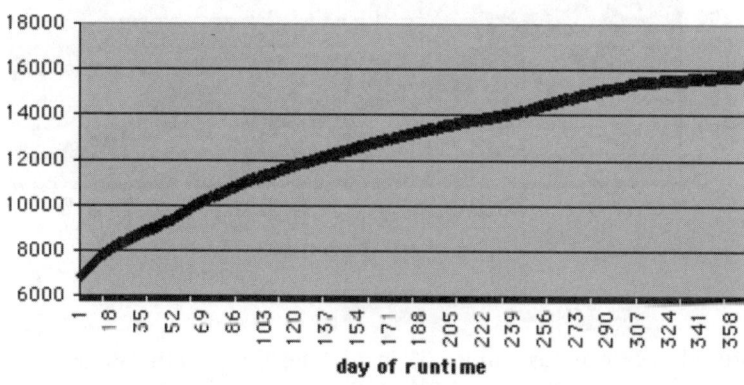

FIG. 9.6. Number of registered users for 365 days in AUT Flirtboat 3.

Flirtboat 3 was launched directly after AUT Flirtboat 2, and all avatars that were visited within 35 days of this launch were taken over, which is why the initial number of avatars in Fig. 9.6 is higher than zero.) On the contrary, daily logins stagnate quickly, see Figs. 9.7 and 9.8.

Visits and Time on Board. (*Note*: Unless stated otherwise, all further analyses are limited to those avatars that were visited by their users for more than one day [> 1d].) The average number of visits to the game per

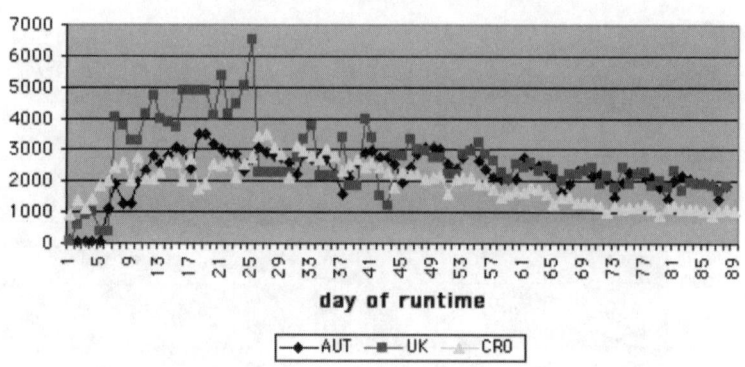

FIG. 9.7. Daily logins for UK Flirtboat, AUT Flirtboat, and CRO Flirtboat.

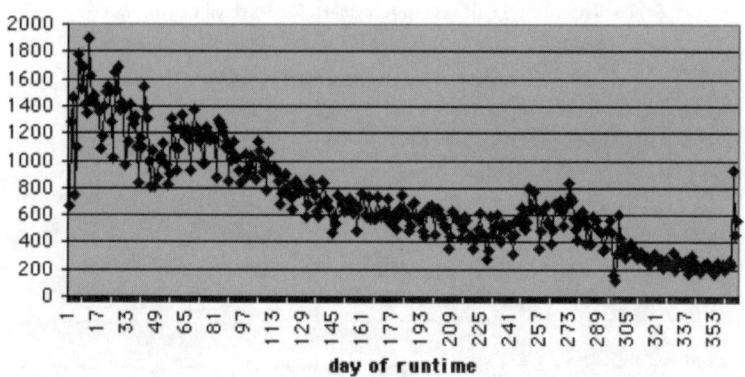

FIG. 9.8. Daily logins in AUT Flirtboat 3 over 1 year.

registered user was just under 35 in AUT Flirtboat, with average time on board amounting to 36 days. In UK Flirtboat, on average an avatar was only visited 21 times over a duration of 27 days. In CRO Flirtboat, avatars were visited nearly 47 times during a period of 37 days on average. In AUT Flirtboat, 25% of the users had a time on board of more than 60 days, visiting their avatar more than 34 times. The duration is the same for the top quartile[5] in CRO Flirtboat, but the number of visits for the top user group is

---

[5]A quartile refers to 25% of the users. All data are split into equally large groups for analysis. This concept is particularly well known for analysis of income groups in a population. The 50% quartile equals the median. It means that 50% of the users show a lower value, 50% a higher value than the median.

more than 52—the highest of all three countries. In the United Kingdom, the quartile with the most visits had over 20 visits and an average duration on board of more than 42 days. The fact that the median (50% quartile) of visits is much lower than the mean, and only the 75% quartile is about as high as the mean, leads to the conclusion that among the top 25% of the users the number of visits is actually well above the mean. In fact the maximum number of visits by an individual user was over 1,000 in AUT Flirtboat, 1,174 in CRO Flirtboat, and over 800 in UK Flirtboat.

As can be seen in Fig. 9.7, after approximately 1½ months of runtime, daily logins were established at a fairly constant level for all three applications. The average number of logins per day amounted to 2,734 in the UK, 1,941 in Croatia, and 2,265 in Austria. As we learn from AUT Flirtboat 3 (the application with the longest runtime up to date), the number of daily logins stagnates at a considerably lower level after a while (see Fig. 9.8).

Based on the development of user numbers over time and the consolidation of daily logins at a range of 2 to 3,000, we conclude that the Flirtboat application has a high potential to attract users over time, and this general potential is not particularly affected by the repetition of a launch in one country or by launches in different countries. In the remainder of this chapter, we discuss the distribution of gender and age in the applications, analyze the personality traits the users have assigned to their avatars, and examine which looks the users have assigned to their avatars.

Regarding gender, there are hardly any differences between the United Kingdom and Austrian applications. In general the average number of visits is slightly lower for males than for females, but the difference is less than one average visit. The difference in the average duration of visits in days to female and male avatars is four in the Austrian sample and three in the United Kingdom sample, and again male avatars are visited less often than female ones. CRO Flirtboat shows a different picture. Here male avatars on average have nine more visits than female ones, although the average time on board is slightly higher for female avatars (39 and 37, respectively).

The average number of visits is highest in all three samples among the 30- to 39-year-olds. In the Austrian data, the difference is more pronounced than in the United Kingdom data (see Fig. 9.9).

Avatar Profiles

In the following, we discuss data reflecting the user assignments of age groups, gender distribution to their avatars, as well as personalities.

Age Groups of Avatars. Figure 9.10 gives an overview of the distribution of age groups of the avatars in the three applications. Age is grouped into five broad classes by the system, and the users assign one of these age groups to their avatars.

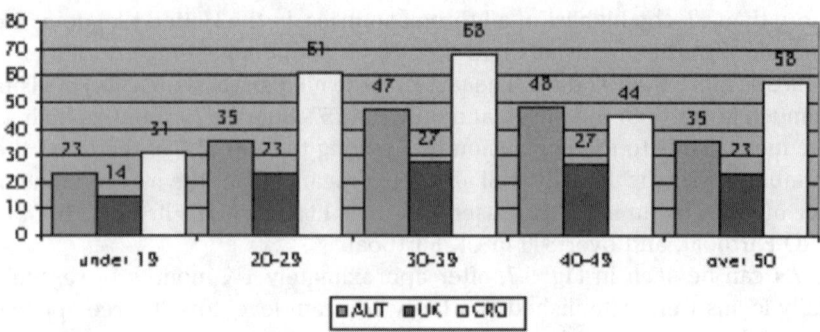

FIG. 9.9. Visits per avatar age in AUT Flirtboat, UK Flirtboat, and CRO Flirtboat.

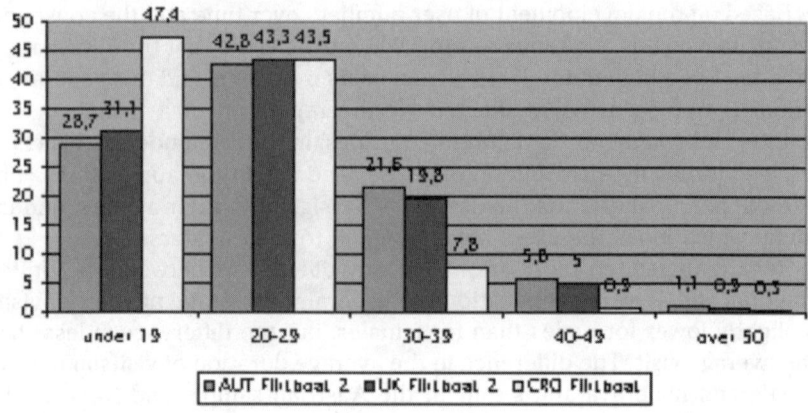

FIG. 9.10. Distribution of age groups in AUT Flirtboat, UK Flirtboat, and CRO Flirtboat.

From the figure we can see that the age distribution of avatars is fairly constant for UK Flirtboat and AUT Flirtboat. The vast majority of avatars belong to the group of under 30s (i.e., 74.4% and 71.5% of the avatars in UK Flirtboat and AUT Flirtboat, respectively). CRO Flirtboat, however, shows the strongest tendency toward the under-30s. In this application, as much as 90.09% belong to this group. However, 40+s are strongly underrepresented (i.e., 5.9% in UK Flirtboat 2, 6.9% in AUT Flirtboat 2, and 1.2% in CRO Flirtboat). In particular, there are few avatars in the age group over 50 (i.e., 0.9% in UK Flirtboat 2, 1.1% in AUT Flirtboat 2, and 0.3% in CRO Flirtboat). Yet again we have to keep in mind that we do not have any reliable means of finding out the true age of the users behind the agents.

If we consider the distribution of Austrian Internet users according to age group (see Table 9.2), we find that the younger age groups (15–20,

TABLE 9.2
Distribution of Age Groups in AUT Flirtboat

Age Group	15–20	20–30	30–40	40–50	over 50
Proportion of Austrian Internet users (%)	11.76	19.83	23.80	18.65	25.97
Proportion of AUT Flirtboat avatars (%)	28.70	42.80	21.60	5.80	1.10

20–30) are overrepresented in the Flirtboat application, whereas the older age groups (40–50, over 50) are increasingly underrepresented.

The age distribution reflected in the avatar ages may be partly due to the theme (i.e., people of the upper age groups are not particularly interested in dating), or partly due to social conventions or set ideas (i.e., net dating is not the proper thing to do at a certain age). Alternatively, it could be that users believe that younger agents will be more successful in dating, and the graphical representations of the agents suggest a specific age range that excludes the upper age groups. Although there are more younger avatars in the game, the older avatars are taken care of more (i.e., are more often visited by their users; see Fig. 9.9). This is particularly the case in AUT Flirtboat and CRO Flirtboat.

Table 9.2 compares the distribution of age groups relative to Austrian Internet users in general and specified avatar ages in AUT Flirtboat. These figures are based on ÖSTAT, census 2001, http://www2.statistik.gv.at/gz/einwohner1.shtml, and AIM, Austrian Internet Monitor, http://www.integral.co.at/Download/ergebnisse.php?level1=6.

Gender Distribution of Avatars. As can be seen in Table 9.3, female avatars outnumber male ones in UK Flirtboat and CRO Flirtboat, with 51.5% versus 48.5% and 53.2% versus 46.8%, respectively. In AUT Flirtboat, however, the distribution is inverse, with 58% male avatars versus 42% female ones. These data refer to avatars visited more often than once (> 1d on board). Considering that in all cases the proportion of females among all registrations is lower than among those staying for more than 1 day, we find that there is obviously a higher dropout of male avatars from the game than of female ones.[6]

The distribution of male (58%) and female (42%) avatars in AUT Flirtboat is comparable with the distribution of male and female Internet users in Austria (i.e., according to AIM, 57.43% of the Internet users are male compared with 42.57% females). We take this as evidence to support the as-

---

[6]For calculation of the dropout rate, the total number of registered avatars was compared with the number of avatars that have been visited more than once. See "Registered Frequency" and "> 1d on Board Frequency."

TABLE 9.3
Gender Distribution

Flirtboat Version	Gender	> 1d on Board		Registered	
		Frequency	Percent	Frequency	Percent
AUT Flirtboat	Female	2,866	42.0	4,134	37.4
	Male	3,954	58.0	6,919	62.6
	Total	6,820	100.0	11,053	100.0
UK Flirtboat	Female	5,279	51.5	10,914	48.1
	Male	4,981	48.5	11,767	51.9
	Total	10,260	100.0	22,681	100.0
CRO Flirtboat	Female	2,442	53.2	3,534	51.5
	Male	2,150	46.8	3,330	48.5
	Total	4,592	100.0	6,864	100.0

sumption of correspondence between the true gender of the user and the gender of the avatar.

Gender and Age. As shown in Fig. 9.11, the distribution of (avatar) gender over (avatar) age groups is comparable for the Flirtboat applications in the United Kingdom, Austria, and Croatia. In all three launches, female avatars clearly outnumber the male ones in the age group of less than 19 years. For all other age groups in all launches, the tendency is reversed (i.e., male avatars outnumber female ones).

Aspects of Personality. The personality model—together with a need model that controls the agents—forms the core of the virtual life in Flirtboat. The model is based on the Jungian theory of personality. In particular,

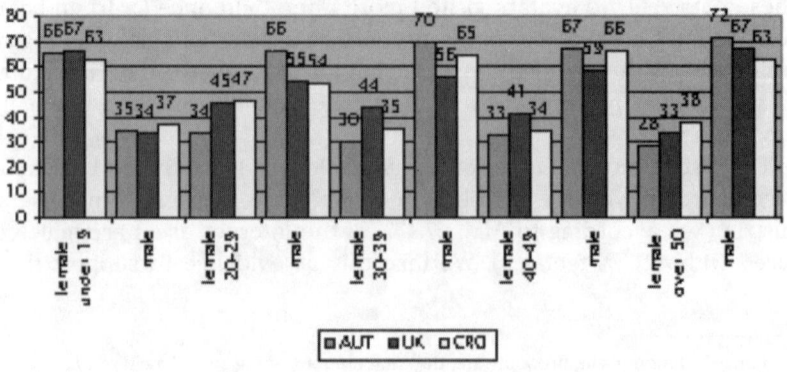

FIG. 9.11. Distribution of gender in age groups.

it is an adaptation of the Myers–Briggs Type Indicator (cf. Keirsey & Bates, 1984), a paper-and-pencil personality test according to which personality is modeled along four dimensions:

- extroversion (E)–introversion (I)
- intuition (N)–sensing (S)
- thinking (T)–feeling (F)
- judgment (J)–perception (P)

Combinations of these dimensions lead to 16 personality types.[7]

This particular approach has been chosen because it is easily operationalizable for the matching and dating mechanism underlying the application, and also for the assignment of personality to the avatars by means of an online questionnaire presented to the user as part of the registration process of a new user/avatar. As regards the former, the approach allows for precise assumptions about the personal relationships each personality type may have with any of the other types and how they are expected to develop over time (see http://www.socionics.com/rel/rel.htm). With respect to the latter, question–answer lists in the style of Flirtboat texts have been designed for each personality type by a psychologist addressing several areas of (the users') life, such as social behavior, partnership, career, and so on. The answers to these questions are used to decide for individual Jungian dimensions whether a person belongs to the one or the other extreme, resulting in profiles like ESFJ, which stands for a personality type characterized as extroverted feeling with sensing (see Boeree, 1997, for an online description of the Myers–Briggs types).[8]

Examples for question–answer pairs as presented in the UK Flirtboat are given in Table 9.4.

Comparing the personality types assigned to the avatars in the Flirtboat applications in Austria, the United Kingdom, and Croatia, we find there is little variation in the most frequently assigned personality types in all three countries. In particular,

- ENFJ (extroverted feeling with intuiting) is the most frequently assigned personality type in UK Flirtboat (10.64% of the avatars) and in CRO Flirtboat (17.51%), and it is the second most frequently assigned personality type in AUT Flirtboat (11.22%).
- INFJ (introverted intuiting with feeling) is the most frequently assigned personality type in AUT Flirtboat (11.41%), the second most frequently

---

[7]See http://www.socionics.com/main/types.htm.
[8]For a description of this particular type, see http://www.socionics.com/advan/prof/esfj.htm.

TABLE 9.4
Examples for Question–Answer Pairings for the
Assignment of Avatar Personality

Imagine your date is keeping you waiting. How long does it take before you start to feel annoyed?	5 minutes
	10 minutes
	15 minutes
	30 minutes
What would you say sounds more like you: "Hello, here I come!" or "Let's wait and see."	"Hello, here I come!"
	"Let's wait and see."
Are you a dreamer or more of a practical type?	Dreamer
	Practical type
Are you a rational or an emotional person?	Rational
	Emotional
Do you like to keep everything in good order or are you inspired by chaos?	Order
	Chaos
How do you feel when you are in a crowd?	Relaxed
	Tense
What do you think about visionaries?	They're tedious
	They're fascinating
Be honest: Are you likely to be impressed by an emotional speech or will only hard facts convince you?	Emotional speech
	Hard facts
Do you like to check things out first or do you act on the spur of the moment?	Investigate first
	I'm impulsive

assigned personality type in CRO Flirtboat (12.18%), and still the third most frequently assigned one in UK Flirtboat (9.48%).

Interestingly, the two types differ only in the dimension extroversion–introversion.

Considering the least frequently assigned personality type, we again find similarities, with ENTP (extroverted intuiting with thinking) being assigned to 2.37% of the avatars in AUT Flirtboat and 2.42% in CRO Flirtboat. ENTP is the third least frequently assigned personality type in UK Flirtboat (4.08%).

When we separate male from female data, we find complete divergence of the most frequently assigned personality types of male avatars (i.e., INFJ in AUT Flirtboat [10.48%], ISTP [introverted thinking with sensing] in UK Flirtboat [10.35%], and the generally high scoring type ENFJ in CRO Flirtboat [14.68%]). Although the most frequently assigned male personality types in AUT Flirtboat and CRO Flirtboat differ only in a single dimension—the introvert Austrian versus the extrovert Croat—there is little or no convergence with the most frequently assigned UK male type.

Looking at female avatars, the similarities between Austrian and Croatian types are even more pronounced, with ENFJ having been assigned as most frequent type in AUT Flirtboat (14.37%) and CRO Flirtboat (20.18%). It is again the UK sample that is distinct, with INFP (introverted feeling with intuiting) the most frequently assigned personality type (13.49%).

More convergence can be found with regard to the least frequently assigned personality type, ENTP, which has been assigned least in five of six clusters—namely, in AUT Flirtboat (1.99% male, 3.00% female), CRO Flirtboat (2.43% male, 2.41% female), and UK Flirtboat (3.34% female).

In summary, with respect to personality assignment, there seems to be a stronger convergence between Austria and Croatia than between Austria and the United Kingdom or between Croatia and the United Kingdom. Note the data discussed refer to all avatars registered.

Choice of Representation. For each gender, there was a choice of 16 representations for the avatar.[9] For AUT Flirtboat and UK Flirtboat, we evaluated which avatars were chosen most and least frequently. Generally, it is interesting to note that there seems to be little difference between the two countries at both ends of the spectrum regarding choice of appearance (i.e., the two most frequently and the two least frequently chosen appearances are the same).[10] Figure 9.14 features pictures of the most and least frequently chosen avatar appearances.[11] The most obvious difference between the two groups is skin color (i.e., all least frequently chosen characters have pale skin, whereas all most frequently chosen characters show a darker complexion). Another uniform characteristic of the most frequently chosen characters is the type of dress (i.e., all characters wear sporty dresses). Dress code is not this uniform for the least frequently chosen characters. To find out why particular optical representations were especially popular or unpopular in AUT and the United Kingdom, a qualitative analysis is necessary. More details in the differences between United Kingdom and Austrian users can be found in Figs. 9.12 and 9.13. Here we find that differences are a bit more pronounced among the female avatars, whereas the selection of appearances of male avatars is fairly similar in both countries. We also see that the difference (in terms of how often they have been chosen) between the three most frequently chosen female appearances in AUT are rather small, whereas the variation in the United Kingdom is quite large. In the United Kingdom, Pic 6 is much more popular than Pic 1, which is the most popular female appearance in AUT. There is no such large difference between AUT and UK regarding the least popular female appearances, Pic 16 and Pic 4. Overall, the data we currently have are a useful resource to derive a number of items based on which initial qualitative investigations shall be directed, such as the connotation of skin color or dress code and their relevance in the context of Flirtboat.

---

[9]We are not able to show the full range of 32 pictures because of space limitations.

[10]Note that comparisons can only be made gender-specifically because male and female avatars have different appearances.

[11]For printing, the original color scheme has been reduced to grey scale.

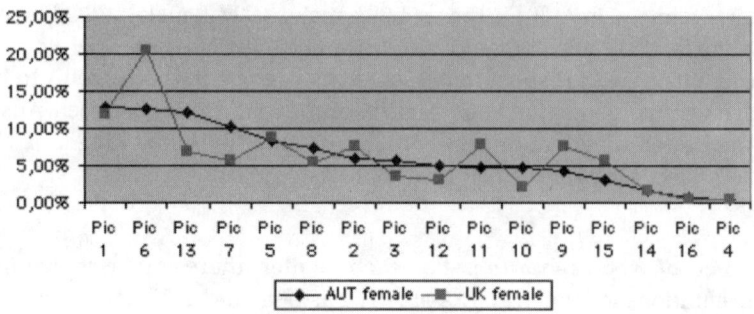

FIG. 9.12. Ranking of avatar appearance, females.

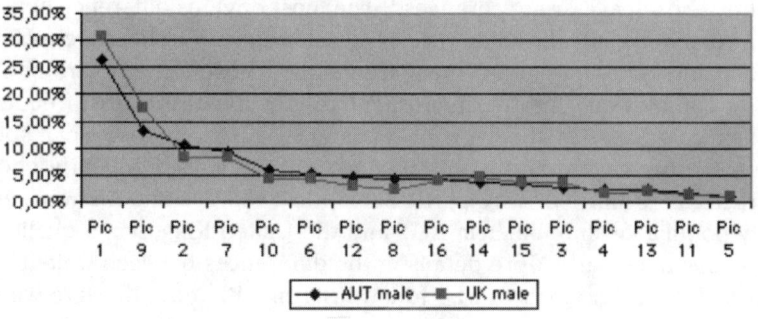

FIG. 9.13. Ranking of avatar appearance, males.

CONCLUSION

The most striking difference across the launches of Flirtboat in three different countries (Austria, the United Kingdom, Croatia) was the assignment of personality traits, the Austrian and Croatian groups being closer to each other than either of them were to the United Kingdom group.

As regards other properties such as the development of user numbers and logins over time, or findings related to avatar characteristics such as age and gender, we found a number of similarities across the launches of Flirtboat in the different countries. This can be summarized as follows.

Based on the development of user numbers over time and the consolidation of daily logins at a range of 2 to 3,000, we can conclude that the Flirtboat application has a high potential to attract users over time, and

9. LIFELIKE AGENTS FOR THE INTERNET

Most frequently chosen			
Pic 1	Pic 9	Pic 1	Pic 6

Least frequently chosen			
Pic 11	Pic 5	Pic 4	Pic 16

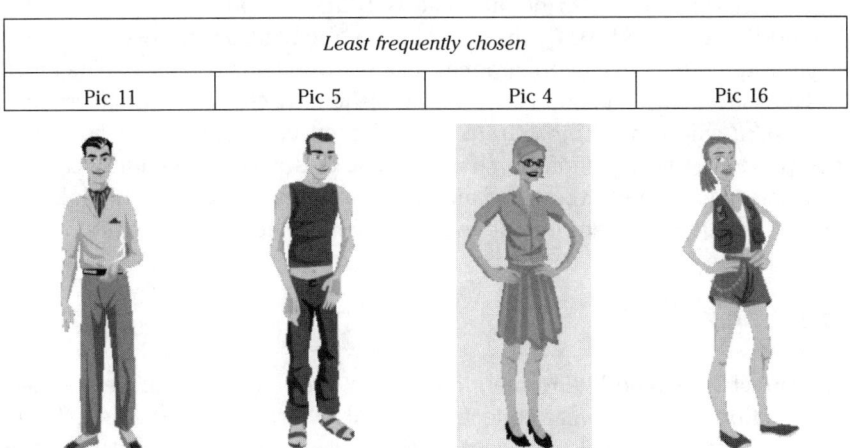

FIG. 9.14. Most frequently and least frequently chosen avatars in AUT Flirtboat and UK Flirtboat.

that this general potential is not particularly affected by the repetition of a launch in one country or by launches in different countries.

Although there are more younger avatars in the game, there is a tendency for older avatars to be taken care of more (i.e., they are more often visited by their users). This is particularly the case in AUT Flirtboat and CRO Flirtboat.

Because we do not have reliable data about user demographics, we cannot make assumptions about the relation between user age or gender and avatar age or gender. As regards gender, we only have slight evidence for a correspondence between true user gender and avatar gender from a correlation between the distribution of male and female avatars in AUT Flirtboat and the distribution of male and female Internet users in Austria. Thus, personal properties must always be understood as avatar properties and not as user properties.

Keeping this in mind, we have further learned from the data that female avatars outnumber male ones in UK Flirtboat, CRO Flirtboat, and AUT Flirtboat, whereas the distribution is inverse for male avatars. These findings hold for avatars visited more often than once. Considering all avatars ever created, there is a higher dropout of male avatars from the game than of female ones. The distribution of (avatar) gender over (avatar) age groups is comparable for the Flirtboat applications in the United Kingdom, Austria, and Croatia. In all three launches, female avatars clearly outnumbered the male ones in the age group of less than 19. For all other age groups in all launches, the tendency is reversed (i.e., male avatars outnumber female ones).

Considering the looks of the avatars (data are only available for the United Kingdom and AUT), there is basic agreement on the two most and least frequently chosen appearances for male and female avatars. The most obvious factor is skin color (i.e., all least frequently chosen characters have pale skin, whereas all most frequently chosen characters show a darker complexion). Another uniform characteristic of the most frequently chosen characters is type of dress (i.e., all characters wear sporty clothes). Dress code is not this uniform for the least frequently chosen characters.

OUTLOOK

An aspect we would like to mention here is derived from a survey of user satisfaction. The data were collected after AUT Flirtboat 1 had gone off-line. Although the data stem from a small set of users, representing 0.9% of the active users,[12] we find the following feedback important with regard to the design of future applications with animated characters.

Asked what they wished for a new release, 39% (by far the largest portion) wished they had more information about the other avatars (users) in the environment, but only 11% requested more animation.[13] What we have found here is a clear interest in social engagement—the wish to get to know the co-inhabitants in the environment of which they or their avatars are part. One reason that the requirement for more animation was quite low might be that character animation in general is rudimentary in the current Flirtboat. The visual representation of the avatar functions more as a mask

---

[12]Users were considered inactive if 35 days elapsed without a login.

[13]The second largest portion was 12% of the respondents who wished for the incorporation of a calendar from which past events could be accessed.

behind which users can conceal/reveal themselves than as an active means of communication. The communication aspect in the current Flirtboat is covered by the actions that can be instigated by the user, free text input, e-mail, and chat facilities.

Nevertheless, we assume that the judgment about the importance of animation will change in applications when the characters become more animated and thus more expressive. This assumption is investigated in the research project NECA (http://www.oefai.at/NECA; Krenn et al., 2002). In the NECA project, a system is developed on the basis of *sysis* NetLife, which enables affective face-to-face, lifelike interaction between agents (i.e., the agents communicate with each other using speech, facial expression, gesture, and posture). The users are presented with movies of agents interacting and performing a clearly defined task (currently two of them discussing an encounter one of them previously had with a third avatar).

User reactions to these clips are collected by allowing the users to rate the appropriateness of the agent's performance in the clip at various levels. Presenting the user with example scenes and dedicated (online) questions, we are able to access users' feelings about the appropriateness of their avatars' behavior of which the user might not be aware. For example, users are doubtless aware that a scowl and a friendly smile are not equally appropriate actions when making friends or rejecting an advance. They might, however, have difficulties pinpointing or even verbalizing how the type of intonation or voice quality used plays a role in signaling the avatars' intentions.

Another aspect that is already available in *sysis* NetLife is that the users can tutor their avatars. In the NECA context, this means after watching the movie, users can tell their avatars (the system) what they did right and where they went wrong, ideally with hints on how to perform better the next time a similar situation arises. Such feedback provides the developers with information as to what users believe to be socially acceptable behavior in a given situation. Even if the technology used in the application is not advanced enough to simulate this behavior, it provides a goal for future research.

Once the target behavior is established across a group of users in a given situation, the same method can be used for more diagnostic testing of the effectiveness of different aspects (facial expression, gesture, posture, intonation, voice quality) of avatar communication across a range of different cultures and subcultures.

REFERENCES

Allbeck, J., & Badler, N. (2001). Consistent communication with control. *In Autonomous Agents 2001 Workshop on Non-Verbal and Verbal Communicative Acts to Achieve Contextual Embodied Agents*, Montreal, Canada.

Austropolis: http://sysis.at/website/web/pages/portfolio/community/21/. Last visited September 8, 2003.

Badler, N., Allbeck, J., Zhao, L., & Byun, M. (2002, June). Representing and parameterizing agent behaviors. In *Proceedings of Computer Animation*, IEEE Computer Society, Geneva, Switzerland.

Boeree, C. G. (1997). Personality theories. Carl Jung. http://www.ship.edu/~cgboeree/jung.html. Last visited September 8, 2003.

Cassell, J. (2000). Nudge nudge wink wink. Elements of face-to-face conversation for embodied conversational agents. In J. Cassell et al. (Eds.), *Embodied conversational agents* (pp. 1–27). Cambridge, MA: MIT Press.

Cassell, J., & Bickmore, T. (2002). Negotiated collusion: Modeling social language and its relationship effects in intelligent agents. In *User Modeling and Adaptive Interfaces*.

Cassell, J., Nakano, Y., Bickmore, T., Sidner, C., & Rich, C. (2001, July). Non-verbal cues for discourse structure. In *Proceedings of the ACL-2001 Conference*, Toulouse, France.

Cassell, J., Sullivan, J., Prevost, S., & Churchill, E. (Eds.). (2000). *Embodied conversational agents*. Cambridge, MA: MIT Press.

Colburn, A., Cohen, M. F., & Drucker, S. (2000). *The role of eye gaze in avatar mediated conversational interfaces* (MSR-TR-2000-81). Microsoft Research.

Cowie, R., Douglas-Cowie, E., Appolloni, B., Taylor, J., Romano, A., & Fellenz, W. (1999). In N. Mastorakis (Ed.), *Computational intelligence and applications*. World Scientific & Engineering Society Press.

De Carolis, B., de Rosis, F., Carofiglio, V., Pelachaud, C., & Poggi, I. (2001, December). *Interactive information presentation by an embodied animated agent*. International Workshop on Information Presentation and Natural Multimodal Dialogue, Verona, Italy. (Online proceedings: http://i3p-class.itc.it/ipnmd-proceedings.html)

Ekman, P. (1972). Universals and cultural differences in facial expressions of emotion. In J. Cole (Ed.), *Nebraska Symposium on Motivation 1971* (Vol. 19, pp. 207–283). Lincoln, NE: University of Nebraska Press.

Eckman, P. (1979). About brows: Emotional and conversational signals. In M. von Cranach, K. Foppa, W. Lepenies, & D. Plog (Eds.), *Human ethology: Claims and limits of a new discipline: Contributions to the Colloquium* (pp. 163–202). Cambridge: Cambridge University Press.

Ekman, P. (1999). Facial expressions. In T. Dalgleish & M. Power (Eds.), *Handbook of cognition and emotion* (pp. 301–320). New York: Wiley.

Ekman, P., & Friesen, W. (1978). *Facial action coding system*. Palo Alto, CA: Consulting Psychologist Press.

Elliott, C. (1992). *The affective reasoner: A process model of emotions in a multi-agent system*. Unpublished doctoral dissertation, Northwestern University, Institute for Learning Sciences (Technical Report).

Flirtboat: http://sysis.at/website/web/pages/portfolio/community/8/. Last visited September 8, 2003.

Gibbon, D. (1998). German intonation. In D. Hirst & A. di Cristo (Eds.), *Intonation systems* (pp. 78–95). Cambridge: Cambridge University Press.

Höök, K., Persson, P., & Sjölinder, M. (2000). Evaluating users' experience of a character-enhanced information space. *Journal of AI Communications, 13*(3).

Hopper, D. I. (2000). Virtual news jock Ananova goes live. (http://www.cnn.com/2000/TECH/computing/04/18/ananova.launch/). Last visited September 8, 2003.

Isbister, K., & Barbara Hayes-Roth, B. (1998). *Social implications of using synthetic characters: An examination of a role-specific intelligent agent* (Technical Report KSL 98-01). Stanford, CA: Knowledge Systems Laboratory.

Järvenpää, E., & Immonen, S. (2002, March). Challenges of cross-cultural management: Case studies in Finnish companies. In *Proceedings of the Project Management Global Conference*, Hong Kong.

Keirsey, D., & Bates, M. (1984). *Please understand me: Character and temperament types* (5th ed.). Del Mar, CA: Prometheus Nemesis.

Krenn, B., Grice, M., Piwek, P., Schröder, M., Klesen, M., Baumann, S., Pirker, H., van Deemter, K., & Gstrein, E. (2002, September 30–October 2). Generation of multi-modal dialogue for net environment. In *Proceedings of KONVENS-02*, Saarbrücken, Germany.

McCarthy, J. (2000). Sprint's virtual newsman more than a talking head (http://www.cnn.com/2000/TECH/computing/05/03/chase.walker.idg/). Last visited September 8, 2003.

Moon, Y., & Nass, C. I. (1996). *Adaptive agents and personality change: Complementarity versus similarity as forms of adaptation*. In Proceedings of the CHI Conference, Vancouver, Canada.

Nass, C., Isbister, K., & Lee, E.-J. (2000). Truth is beauty: Researching embodied conversational agents. In J. Cassell et al. (Eds.), *Embodied conversational agents* (pp. 374–402). Cambridge, MA: MIT Press.

Nass, C., Steuer, J., & Tauber, E. (1994). *Computers are social actors*. Proceedings of the CHI conference on human factors in computing systems: Celebrating interdependence, Boston, MA.

Oberlander, J., & Brew, C. (2000). Stochastic text generation. *Philosophical Transactions of the Royal Society of London* (Series A), *358*, 1373–1385.

Ortony, A. (2003). On making believable emotional agents believable. In R. Trappl, P. Petta, & S. Payr (Eds.), *Emotions in humans and artefacts* (pp. 183–211). Cambridge, MA: MIT Press.

Ortony, A., Clore, G., & Collins, A. (1988). *The cognitive structure of emotions*. Cambridge: Cambridge University Press.

Ostermann, J., Beutnagel, M., Fischer, A., & Wang, Y. (1998, December). Integration of talking heads and text-to-speech synthesizers for visual TTS. In *Proceedings of ICSLP 98*, Sydney, Australia.

Pasquariello, S., & Pelachaud, C. (2001, September). Greta: A simple facial animation engine. 6th Online World Conference on Soft Computing in Industrial Applications, Session on Soft Computing for Intelligent 3D Agents. (Available online at http://www.dis.uniroma1.it/~pelachau/#publications). Last visited September 8, 2003.

Pelachaud, C. (2001). Contextually embodied agents. In N. Magnenat-Thalmann & D. Thalmann (Eds.), *Deformable avatars* (pp. 98–108). New York: Kluwer.

Pelachaud, C., Badler, N., & Steedman, M. (1996). Generating facial expressions for speech. *Cognitive Science*, *20*, 1–46.

Persson, P. (1999). *AGNETA & FRIDA: A narrative experience of the web?* AAAI Fall Symposium on Narrative Intelligence, North Falmouth, Massachusetts.

Ruttkay, Z., Dormann, C., & Noot, H. (2002). *Evaluating ECAs—What and how?* Proceedings of the AAMAS'2002 Workshop on Embodied Conversational Agents—let's specify and compare them!, Bologna, Italy.

Schröder, M., Cowie, R., Douglas-Cowie, E., Westerdijk, M., & Gielen, S. (2001). Acoustic correlates of emotion dimensions in view of speech synthesis. *Proc. Eurospeech*, *1*, 87–90.

Stronks, J. J. S. (2002). *Friendship relations with embodied conversational agents: Integrating social psychology in ECA design*. Unpublished master's thesis, Faculty of Computer Science, University of Twente, Enschede, The Netherlands.

Weizenbaum, J. (1966). ELIZA—A computer program for the study of natural language communication between man and machine. *Communications of the Association for Computing Machinery*, *9*, 36–45.

Zong, Y., Dohi, H., Prendinger, H., & Ishizuka, M. (2000). *Emotion expression function in multimodal presentation. Advances in Multimodal Interfaces—ICMI2000*. Proceedings of the 3rd Int'l Conference on Multimodal Interfaces, Beijing, China. (Available online at http://www.miv.t.u-tokyo.ac.jp/papers/yzong-beijing-ICMI2000.pdf). Last visited September 8, 2003.

PART

III

AGENTS FOR INTERCULTURAL COMMUNICATION

CHAPTER

10

Building Bridges Through the Unspoken: Embodied Agents to Facilitate Intercultural Communication

Katherine Isbister
Stanford University

THE CHALLENGE: DEVELOPING INTERCULTURAL NONVERBAL COMMUNICATION SKILLS

Increasingly, living and working in the modern world requires engaging with people from cultures other than one's own. Collaboration with colleagues from other nations, visits to other countries for conferences or work meetings, and everyday encounters in one's hometown with people from other cultures are all more frequent events (see Ting-Toomey, 1999, for an excellent summary of statistics related to these changes).

Knowing a common language is necessary, but not sufficient, for successful communication. Much of the emotional and interpersonal content of conversations happens nonverbally (Leathers, 1986). A raised brow, a lean forward, a glance away, a long pause—these reveal how the speaker feels about the conversation and the other participants. As Ting-Toomey (1999) pointed out, "nonverbal messages are often the primary means of signaling our emotions, attitudes, and the nature of our relationships with others" (p. 115). Becoming fluent in *reading* these cues must be part of a complete education in intercultural communication. For the goal of communication is not just to transfer information, but also to develop relationships, which depend on emotions and attitudes that are expressed nonverbally. We determine, for example, whether someone is trustworthy from their nonverbal behavior (Ekman et al., 1980), and trustworthy behavior is culturally specific (Leathers, 1986). Thus, to be effectively trustworthy, a person must learn how to convey this nonverbally in the local culture.

It is challenging to learn these skills. Few of us ruminate on, analyze, or discuss our nonverbal communication. In fact studies have shown that people react to nonverbal cues, altering their subjective opinions as well as their actions, but stating that they did not observe differences (Colburn et al., 2000; Isbister, 1998). This indicates that we are affected by, although not necessarily consciously aware of, nonverbal cues. Language teachers typically do not coach students in this aspect of communicating. If they are native speakers, they may unconsciously model these cues, but may not be self-aware enough to know this, much less consciously readjust their students' movements, eye-gaze patterns, and interpersonal distance. In fact students with little knowledge of other cultures may misattribute nonverbal communication their teacher displays to personality rather than culture (Ting-Toomey pointed out that this frequently happens in intercultural communication), and may not realize that these cues and rhythms are part of participating fully in conversation like a native.

Within intercultural contexts, it is unlikely that the other person will verbalize what he or she is reading from our nonverbal communication to check in on its accuracy. More likely, he or she will simply respond nonverbally to cues we may not be aware we are sending. This can result in some confusing and awkward encounters, with little hope for clarification.

WHY AGENTS ARE SUITED FOR TRAINING INTERCULTURAL NONVERBAL COMMUNICATION

It is this author's belief that we can use social agent technology to help people understand, practice, and refine nonverbal communication in intercultural contexts. It has been demonstrated that people do read nonverbal cues in agents and make use of them to determine the worthiness of the agents' advice (Isbister & Nass, 2000). Agents with nonverbal skills have been shown to be effective in pedagogical contexts (Lester et al., 1997). There are several reasons that agents could be of particular value in training nonverbal communication:

> Repetition: As with a video or any other canned medium, one can play and replay an interaction with an agent. The agent never tires of the interaction, and there is no fear of negative social impact of repetition. This allows the person to focus on different aspects of what is being communicated nonverbally to understand the nuances of the various layers of what is happening.

Low social embarrassment: One need not fear looking stupid to an agent—one can practice and work through the awkward phase essentially alone.

Distillation of nonverbal cues with little noise: In designing an agent's body language, we can distill and isolate cues in ways that are not as easy with video, which also picks up many other nuances that are not necessarily of interest (a dialect, a characteristic way of doing things). Cartoonists use well-documented effects to draw the eye's attention (Lasseter, 1987). We can use these techniques to draw broad strokes in cultural cues. This also means we can educate people about cultural types (e.g., stereotypical teacher, mother, student) through choice of the characters that we develop.

Demonstration of culture-hopping: If we use the same character and show him or her acting in two culturally distinct ways, this can help the learner project him or herself into the same fluency and view it as a possibility. It may also be easier to see, through the contrast, where one may need to focus in developing one's own efforts.

Learner behavior tracking and feedback: Software agents can make use of user behavior detection to help correct their nonverbal performances in real time. Audio detection, gaze, and motion tracking are already in use in agent projects; as we get better at tracking facial expressions and gestures, these cues can be read and evaluated by the agent. Having a bodily representation, the agent can give feedback in the appropriate nonverbal ways to the learner as well as with direct verbal feedback.

DIMENSIONS OF NONVERBAL COMMUNICATION DIFFERENCES AMONG CULTURES

To illustrate where agents can be helpful in training intercultural nonverbal skills, I include a brief taxonomy of the kinds of cues that have been observed to differ and matter interculturally. It is important to note that these dimensions can also vary widely within cultures—gender, age, subculture, and other factors can lead to differences in how members of any given culture behave. This chapter focuses on the broad variances between cultures that have been observed, although certainly subculture communication might also be enhanced with practice and training using social agents.

Interpersonal Distance

Interpersonal distance (how close together we are when we talk) varies widely among cultures. Hall (1966) pointed out that Arabs tend to stand or sit much closer together when talking (in fact within scent range) than

Western Europeans or Americans. To not do so is to be cold or suspicious. He distinguished between *high-contact* and *low-contact* cultures. According to Ting-Toomey (1999), French, Italians, Latin Americans, Russians, Arabs, and Africans belong to the high-contact set, whereas Americans, Canadians, Northern Europeans, and New Zealanders are moderate contact. East Asians are low contact (Chinese, Japanese, Koreans). She said,

> In a high-contact culture communicators face one another directly, often look one another in the eye, interact closely with one another, and speak in a rather loud voice. In a low-contact culture, in contrast, interactants face one another more indirectly, interact with a wider space between them, engage in little or no touching, prefer indirect eye glances, and speak in a soft-to-moderate tone of voice. (p. 129)

Interpersonal distance shifts during conversation, working within the acceptable range, to indicate things like agreement/disagreement (leaning toward/away) or increased intimacy. Thus, when people of different cultures are constantly moving within intimate range unwittingly or staying too far away while floating within their own culture's acceptable range, they may be signaling something they do not intend (see Hall, 1966, for a discussion of the different zones of interpersonal distance and their meaning).

Eye Gaze

Eye gaze, as mentioned earlier, seems to be culturally interrelated with accepted interpersonal distance and contact levels. Eye gaze is used to establish that one is paying attention (Cassell, Bickmore, Vilhjálmsson, & Yan, 2000), to indicate one's relative status (Johnstone, 1981), to help signal when one is ready to take a turn in a conversation, or to yield the floor, as well as to gesture at people or things one is discussing (Clark, 1996; Colburn et al., 2000; Garau et al., 2001). Leathers (1986) pointed out that Japanese and Arab cultures differ quite a bit in the accepted eye contact practice. In Arab culture, sustained eye contact is a sign of engagement and sincerity, whereas in Japanese culture, it is more polite to use direct eye contact only sparingly (Leathers, 1986). Ting-Toomey (1999) observed that even within the United States different subcultural groups have different turn-taking norms for gaze. Euro-Americans "tend to break off eye contact when speaking and maintain eye contact when listening," whereas African Americans "tend to maintain eye contact when speaking and break off eye contact when listening" (p. 126). Thus, African Americans may perceive Euro-Americans as nonresponsive or indifferent, whereas African Americans may be perceived as confrontational or aggressive when speaking, by European Americans.

Gestures and Range of Movement

Gestural differences between cultures are well documented. There are emblematic gestures (such as how to signal "everything is OK") even within Europe that mean disastrously different things (Morris et al., 1979). A simple gesture such as a nod means "yes" in the United States, but simply means "I understand and hear what you're saying" in Japan, which can lead to misunderstandings (Leathers, 1986). The gesture that means "come here" in many Asian cultures (palm down, fingers waving toward the body) may be read as "go away" by people from the United States (Ting-Toomey, 1999). How widely one moves one's arms while gesturing varies by culture as well. What might seem like violent gesticulating to someone from Japan would seem quite normal and usual to someone from a Latin culture.

Emotional Expression

People convey their emotions nonverbally in various ways—with facial expressions (raised or furrowed brow, smile or frown, widened or narrowed eyes), with tone of voice during speaking, with eye gaze, and with overall posture. Although there are widely accepted results that show some emotions are universally legible (Ekman, 1984; Ekman et al., 1980), the display of emotions during conversation is not universal. Members of cultures that devalue an emotion suppress the expression of that emotion. For example, Japanese people learn to suppress the automatic eyebrow flash that occurs when one is greeting someone because it is considered indecent (Leathers, 1986). In addition, many emotional expressions during conversation are more for demonstration purposes than emergent from the person's innate reaction (Clark, 1996; Ting-Toomey, 1999). Display emotions reflect cultural values, and some expressions may only appear in certain cultures. For example, the *wry smile* in Britain and the *smug* look in America seem to be unique to those cultures (Leathers, 1986).

Status Awareness and Respect

Nonverbal status markers include how still one holds one's head, how much one fidgets and touches one's own body, patterns of eye contact, downward turn of gaze, posture, and the amount one smiles (Johnstone, 1981). Status is also expressed in who stands when greeting, who goes first through a door, and who sits in what place at the table. Cultures have different rules about who gets status when and how deference should be shown—the bowing ritual in Japan is a well-known example of a culturally specific nonverbal way to convey status. Not showing the culturally proper degree of deference for one's perceived social status can cause serious problems.

Physical Contact and Intimacy

Physical contact means different things in different cultures. Japanese rarely make contact at all in public even with close friends; other cultures have greeting rituals that include hugs and kisses even for those whom one has never met. In some cultures, amount of public physical contact increases among friends and intimates; in other cultures, this is not a socially sanctioned transition.

Displaying Understanding, Agreement, Disagreement

During the course of a conversation, there are various nonverbal cues used to indicate "I am listening" as well as "I agree with what you say" or "I disagree" (Cassell et al., 2000; Clark, 1996). These include nods, empathetic emotional gestures that indicate understanding of the emotion being described, as well as sounds like *Unh-hunh* or *ummm*. As mentioned previously, the nonverbal signals for comprehension vary across cultures—the nod is agreement in the United States, but merely a sign of understanding in Japan. In Japan, people work to muffle initial expressions of disagreement, carefully framing them to minimize embarrassment for the speaker. Disagreement is typically much more plainly displayed on the face of a U.S. listener and may help the speaker shape his or her discourse accordingly.

Conversational Mechanics

Much of how we handle synchronizing with one another in conversation is based on nonverbal communication. Knowing when a conversation is about to begin or end, knowing when someone wants to take a turn, the negotiation of giving up the conversational floor to another, and confirming understanding (as mentioned earlier) are all indicated mostly nonverbally. There are cultural differences in how these activities take place. One example in turn taking is whether it is acceptable to overlap another person's turn or not—to break in. In France, frequent overlaps are considered the sign of a lively and fun exchange (Clark, 1996). This is not the case in Scandinavian cultures, for example, where one expects a healthy pause between turns, in which each person thinks about what was said.

CURRENT USES OF AGENTS TO MODEL AND DETECT NONVERBAL BEHAVIOR

The agent design community has grasped the value of nonverbal communication and, as graphics and computer power have grown, has been working to incorporate perception and generation of appropriate nonverbal cues into social agents. The following is an overview of some of the ongoing projects in this domain.

Justine Cassell and her students at the MIT Media Lab have developed several iterations of systems that generate nonverbal cues based on a person's statements and nonverbal responses. Cassell collaborated with Norm Badler and others at the University of Pennsylvania on a system called *Animated Conversation*. This system "automatically generated context-appropriate gestures, facial movements, and intonational patterns" (Cassell et al., 2000, p. 53). However, the interaction was a conversation between two agents and was not generated in real time. Gandalf, created at MIT, was an agent that "recognized and displayed interactional information such as gaze, simple gesture and canned speech events" (Cassell et al., 2000, p. 53). She and her students demonstrated with experiments that people found the interaction smoother and thought the agent's language skills were higher when Gandalf used these nonverbal abilities, as opposed to when they were disabled. More recently, Cassell's group worked on REA, a virtual real estate agent. REA tracks a person's gaze and listens for speech. She "can sense the user passively through cameras and audio input, and is capable of speech with intonation, facial display, and gestural output" (Cassell et al., 2000, p. 55). REA uses audio sensing to guess when a person wants to take a conversational turn and yields the floor. She also uses sensing of the person's position to adjust her gaze to follow and will turn to face the person when beginning a conversation. She nods to acknowledge the person's comments, at the end of his or her speech, and does the proper look-away behavior when she is planning what to say to hold the conversational floor. REA also uses gestures to underscore her points and help clarify what she is talking about.

Churchill and colleagues (2000) developed an agent to assist office workers with the operation of a multimedia presentation room using a similar base architecture to REA's for conversational gesture generation.

Elisabeth André and Thomas Rist at DFKI developed several software agents that make use of nonverbal cues to aid in communication. Their earlier work, on the AiA and PPP Personas, focused on use of gestures such as pointing to help make information clearer to people browsing desktop or Web-based information (André, Rist, & Müller, 1999). More recently, they created a system with multiple characters that uses gesture and eye gaze in conversation with one another, demonstrating multiple roles, personalities, and attitudes toward the topic under discussion (André & Rist, 2000).

James Lester's group at North Carolina State University developed several iterations of pedagogical agents, which use gestures, eye gaze, and dynamic speech to engage learners. Herman the Bug included a real-time behavior sequencing engine, which allowed it to assist students in a flexible manner as they worked to learn about plants by designing a plant that could thrive in a given environment. Lester's group found that the mere presence of a lifelike character in an interactive learning environment "can have a strong positive effect on students' perception of their learning experience" (Lester et al., 1997, p. 359).

Rickel and Johnson at USC spent several years developing and refining agents that can collaborate with and educate people who are learning tasks in a virtual world. They've taken advantage of the fact that an agent can more easily track the actions and gaze of a person in a virtual world, to model reasonable and helpful responses in the agent. The agent can point out or demonstrate steps to take or places to find things, using gaze and gestures, as well as bodily orientation. The agent can also give nonverbal tutorial feedback, such as a nod of approval, or a look of puzzlement if someone is on the wrong track (Rickel & Johnson, 2000).

Colburn, Cohen, and Drucker (2000) constructed avatars for virtual conferencing that have eye gaze patterns generated to resemble normal conversation patterns—specifically, looking away and looking back to control turn taking and attention to who has the floor. They found that people in one-on-one interactions with the avatar tended to look at it more when listening (as they would look at a real person) than at an avatar without gaze modeling.

Finally, Isbister and Nakanishi designed an agent to support conversations in a virtual world that detected awkward silences during conversation and offered topics to talk about. The agent used body orientation, facial expression, and emblems such as nods to get the conversational partners to take up the topic (Isbister et al., 2000).

CURRENT USE OF AGENTS IN INTERCULTURAL COMMUNICATION SITUATIONS

A team at USC is currently building a mission rehearsal environment in which soldiers can practice difficult and potentially emotional interactions with locals in a foreign setting while trying to accomplish their mission. The project makes use of the agents that act out roles in the drama, including a mother grieving over a child, and other soldiers who can provide advice as the soldier goes through the exercise (Hill et al., 2001). These agents make use of nonverbal cues, such as emotional expression and tone of voice, as well as gestures to make the simulation feel more real to the participant.

Also the agent mentioned in the previous section, who tries to get two people to take up a new conversational topic, was designed expressly to facilitate conversation between Japanese and U.S. students. The agent's animations and behaviors were designed to be legible to people from either culture. However, this work did not attempt to exploit the differences by having the agent act more Japanese or more American depending on who it addressed (Isbister et al., 2000).

Most of the work being done on nonverbal cues in agents is local to each research group's own country. As Badler and Albeck pointed out, this may be because it is difficult enough to take into account all the nuances of creating workable gestures for one's own culture (Badler & Allbeck, 2001, p. 9). However, this author believes we do not need to model everything at once to create helpful applications in this space.

SUGGESTED DIRECTIONS FOR DEVELOPING AGENTS TO SUPPORT LEARNING NONVERBAL INTERCULTURAL COMMUNICATION SKILLS

Faces

Because so much of the crucial nonverbal communication during conversation happens in the face—gaze, emotional expression, signs of comprehension—it would be quite valuable to have a one-on-one conversational partner even if it was simply a face. Although the ever-improving realistic style facial models could be used (such as the system described in Colburn et al., 2000), this is not a prerequisite. As mentioned in the section on the benefits of using agents, one could take a distilled and exaggerated approach to displaying the relevant cues. One example of this method is Elliot's demonstration of his affective reasoner. He created a simple, line-drawing style face that transitioned between emotions. Those who saw the demo were highly entertained and amused by it—the emotions were very readable (see Elliot, 1992, for a discussion of the affective reasoner).

To provide meaningful feedback with its own face, the agent would need to have ways to track the learner's face. Gaze-tracking techniques already exist that could be used to help teach a person when their gaze patterns are conforming to acceptable norms for turn taking, deference, and attentiveness. Presuming there was some way to monitor the learner's facial expression, the agent could provide real-time emotional feedback of its own in response to changing expressions. For example, one could be coached not to raise one's eyebrows when greeting someone from Japan, by getting disapproving verbal and nonverbal feedback from the agent, based on one's real-time reactions. The agent might also note expressions that are too exaggerated for the Japanese cultural context.

Full Bodies

With full-bodied agents, particularly life-size ones like REA (Cassell et al., 2000), one could monitor gestural range and interpersonal distance in the learner and provide emotional and other nonverbal feedback (such as backing up) in the agent in response to inappropriate closeness, distance, or excessive or overly minimal gesticulation.

Multiple Agents

It would be beneficial to make use of multiple agents, where possible, to help learners by giving them the chance to observe appropriate and inappropriate interactions and to model on the agents. Agents could use nonverbal cues in interaction with one another (such as a sidelong glance of disbelief) to help give the learner a visceral impression of what sort of impression he or she is making at present.

Settings and Scripts

Many of the agent environments developed in other contexts by the research community could be repurposed to help a person practice culturally appropriate rituals, such as greetings and farewells, hosting someone at one's home, going to a restaurant, and other social situations. Agent behaviors could be scripted and one could be coached by example, then asked to make decisions about how to behave, to practice the right responses.

CONCLUSION

Nonverbal skills are an important part of intercultural communication fluency. This chapter outlined some of the ways that nonverbal behaviors differ between cultures and ways that the social agent community might apply existing technologies to the challenge of training these skills. Social agents are well suited to this application—offering the opportunity for repetition, low social anxiety, and to clearly isolate and model the right cues for the student.

One challenge in the development of this kind of training assistance is the further evolution of our abilities to track the learner's nonverbal cues. Real-time tracking and response are an important part of the quality and effectiveness of the learning experience. Some of the cues outlined in the overview of cross-cultural nonverbal differences earlier in this chapter are currently beyond our ability to detect. However, detection techniques are continually improving, and this could be an excellent testbed for refining and applying them.

I hope this overview serves as an inspiration to other members of the social agent research and design community to begin exploring this application area. It seems likely that creating pilot projects will further some of the basic goals of the community, evolving our understanding of realistic and engaging behavior in real social contexts.

REFERENCES

André, E., & Rist, T. (2000). Presenting through performing: On the use of multiple lifelike characters in knowledge-based presentation systems. *IUI (Intelligent User Interfaces) 2000 Proceedings*, pp. 1–8.

André, E., Rist, T., & Müller, J. (1999). Employing AI methods to control the behavior of animated interface agents. *Applied Artificial Intelligence, 13*, 415–448.

Badler, N., & Allbeck, J. (2001). Towards behavioral consistency in animated agents. In N. Magnenat-Thalmann & D. Thalmann (Eds.), *Deformable avatars* (pp. 191–205). Dordrecht: Kluwer Academic Publishers.

Cassell, J., Bickmore, T., Vilhjálmsson, H., & Yan, H. (2000). More than just a pretty face: Affordances of embodiment. *CHI 2000 Proceedings*, pp. 52–59.

Churchill, E. F., Cook, L., Hodgson, P., Prevost, S., & Sullivan, J. W. (2000). "May I help you?": Designing embodied conversational agent allies. In J. Cassell, J. Sullivan, S. Prevost, & E. Churchill (Eds.), *Embodied conversational agents* (pp. 64–94). Cambridge, MA: MIT Press.

Clark, H. (1996). *Using language*. Cambridge, United Kingdom: Cambridge University Press.

Colburn, A. R., Cohen, M. F., & Drucker, S. M. (2000). *The role of eye gaze in avatar mediated conversational interfaces* (Technical Report MSR-TR-2000-81). Redmond, WA: Microsoft Research.

Ekman, P. (1984). Expression and the nature of emotion. In K. Scherer & P. Ekman (Eds.), *Approaches to emotion* (pp. 319–343). Hillsdale, NJ: Lawrence Erlbaum Associates.

Ekman, P., Friesen, W., O'Sullivan, M., & Scherer, K. (1980). Relative importance of face, body, and speech in judgments of personality and affect. *Journal of Personality and Social Psychology, 38*(2), 270–277.

Elliot, C. (1992). *The affective reasoner: A process model of emotions in a multi-agent system*. Unpublished doctoral dissertation, Northwestern University, Evanston, IL.

Garau, M., Slater, M., Bee, S., & Sasse, M. A. (2001). The impact of eye gaze on communication using humanoid avatars. *CHI 2001 Proceedings*, pp. 309–316.

Hall, E. T. (1966). *The hidden dimension*. New York: Anchor Books, Doubleday.

Hill, R., Morie, J., Rickel, J., Thiébaux, M., Tuch, L., Whitney, R., Douglas, J., Swartout, W., Gratch, J., Johnson, W. L., Kyriakakis, C., LaBore, C., Lindheim, R., Marsella, S., Miraglia, D., & Moore, B. (2001). Toward the Holodeck: Integrating graphics, sound, character, and story. *Proceedings of the Fifth International Conference on Autonomous Agents*, pp. 409–416.

Isbister, K. (1998). *Reading personality in onscreen interactive characters: An examination of social psychological principles of consistency, personality match, and situational attribution applied to interaction with characters*. Unpublished doctoral dissertation, Stanford University.

Isbister, K., Nakanishi, H., Ishida, T., & Nass, C. (2000). Helper agent: Designing an assistant for human–human interaction in a virtual meeting space. *CHI 2000 Conference Proceedings*, pp. 57–64.

Isbister, K., & Nass, C. (2000). Consistency of personality in interactive characters: Verbal cues, non-verbal cues, and user characteristics. *International Journal of Human-Computer Studies, 53*(2), 251–267.

Johnstone, K. (1981). *Impro: Improvisation and the theatre*. New York: Routledge.

Lasseter, B. (1987). Principles of traditional animation applied to 3D computer animation. *Computer Graphics (SIGGRAPH '87 Proceedings), 21*(4), 35–44.
Leathers, D. G. (1986). *Successful nonverbal communication: Principles and applications.* New York: Macmillan.
Lester, J. C., Barlow, S. T., Converse, S. A., Stone, B. A., Kahler, S. E., & Bhogal, R. S. (1997). The persona effect: Affective impact of animated pedagogical agents. *CHI '97 Proceedings,* pp. 359–366.
Morris, D., Collett, P., Marsh, P., & O'Shaughnessy, M. (1979). *Gestures: Their origins and distribution.* New York: Stein & Day.
Rickel, J., & Johnson, W. L. (2000). Task-oriented collaboration with embodied agents in virtual worlds. In J. Cassell, J. Sullivan, S. Prevost, & E. Churchill (Eds.), *Embodied conversational agents* (pp. 95–122). Cambridge, MA: MIT Press.
Ting-Toomey, S. (1999). *Communicating across cultures.* New York: Guilford.

CHAPTER
11

Designing a Social Agent for Virtual Meeting Space

Hideyuki Nakanishi
Toru Ishida
Kyoto University

Katherine Isbister
San Francisco, CA

Clifford Nass
Stanford University

Virtual meeting spaces enable people to meet accidentally and have multiple conversations simultaneously. There are many virtual meeting spaces such as Community Place (Lea et al., 1997), InterSpace (Sugawara et al., 1994), Diamond Park (Waters & Barrus, 1997), DIVE (Hagsand, 1996), Massive (Greenhalgh & Benford, 1995), and CU-SeeMe VR (Han & Smith, 1996). Virtual meeting spaces make it easy to have casual meetings between strangers from across town or even across the world. Virtual meeting spaces usually provide little socially meaningful context to use as a basis for finding common ground with each other. Because it is easy to arrive at a virtual meeting space from many entry points, it is often hard for visitors to assume much about one another's cultural backgrounds, group memberships, and other aspects of social identity. People need this sort of common context to build new human relationships (Clark, 1996).

We believe social agents could provide ongoing, in-context help in forming social relationships and building common ground between visitors to virtual meeting spaces. We developed a social agent playing a role of party host in a virtual space. This agent was applied to support cross-cultural communication in our experiment.

SOCIAL AGENTS

Related Work

Previous studies have discussed and demonstrated some benefits of interface agents in one-on-one task settings, such as taking an educational tutorial, going on a tour (Isbister & Doyle, 1999), or looking at real estate. Lester et al. (1997) noted that the presence of an agent can lead to a strong positive effect on students' perceptions of their learning experience. Cassell et al. (1999) discussed the value of an embodied real estate agent with the proper human verbal and nonverbal communication skills. However, these findings concern task-support agents interacting with a single user.

There are projects that have created agent-based social support through text-based conversation. Julia (Foner, 1997) plays the role of a guide in virtual worlds of MUD; the Extempo bartender agent converses with chat visitors and is designed to enhance the social atmosphere (Isbister & Hayes-Roth, 1997). There are bots for Web sites that answer questions and direct visitors in a friendly way (e.g., http://www.artificial-life.com). However, these agents are designed to engage in one-on-one social interactions, rather than facilitating human–human interaction. There are few studies about agents who interact with multiple people. Takeuchi and Naito (1995) proposed a synthesized facial display as a conversational partner in multi-party meetings.

The social agent we developed differs from the agents described earlier, which support specific tasks or play the role of a conversation partner. Our agent aims to work as an in-between of human–human interaction. Our agent is designed to conduct simple question-and-answer sessions so that people whose conversation is faltering can find a common topic to talk about. Another possible solution for such an awkward situation is providing an information search tool to find a common topic based on the retrieved data about the social identities of conversation partners. However, that tool does not help the process to start a conversation. There is a gap between finding topics and beginning conversations. Through question and answer, people can share one another's answer to the same question. That is an opportunity to start a conversation based on the answers. Furthermore, it may be invasive for the participants' privacy to collect personal information about conversation partners.

Cross-Cultural Communication

For testing our agent, we focused on an extreme case of low social context in a virtual meeting space: strangers from different national cultures meeting for the first time. Even when people can use a common language with reasonable fluency, they do not necessarily have a common context for their conversation. Different cultures have different notions of how to begin and develop conversations. What is a safe topic that is unlikely to harm the conversation and destroy the relationship in one culture may be unsafe in another culture. For example, in some cultures it is appropriate to ask about family members right away, whereas in other cultures this is private (Clark, 1996; Hall & Hall, 1990). Because it is hard to establish a common ground in this sort of meetings, we thought we could find the clear effect of our agent's assistance in conversations.

We focused on conversations between Japanese and Americans. These two national groups are known to have different interaction styles and cultural norms (Hall & Hall, 1990), and so we felt this was a good test case.

DESIGN OF THE AGENT

In a virtual meeting space called FreeWalk that we developed (Nakanishi et al., 1994) our agent basically acts like a busy party host looking for clues that the guests' conversations are going badly. The agent tracks audio from a two-person conversation to look for longer silences that will trigger its conversation aid. Pauses are a powerful cue for what is happening in a conversation (Clark, 1996). When the agent finds the pause, it approaches the conversation pair. The agent then directs a series of yes/no questions to both conversation partners in turn and uses their answers to guide its suggestion for a new topic. Then the agent retreats until it is needed again.

Design of the Virtual Meeting Space

Figure 11.1 shows an image of a FreeWalk screen. FreeWalk provides a three-dimensional community common area where people can meet. Participants move and turn freely in the space using their mouse (just as in a videogame). Locations and view directions of participants in the space determine which pictures and voices get transmitted.

In this three-dimensional space, a pyramid of three-dimensional polygons represents each participant. The system maps live video of each participant on one rectangular plane of the pyramid, and the participant's viewpoint lies at the center of this rectangle. The view of the community common area from a participant's particular viewpoint appears in the FreeWalk screen.

Participants standing far away in the three-dimensional environment appear smaller and those closer appear larger. FreeWalk does not display participants located beyond a predefined distance. The system also transfers voices under the same policy—that is, voice volume changes in proportion

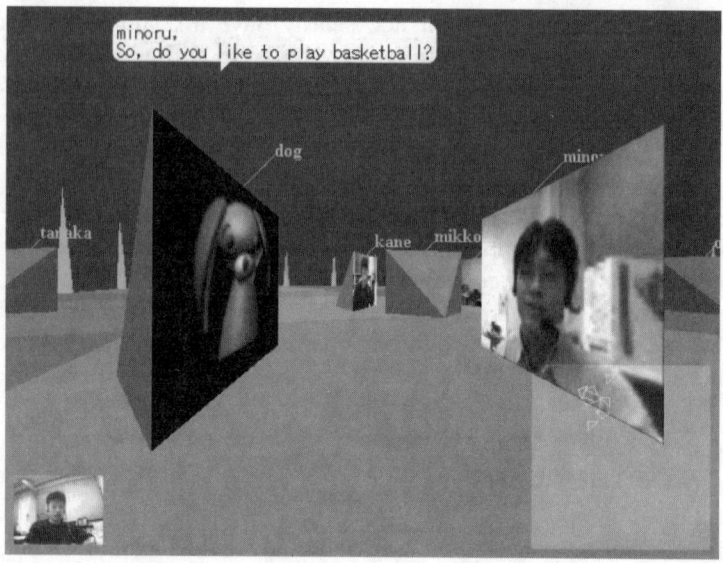

FIG. 11.1. Social agent in a virtual meeting space.

to the distance between sender and receiver. Moreover, a participant hears others' voices in stereo so that he or she can easily recognize the speaker.

Because distance attenuates voice, a participant must approach the others to talk to them. However, not only can the participants in the conversation hear the speaker's voice, but also anyone in the neighborhood can listen. This mechanism forces people to combine actions and conversations in the space. People can smoothly join the conversation that attracts their interest because they can guess the subject by listening to the conversation beforehand. People can exit a conversation by leaving a group and join a conversation by approaching another group.

Because the participants' locations and view directions reflect a pyramid orientation, each participant can grasp the locations and view directions of other participants, and observe what other people are doing from a distance. Participants can also observe others around them by turning their body. The volume and direction of one's voice is determined by its position, therefore participants can easily identify the others in the same conversation group. Comprehending the correspondence between each face and voice is not as difficult as that in conventional videoconferencing systems. The speaker can assume that the listeners recognize who is speaking. Furthermore, the speaker can turn to face the person to whom he or she is talking. These natures may transmit nonverbal information, which makes it unnecessary to call the name of the person whom you talk to before you begin speaking.

Nonverbal Communication Abilities

In virtual spaces, our social agent is embodied the same way as users (see Fig. 11.1). This allowed us to take advantage of nonverbal cues in designing the agent, such as a spatial position and direction for turning to face users, and animated pictures to present facial expressions and gestures.

The agent approaches the conversation pair to direct yes/no questions when it detects an awkward silence in their conversation. After concluding a suggestion cycle, the agent departs from the conversation zone and wanders at a distance until it detects another awkward silence. This makes it easy for the conversation pair to know whether the agent joins their discussion (Hall, 1966). The agent orients its face toward the conversation partner it is addressing so that the pair can intuitively recognize whom the agent asks.

On the rectangle of the agent's embodiment, an animated dog is pasted. We chose a dog because we wanted users to think of the agent as subservient, friendly, and reasonably socially intelligent (Reeves & Nass, 1996). The agent has a set of animations for asking questions, reacting to affirmative or negative responses, and making suggestions. Each of them corresponds to a phase in the process of question and answer. We crafted these animations as a supplement to the agent's speech (Cassell et al., 1999).

Topic-Suggestion Mechanisms

Silence Detection. The agent decides there is silence when the sum of the voice volumes of both participants is below a fixed threshold value. When the agent detects a silence that lasts for more than a certain period of time, it decides the participants are in an awkward pause.

Positioning. The agent decides how to position itself based on the location and orientation of each participant. The agent turns toward the participant it is currently addressing. If the participants move while the agent is talking, the agent adjusts its location and orientation. The agent tries to pick a place where it can be seen well by both people, but also tries to avoid blocking the view between them. If it is hard to find an optimal position, the agent stands so that it can at least be seen by the participant to whom it is addressing the question.

State Transitions. The agent has three states—namely, idling, approaching, and talking. When idling, the agent strolls at the corner of the virtual space farther away than the normal conversation zone (Hall, 1966). When the agent detects an awkward pause in the participants' conversation, it begins an approach. On reaching the participants, the agent goes into the talking state. However, if the participants start talking again before the agent reaches them, it stops the approach and goes back to idling. This behavior is strikingly similar to the actions of a hesitant subordinate trying to approach a superior who is engaged in a conversation with another dominant person. The agent also remains in an idling state if the participants are standing far apart from each other (out of conversation range) or are not facing each other. If the participants turn away from each other during the agent's approach or while it is talking, it returns to idling state as well.

Topic Knowledge

We gathered safe and unsafe topics for the first-time meeting using a Web survey filled out by Japanese and U.S. university students. We used the collected pool of topics to select common safe and unsafe topics for people from both countries. From these topics, we crafted a set of questions that the agent could ask in the question-and-answer process. Safe topics included: movies, music, weather, sports, and what you did yesterday. Unsafe topics included: money, politics, and religion. A sample safe question is: "Is the weather nice where you are right now?" A sample unsafe question is: "So, do you think it is all right for a country to fish for and eat whales?"

Conversation Model and Interface

The user interface for communicating with the agent is easy to learn. The agent presents questions to the participants in a text balloon above its head. We did not use a synthesized voice because we were afraid that unnatural utterances may affect participants, and participants may fail to catch what the agent says. The participant indicates yes or no by clicking the mouse on his or her answer displayed under the question in the text balloon. We did not use natural language as an input interface to prevent participants from expecting too much intelligence from the agent because they might be frustrated by rough conversation with the agent. Both participants see all questions, but only the addressed person sees the yes/no options. When the person answers the question, his or her answer is displayed in a text balloon above his or her own embodiment (see Fig. 11.2).

Each topic has a tree structure with nodes that are: a question for a participant, possible answers by participants, agent's reply to each answer, and flags indicating whether the agent will address its next question to the other person or to the same person (see Fig. 11.3). Topics were designed to draw participants into a dialog, so each turn is tailored for this purpose. Basically, the agent asks both participants the same question to draw shared or conflicting points from the interaction. The cycle always concludes with a recommendation for how the participants could make use of the particular topic area given their own answers to the agent.

(1)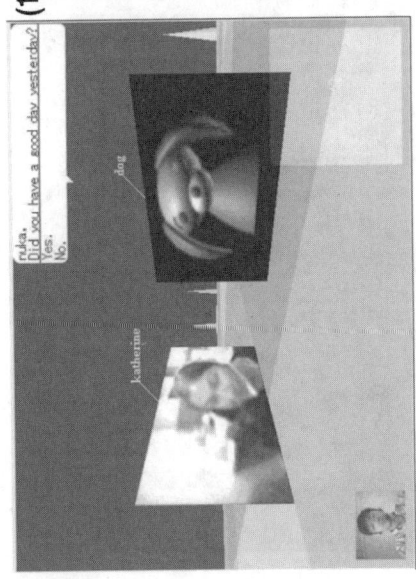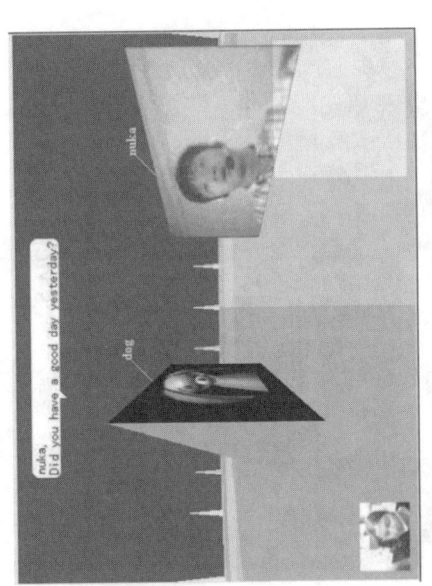

(2)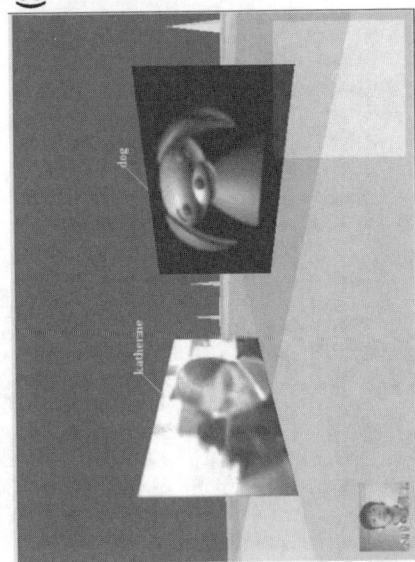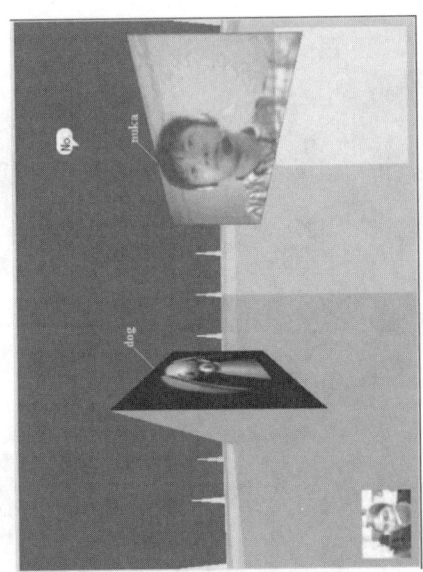

FIG. 11.2. *(Continued)*

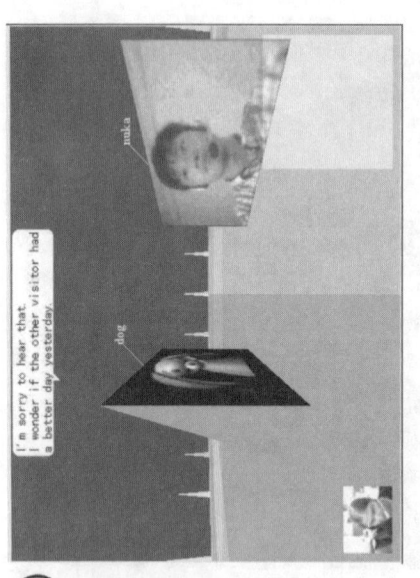
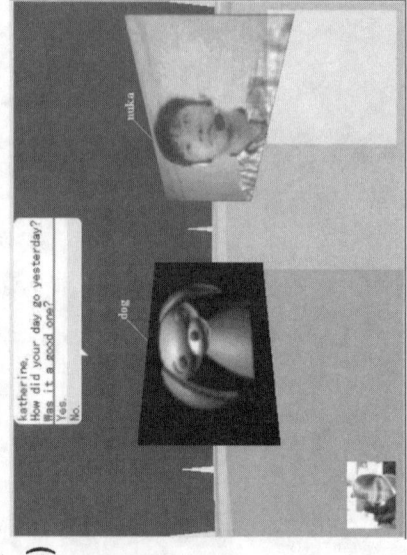
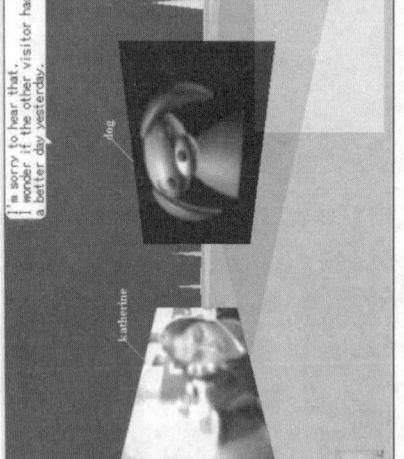

A's view   B's view

FIG. 11.2. Conversation from both participants' points of view.

11. SOCIAL AGENT FOR VIRTUAL MEETING SPACE

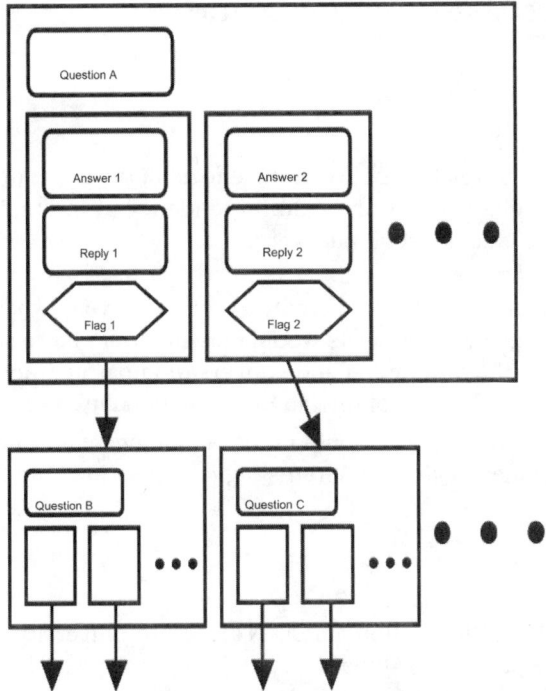

FIG. 11.3. Tree structure of topic data.

When the agent approaches to start a cycle, it selects a topic from its repertoire of safe (or unsafe) topics randomly out of those not yet used. Then it randomly chooses one of the two participants as the target for the first question. Let us call this person A. When A answers, the agent replies to A's answer. Based on what A answered, the agent then chooses a follow-up question. This question might be directed at A or B. If it is directed at B, the agent turns to B to pose the question. When B answers, the agent makes a general comment meant to guide the participants into using this topic. This general comment is selected based on the previous answers from the participants. Figure 11.2 shows part of this cycle from both participants' points of view. In this figure, (1) person A is asked the first question (2) and responds, (3) then the agent comments. (4) Next person B is asked a question. As described earlier, the agent faces the person it is addressing.

After making this comment, the agent departs. If at any time a user does not respond to the agent's question, the agent waits for an interval and then goes back into idling mode without trying to continue its question cycle. This makes it clear that the agent is an in-between only to lead a question cycle. Therefore, participants can intuitively understand they do not need to include the agent in their conversation.

CROSS-CULTURAL COMMUNICATION EXPERIMENT

Hypotheses

We focused on the relation between the effects of the agent on social interaction and the difference in the cultural tendency of the topics the agent provided. Our initial expectations were:

1. The safe-topic agent would create a more satisfying experience than if there were no agent. Participants would feel they were more similar, would be happier with the interaction and conversation partner, and would form more positive impressions of one another's nationality.

2. The unsafe-topic agent would make people uncomfortable, but might lead to a more meaningful and interesting conversation than the safe-topic agent.

Procedure

The study was a collaboration among NTT, Kyoto University, and Stanford University. We used a 1.5-Mbps dedicated line to connect both PCs in the two universities. The two research teams used chat software to communicate while running the study. We set up a PC with a camera and headset at each location (see Fig. 11.4).

We designed a three-condition experiment using pairs of students who were located in the United States and Japan. Pairs either interacted one on one or had the help of the safe-topic or unsafe-topic agent. We divided the 20-minute conversation session into five segments and forced the agent to display a topic within each four-minute segment. The agent looked for an awkward pause during a minute in each time segment. The agent introduced topics immediately if it could not find a pause. Thus, in the safe-agent condition, the agent introduced all five safe topics in random order. In the unsafe-agent condition, the agent introduced all five unsafe topics in random order.

Each research team recruited students for the study. The Stanford students were part of an undergraduate class that required participation in experiments for credit. The Japanese students were undergraduates from Kyoto University and other nearby universities who were paid for their participation. Because the study would be held in English, we screened Japanese students and selected those who scored at a reasonably high level on English proficiency tests. Both sets of students were screened for a high level of familiarity with one another's culture, and those with high experience were not asked to participate. We had 90 participating students. Stu-

11. SOCIAL AGENT FOR VIRTUAL MEETING SPACE 255

FIG. 11.4. Setup for the experiment (Stanford Side).

dents were assigned randomly to same-gender pairs. Each pair was randomly assigned to one of the three conditions.

Students were told that they would be testing out a new communication environment with a student from the other country. They were asked to talk about anything they liked. They were trained in how to use the system, then left alone to talk for 20 minutes. We made video recordings of all sessions by capturing what was on the screen on the Kyoto side onto videotapes. After their conversation, participants filled out a survey presented on a Web browser.

We prepared the survey in their native language. Questionnaire items of the survey were translated and then reverse translated for accuracy. The questionnaire included questions about the conversation, their conversation partner, and the agent (in agent conditions). We also asked them to make assessments of themselves and the typical person of both participants' cultures on some commonly used stereotypic adjectives. We ended up with data from 45 Japanese students and 43 American students (because we could not get the questionnaire results from two subjects in the unsafe agent condition) for our analysis.

Results

Safe Agent Versus No Agent

We summarized the values of only the questionnaire items that had significant differences.

TABLE 11.1
Summary of *T* test Comparisons of American Students'
Ratings, Safe Agent versus No Agent

Variable	Mean (safe agent)	Mean (no agent)	t value (n = 13)
Confident	6.46	5.54	−2.33**
Domineering	4.00	4.92	2.03*
Restrained	3.61	5.00	2.52**
Partner trustworthy	6.54	5.91	−2.46**
Japanese creative	5.38	4.54	−2.06*
Japanese friendly	5.92	5.23	−2.08**
Japanese emotionally expressive	3.15	4.23	2.75***

*$p = .05$. **$p < .05$. ***$p = .01$.

American Reaction. The safe agent had positive effects for American participants as we expected (see Table 11.1; all items on an 8-point scale, 8 highest):

- *opinion of their own behavior higher*

    They rated themselves as more confident, less domineering, and less restrained in the safe agent condition.

- *opinion of partner higher*

    They rated their partner as significantly more trustworthy in the safe-agent condition.

- *opinion of the typical Japanese person higher*

    The safe-agent condition had a positive effect on impression of typical Japanese people. Those in the safe-agent condition rated the typical Japanese person as more creative and friendly. However, they rated the typical Japanese person as less emotionally expressive. (Americans typically stereotype Japanese people as less creative, less friendly, and less emotionally expressive.)

Japanese Reaction. The Japanese participants had a different response to the safe agent's presence. The safe agent did not improve their experience. However, it made them think their partner was more like themselves as we expected (see Table 11.2).

- *opinion of the experience lower*

    Japanese in the safe-agent condition rated the experience as less safe and more uncomfortable. They were less interested in continuing such a conversation and were less satisfied afterward.

TABLE 11.2
Summary of T test Comparisons of Japanese Students'
Ratings, Safe Agent versus No Agent

Variable	Mean (safe agent)	Mean (no agent)	t value (n = 15)
Unsafe	3.29	2.24	−2.05*
Uncomfortable	5.14	2.71	−3.9*****
Desire to continue	4.86	7.07	3.55****
Satisfying	4.79	6.14	2.32**
Self-evasive	5.86	4.71	−2.09**
Self-quieter	4.68	3.36	−2.08**
Partner talkative	5.00	4.07	−2.06**
Partner effusive	2.21	1.50	−2.06**
Partner engaging	6.00	6.78	2.47**
Partner typically American	5.22	6.14	2.26**
Partner similar to self	5.28	4.21	−2.20**
Americans competitive	6.57	5.00	−2.40**
Americans domineering	6.07	3.36	−4.44******
Americans selfish	6.14	4.93	−2.26**
Americans effusive	6.79	5.86	−2.21**

*$p = .05$. **$p < .05$. ***$p = .01$. ****$p < .01$. *****$p = .001$. ******$p < .001$.

- *opinion of their own behavior lower*

  They also rated themselves in a more negative light in the safe-agent condition. In the safe-agent condition, participants rated themselves as more evasive and quieter than participants did in the no-agent condition.

- *opinion of their partner mixed*

  Their ratings of their American partners were mixed. In the safe-agent condition, participants found their partners more talkative, more effusive, and less engaging. (Japanese tend to stereotype Americans as talkative and emotionally effusive.) Yet they rated their partners as less typically American and more similar to themselves.

- *opinion of the typical American person lower*

  The safe-agent condition seemed to exacerbate negative views of Americans for Japanese participants. In the safe-agent condition, they rated the typical American as more competitive, more domineering, more selfish, and more effusive than those in the no-agent condition. (All of these are stereotypical American traits from the Japanese point of view.)

We cannot be sure why the two groups had such different reactions. One reason may be that the agent's questions were implemented in English. It is possible that Japanese subjects felt it was a two-against-one situation. This might explain why they disliked the interaction, although it seemed to make them rate their partner as more similar to themselves. Another reason may be that Japanese subjects disliked the sudden interruptions by the agent that failed to find an awkward pause. Most of Japanese subjects seemed to be interested in talking with Americans.

Safe Agent Versus Unsafe Agent

- *Awkward is not necessarily bad*

As we expected, the unsafe agent made things more awkward, but also more interesting. We counted awkward pauses by observing the videotapes and found a higher number of awkward pauses in the unsafe-agent condition versus the safe-agent condition ($M = 4.34$(unsafe) and 3.09(safe), $t(56) = -3.06, p < .01$).

Despite the higher level of awkwardness in these conversations, both Japanese and American participants found the conversation that included the unsafe agent more interesting. Americans rated the unsafe-agent interaction more interesting, whereas Japanese rated the unsafe-agent experience more desirable to continue and more comfortable (see Tables 11.3 and 11.4).

TABLE 11.3
Summary of T test Comparisons of Japanese Students' Ratings, Safe Agent versus Unsafe Agent

Variable	Mean (safe agent)	Mean (unsafe agent)	t value (n = 15)
Desire to continue	4.86	6.21	−2.00^
Uncomfortable	5.14	3.57	2.41**
Self-evasive	5.86	4.64	2.03*
Self-restrained	5.43	3.79	2.37**
Self self-abasing	5.54	4.07	2.33**
Self team-oriented	4.00	2.64	2.34**
Partner similar to self	5.29	3.64	2.58**
Partner considerate	7.31	5.93	3.02****
Partner domineering	1.14	2.07	−2.43**
Partner friendly	7.29	6.14	2.31**
Partner talkative	5.00	3.79	2.14**
Americans domineering	6.07	5.00	2.26**
Agent nice	3.43	5.29	−2.25**
Agent competent	4.29	5.57	−2.04*
Agent typically Japanese	4.50	3.43	2.71**
Agent talkative	5.61	4.36	2.66***
Agent nationalistic	1.43	3.29	−2.87***

$*p = .05.$ $**p < .05.$ $***p = .01.$ $****p < .01.$ $^p = .056.$

TABLE 11.4
Summary of T test Comparisons of American Students' Ratings, Safe Agent versus Unsafe Agent

Variable	Mean (safe agent)	Mean (unsafe agent)	t value ($n = 13$)
Interesting	5.85	6.77	−2.18**
Partner similar to self	3.31	4.77	−2.55**
Japanese emotionally expressive	3.15	4.15	−2.16**
Japanese outgoing	4.08	4.77	−2.04*
Japanese talkative	3.77	4.85	−2.30**
Japanese evasive	3.85	4.85	−2.39**
Japanese quiet	6.00	4.38	2.82***
Agent blunt	4.69	7.36	−3.84*****
Agent domineering	3.38	5.25	−2.31**
Agent restrained	3.15	1.92	2.52**
Agent friendly	5.46	4.08	2.03*
Agent typically American	6.62	4.92	2.40**

*$p = .05$. **$p < .05$. ***$p = .01$. ****$p < .01$. *****$p = .001$.

- *American partner seemed worse in the unsafe-agent condition*

  Japanese participants rated their partner as less similar to themselves, less considerate, more domineering, less friendly, and less talkative in the unsafe-agent condition. These rankings suggest that the unsafe agent led to more negative impressions of the partner for Japanese participants.

- *Unsafe topics made Japanese act more American*

  Japanese rated themselves as less evasive, less restrained, less self-abasing, and less team-oriented in the unsafe-agent condition. These are all stereotypically American traits for Japanese. It seems they thought they acted more American than those in the safe-agent condition. Americans rated their partner as more similar to themselves in the unsafe-agent condition. This seems to corroborate the Japanese self-ratings.

- *Safe/unsafe topic choice affected stereotyping in complicated ways*

  Japanese participants in the unsafe-agent condition thought the typical American was less domineering. This conflicts with their ranking of their own partner.

  American participants rated the typical Japanese in conflicting ways: After the unsafe-agent condition, they thought the typical Japanese person was more emotionally expressive, more outgoing, and more talkative, but also more evasive and quieter.

- *Safe/unsafe agents were perceived differently by Japanese and Americans*

The two groups differed in their impressions of the safe and unsafe agents. The Americans formed the intended impression: They rated the unsafe agent's topics as less appropriate and thought it acted more blunt, more domineering, less restrained, and less friendly. They also said it was less typically American, distancing it from their own in-group's behavior. The Japanese thought the unsafe agent was nicer and more competent than the safe agent. They rated the unsafe agent as less typically Japanese and less talkative. They found the unsafe agent more nationalistic probably because it brought up more political topics than the safe agent.

LESSONS LEARNED

The agent's behavior strongly influenced participants' impressions of the agent, their conversation partners, and even stereotypes about their partner's nationality.

Provocative Help Can Be Good

Our evaluation suggested that a communication assistant can be helpful both when it offers safe topics to talk about and when it steers the conversation in less safe directions. In fact the Japanese participants seemed to prefer the unsafe-topic agent, and both groups found it more interesting than the safe-topic agent. We suspect that an agent with a model for offering both kinds of topics, depending on the conversation flow, would be the most desirable.

User Adaptation Would Make the Agent More Effective

The two cultural groups had different impressions of the same agent behaviors and reacted in different ways. For example, behavior perceived as blunt and unfriendly by Americans was seen as nice and competent by Japanese. An effective agent for different types of people probably needs to adapt its behaviors to user subgroups or perhaps to individuals' own interaction styles and preferences. We believe we created a more American identity for our agent by delivering its topic help in English. It is an interesting challenge to create an agent whose presentation is adapted to different user styles and preferences.

Agent Behavior May Shift User Behavior

Both the Japanese and American participants noted that Japanese seemed to act more American in the unsafe-agent condition. This result indicates that it may be possible to mold user behavior with the choices one makes about how the agent behaves and what it talks. This could have interesting implications for those interested in setting a specific group conversational tone or style in a virtual meeting space.

The results suggest that social agents should control cultural effects of interaction style, behavior, and what they talk. Studies on embodied conversational agents and virtual humans are good resources for designing this control mechanism. It is also suggested that social agents are required to recognize cultural characteristics of ongoing conversation, user preferences, and user behavior. The capability of culture recognition will become an attractive and important issue in the development of social agents.

There are several technologies to overcome geographical and language barriers of cross-cultural communication, although few technologies to eliminate cultural barriers have been studied. Our evaluation showed the potential ability of social agents to help people compensate cultural gaps between them. Our evaluation also showed the danger of widening the gap. An ethical code for developers and users of social agents may be necessary in the future.

Social agents need sophisticated cultural knowledge to serve as cultural assistants. Because cultural knowledge of our agent was the only common safe and unsafe topics in both countries, its capability was limited. We continue to conduct cross-cultural experiments to convert existing sociological findings into our agent's knowledge.

EXTENSION OF THE AGENT'S ABILITY

In the experiment described previously, our agent provided a topic chosen randomly from the topic pool prepared beforehand when an awkward pause occurred in the subjects' conversation. Hence, the agent often provided a topic that already had been talked about or that was not related to their ongoing conversation at all. After the experiment, we developed the agent's additional function to solve that problem. In this section, we describe *topic derivation function* to provide a topic related to the ongoing conversation when a pause occurs. To prevent the resumed conversation from shrinking quickly, the agent provides a topic that has not yet appeared in the conversation. This function consists of the following three modules.

Conversation Monitoring

This module has a speech-recognition engine to monitor a conversation and detect keywords included in the conversation. Those keywords are typically proper nouns of a particular category, and the agent provides topics

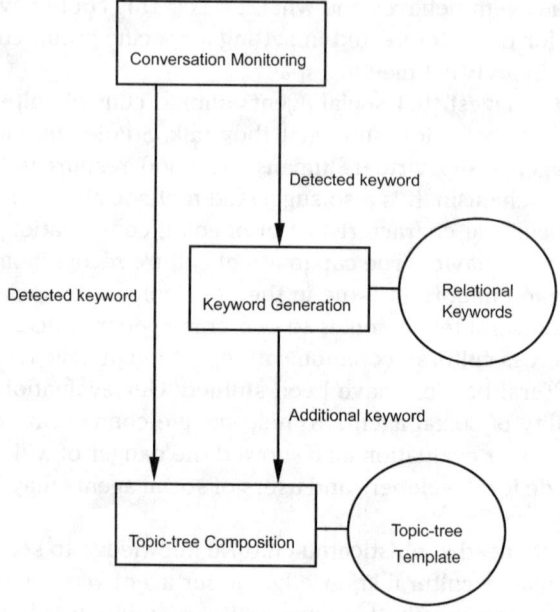

FIG. 11.5. Topic derivation function.

related only to that category. This constraint raises the accuracy of speech recognition, and clarifies the agent's characteristics simultaneously.

Keyword Generation

To provide a fresh topic, this module produces an additional keyword related to the detected keyword. An example of this relation is inclusion. If the detected keyword is *food* or *place*, this module selects a more specific one as the additional keyword.

Topic-Tree Composition

Topic tree is a tree structure of a question and answers that the agent utters (see Fig. 11.3). To construct it, this module uses two keywords, one of which is detected by the conversation monitoring module and another of which is produced by the keyword-generation module. Those two keywords are set in the chosen template of topic tree prepared beforehand. This method can produce a topic closely related to an ongoing conversation and facilitate it at once. The module chooses the template according to the types of keywords to be set. The examples of the type of keyword are *place, person*, and *food*. The designer of the agent has to prepare each template corresponding to each combination of those types.

Figure 11.5 shows the architecture of the topic derivation function described before. We integrated this function with the agent providing topics of sightseeing in Kyoto. We selected *region, temple*, and *shrine* as the types of keywords. The following script is one of the recorded conversations when we tested this agent. A and B are a talking couple, and X is the agent. A_B means A talks to B.

A_B: Have you been to Arashiyama?
B_A: No, I have not.
A_B: Have you been to Higashiyama, then?
B_A: I think I have been to . . .
(A silent pause occurs, and then the agent approaches)
X_B: Have you been to Ginkakuji temple in Higashiyama?
B_X: No.
X_B: You should visit there.
X_A: Have you been to Ginkakuji temple?
A_X: Yes.
X_A: You can tell your partner about Ginkakuji temple.

In this example, the agent's question includes Higashiyama, which appeared in the conversation, and Ginkakuji, which is a temple located in Higashiyama region in Kyoto and has not appeared in the conversation.

SUMMARY

Virtual meeting spaces usually provide little social context to find a common ground for communication with each other. To eliminate this difficulty, we developed a social agent supporting human–human communication in virtual meeting spaces. This social agent mimics a party host and tries to find a common topic for two meeting participants whose conversation has lagged.

Previous studies mainly focused on the autonomous agents to assist users in specific tasks or engage in one-on-one interaction as a conversation partner. Our social agent aims to work as an in-between of human–human interaction in casual meetings. The agent provides opportunities to start a conversation.

We selected the cross-cultural first-time meeting as a good test case for the agent because it seems hard to establish a common ground in such a situation. We focused on conversations between Japanese and Americans because they are known to have different interaction styles and cultural norms.

In the FreeWalk space, the agent tracks audio from a two-person conversation to look for a longer silence that means an awkward pause. When the agent detects a pause, it begins an approach. On reaching the participants, the agent goes into the topic-suggestion cycle. In the suggestion cycle, the agent conducts a series of yes/no questions to both participants to draw shared or conflicted points. After the cycle, the agent makes a general comment based on the previous answers from the participants to recommend how they could make use of the particular topic area. After making this comment, the agent departs from the conversation zone and wanders at a distance until it detects another awkward silence.

In virtual spaces, our social agent is embodied the same way as users. Such moving behaviors of the agent as approaching and retreating make it intuitive for the conversation pair to understand that the agent is an in-between only to lead a question cycle, and that they do not need to include the agent in their conversation. The agent turns toward the participant it is currently addressing so the pair can intuitively recognize who is asked.

On the rectangle of the agent's embodiment, an animated dog is pasted. We wanted users to think of the agent as subservient, friendly, and reasonably socially intelligent. The agent has a set of animations, each of which corresponds to a phase in the process of question and answer. These animations are a supplement to the agent's speech.

We gathered safe and unsafe topics for the first-time meeting by using a Web survey to craft a set of questions the agent could ask in question and answer. Each topic has a tree structure with these nodes: a question, possible answers, agent's reply to each answer, and so on.

We performed an experimental evaluation of the agent's ability to assist in cross-cultural communication between Japanese and American undergraduates. We designed two kinds of agents to introduce culturally common safe or unsafe topics to conversation pairs through a series of question and answer.

In the experiment, the safe agent had positive effects for American students. In contrast, it had negative effects for Japanese students, but simultaneously it made them think their partner was more similar to themselves. In the unsafe-agent condition, both Japanese and American students thought their conversations were more interesting, and Japanese students acted more American.

As a result, we found that the agent's behavior strongly influenced participants' impressions of the agent, their conversation partners, and even stereotypes about their partner's nationality. Provocative topics are useful, an agent adaptive to participants is good, and an agent's presence affects participants' style of behavior.

In the experiment, the agent often provided a topic that was not related to their ongoing conversation at all. To solve this problem, we developed a topic derivation function. The conversation monitoring module detects keywords included in the conversation. The keyword-generation module produces an additional keyword related to the detected one. Those two keywords are set in the topic-tree template chosen by the topic-tree composition module. This method can produce a topic closely related to an ongoing conversation and facilitate it at once.

ACKNOWLEDGMENTS

Thanks to NTT GEMNet project for providing the broadband line, to Eva Jettmar for her assistance in the experiment, and to Satoshi Nakazawa for his efforts to extend the agent.

REFERENCES

Cassell, J., Bickmore, T., Billinghurst, M., Campbell, L., Chang, K., Vilhjalmsson, H., & Yan, H. (1999). Embodiment in conversational interfaces: Rea. *International Conference on Human Factors in Computing Systems (CHI99)*, pp. 520–527.

Clark, H. H. (1996). *Using language*. Cambridge, England: Cambridge University Press.

Foner, L. (1997). Entertaining agents: A sociological case study. *International Conference on Autonomous Agents (AGENTS97)*, pp. 122–129.

Greenhalgh, C., & Benford, S. (1995). Massive: A collaborative virtual environment for teleconferencing. *ACM Transactions on Computer-Human Interaction, 2*(3), 239–261.

Hagsand, O. (1996). Interactive multiuser VEs in the DIVE system. *IEEE MultiMedia, 3*(1), 30–39.

Hall, E. T. (1966). *The hidden dimension*. New York: Doubleday.

Hall, E. T., & Hall, M. R. (1990). *Hidden differences: Doing business with the Japanese* (Reprint ed.). New York: Anchor Books.

Han, J., & Smith, B. (1996). CU-SeeMe VR immersive desktop teleconferencing. *International Conference on Multimedia*, pp. 199–207.

Isbister, K., & Doyle, P. (1999). Touring machines: Guide agents for sharing stories about digital places. *Workshop on Narrative and Artificial Intelligence in AAAI99*.

Isbister, K., & Hayes-Roth, B. (1997). Social implications of using synthetic characters. *Workshop on Animated Interface Agents in IJCAI97*, pp. 19–20.

Lea, R., Honda, Y., Matsuda, K., & Matsuda, S. (1997). Community place: Architecture and performance. *Symposium on Virtual Reality Modeling Language (VRML97)*, pp. 41–50.

Lester, J. C., Converse, S. A., Kahler, S. E., Barlow, S. T., Stone, B. A., & Bhogal, R. S. (1997). The persona effect: Affective impact of animated pedagogical agents. *International Conference on Human Factors in Computing Systems (CHI97)*, pp. 359–366.

Nakanishi, H., Yoshida, C., Nishimura, T., & Ishida, T. (1999). FreeWalk: A 3D virtual space for casual meetings. *IEEE MultiMedia, 6*(2), 20–28.

Reeves, B., & Nass, C. (1996). *The media equation: How people treat computers, television, and new media like real people and places.* Cambridge, England: Cambridge University Press.

Sugawara, S., Suzuki, G., Nagashima, Y., Matsuura, M., Tanigawa, H., & Moriuchi, M. (1994). InterSpace: Networked virtual world for visual communication. *IEICE Transactions on Information and Systems, E77-D*(12), 1344–1349.

Takeuchi, A., & Naito, T. (1995). Situated facial displays: Towards social interaction. *International Conference on Human Factors in Computing Systems (CHI95)*, pp. 450–455.

Waters, R. C., & Barrus, J. W. (1997). The rise of shared virtual environments. *IEEE Spectrum, 34*(3), 20–25.

CHAPTER

12

Designing Intercultural Agents for Multicultural Interactions

Elaine M. Raybourn
Sandia National Laboratories*, University of New Mexico

In the present chapter, I propose that designing large-scale, agent-based community systems for multinational organizations requires fostering a shared sense of community belonging that transcends any one cultural orientation and is thus truly multicultural. I introduce the notion of intercultural agents for multicultural interactions and discuss how agent-generated cultural cues in virtual environments can motivate cross-cultural discovery among users. I introduce why we need agent-based systems that communicate cultural cues and propose using a model called *Designing From the Interaction Out* as a guide for designing intercultural agents. In discussing the model, examples are offered from previous work on an adaptive, agent-based community system and an embodied agent whose purpose is to facilitate organizational change and adoption of new technologies. A scenario is advanced in which humans and intercultural agents work in concert to co-create equitable multicultural environments. I conclude by offering that multicultural interactions need diverse agent solutions that work together to honor the contributions of users' multilevel cultural identities that make up the vibrant multicultural interactions in community-based systems.

---

*Sandia is a multiprogram laboratory operated by Sandia Corporation, a Lockheed Martin Company, for the U.S. Department of Energy under Contract DE-AC04-94A685000.

CULTURE AS INTERACTION

In his seminal book *Silent Language*, anthropologist Edward T. Hall (1959) indicated that "culture is communication" (p. 97) and thus comprised of words; actions; nonverbal behaviors; the handling of time, space, and materials; and worldview, beliefs, and attitudes passed over time from generation to generation. Culture is a set of experiences that are deeply rooted in interactions with our physical and social realities, which often go unsaid but are nevertheless communicated without one's knowing (Hall, 1959). Our physical environment provides the experiential backdrop in which our interactions over time communicate culture to ourselves, as well as those around us. According to Hall (1959), "to interact with the (physical) environment is to be alive. . . . Ultimately everything man does involves interaction with something else." It is through these interactions, and the relevance we place on the feedback we perceive, that culture is formed and expressed to others. The consistent patterns of our interactions with artifacts, our physical environment, and other individuals over time provide cues that others may use to interpret our culture.

For example, as we walk across the lawn during a winter morning in Germany, we may hear the crunching of frost under our feet, feel the frozen grass through the soles of our shoes, and see our footprints in newly fallen snow. That is, when we interact with our physical environment we usually receive feedback on how our presence has influenced or changed the physical reality. Because we receive feedback, we are more aware of the physical world. The physical environment helps us understand more about ourselves, and therefore more about what it means to be alive. One might say that the physical environment interacts back with us as it registers our actions and issues a reaction or feedback. To receive feedback is to receive important information that is necessary to discern cultural phenomena. To receive feedback then is to feel alive and feel that the physical environment is alive.

Similarly, in a virtual or electronic setting, we need to receive feedback from our interactions. We want to know how our actions and presence in the virtual environment influence or change the electronic setting. A virtual environment should help us understand more about ourselves by giving us feedback about what it means to be alive in an electronic setting. When we browse the Internet, we might want to know who else has viewed the same Web page, or we may want to see old read email messages in our inbox fade over time as their saliency diminishes. Hall (1959) indicated, "Experience is something man projects on the outside world as he gains it in its culturally determined form" (p. 192). That is, through communication and feedback from our interactions, we learn how to culturally situate our experiences.

To culturally situate our experiences in electronic settings, we need to sense how our actions affect the virtual setting as we interact with it; we need to be aware of our electronic footprints. Just as in physical settings, we communicate and interpret rich cultural information about ourselves as we interact with artifacts or others in virtual environments. To interact with the virtual environment and have it interact back with you is to feel alive and feel that the virtual environment is alive.

As a designer of culturally situated, social computing environments, I have paid particular attention to the types of cultural footprints we may leave in electronic environments. Cultural information left in the form of an electronic footprint may include users' current activities, electronic artifacts, narratives, assumptions, values, goals, meanings, and history. Whether our interactions are expressed explicitly (we are aware of our interactions) or implicitly (we are unaware of our interactions), intelligent, adaptive systems have the ability to capture and subsequently communicate the electronic footprints users leave in community-based environments. These footprints can serve as the cultural backdrop or context for the system-generated cues that may prompt the multicultural interactions that are then co-created by users and intercultural agents in communication. However, unlike physical environments, most community-based systems today lack these subtle cultural prompts, cues, signposts, or environmental markers that provide us with rich cultural feedback about other users or the environment's current or past state (Raybourn, 2003b). This is one of the reasons that so many community-based systems do not feel alive and we do not feel alive interacting in them.

CULTURE AS THE COMMUNICATION OF PERCEPTION

Culture influences the meanings we attach to mutually agreed upon symbols that are co-created and co-interpreted by interlocutors in communication. In the 1976 edition of *Beyond Culture*, Hall characterized culture as comprised of deep, commonly held, unstated experiences that members of a given culture share, communicate without realizing, and use as a backdrop for judging all other events. Hall (1976) described culture as the communication of deeply embedded shared experiences that influence our perceptions, interpretations, and sense making.

Perception and interpretation largely dictate the manner in which we process information. Each of us makes sense of incoming stimuli by interpreting their meaning and subsequently categorizing the information in accordance with cognitive schema we have developed to best understand our world (Varner & Beaner, 1995). Our culture serves as a perceptual filter with which we determine the cues that are important to us (to which we attend) and those cues that are not important (thus ignored). Culture also provides a framework for grouping information, matching information, and creating new reference frames. Needless to say, we best understand a sociocultural phenomenon when our interpretations of the physical and social environments accurately align with the collective cultural meaning assigned to events and cues over time. When our interpretations and perceptions do not align closely with the collective understanding of meanings assigned to pertinent cues and events, we become more aware of our cultural filter. The process of communication in which culture is salient to the mutual understanding or misunderstandings that ensue among culturally diverse interlocutors has been defined as intercultural communication.

Our physical environment is ripe with cultural cues. In contrast, virtual environments are notorious for providing communication opportunities that lack the presence of rich cultural cues that indicate to interlocutors the saliency of the culture in communication. The absence of such cues in computer-mediated communication may lead to increased cultural misunderstandings among interlocutors as well as a decreased awareness of the salience of culture in intercultural interactions. Communication cannot be as effective in the absence of cultural cues and feedback. As designers of adaptive and intelligent environments, it behooves us to create computer-mediated communication that is capable of issuing rich cultural cues to support intercultural communication.

WHAT ARE INTERCULTURAL AGENTS?

When designing intelligent or adaptive agent-based environments intended to support communication, we need more descriptive definitions of intercultural communication than the one provided in the previous section. It no longer suffices to define intercultural communication as an exchange in which culture is salient to mutual understanding among culturally diverse individuals because in an electronic setting cultural cues may not be present at all or may be so embedded that they go unnoticed by interlocutors. In the virtual environment, intercultural communication may lack the necessary cues and the emergence of patterns that are indicative of culture. Therefore, I define *intercultural communication* as the exchange and co-creation of information and meanings by entities, individuals, or groups when at least one party perceives a difference among entities or perceives itself to be different from others (Raybourn 1998a, 2001). This definition of intercultural communication is slightly different from most other definitions in a fundamental way: Only one entity needs to be aware of the salience of culture in any given interaction, and this awareness does not need to be communicated, perceived by, nor expressed to other entities participating in the communication event. That is, the salience of culture as affecting the process of mutual understanding (or misunderstanding) may go unnoticed by one or more entities during a communication interaction.

Using the present definition of intercultural communication as a framework, I define *intercultural agents* as entities that participate on behalf of users in communication, the exchange of information, and co-creation of meanings by software agents, entities, individuals, or groups when at least one party perceives a difference among entities or perceives itself to be different from others. An intercultural agent's goal should be to assist intercultural computer-mediated communication (ICMC) and social computing by managing the awareness of the presence of users' cultural footprints and cultural cues in electronic interactions. I define cultural cues or signposts in agent-based systems as (a) subtle prompts enacted by individual or collections of agents, or the agent-based environment to encourage a user to action; or (b) subtle (symbolic) markers left by agents or humans in an environment that issue feedback about the users' or environment's current or past state.

Intercultural agents may have several modes of interaction with interlocutors or other agents. First, in an interaction among two or more individuals and software agents, the present definition suggests that software agents may assist the user in achieving intercultural communication competence by direct intervention (i.e., with a direct prompt or interaction with an embodied agent).

Alternatively, intercultural agents may act on the users' behalf by generating subtle cultural prompts in the form of cues in the environment that guide an interlocutor to be more aware of her intercultural communication competence or increase her awareness of the saliency of culture in the interaction. Cultural cues or prompts may be issued in the form of narratives co-created with users, in the graphical user interface, or through adaptive environments that respond to users' cultural footprints (Raybourn, 1998a, 2000).

In some cases, it may be software agents working in concert that first know whether interlocutors are operating from different cultural orientations based on user models, profiles, current context, history, and so on. That is, intercultural agents may be privy to user profiles gathered explicitly through a user's feedback to postings in the community-based system that reflect a user's long-term interests or implicitly through the user's activities, history, and current context of work. However, agents do not need to overtly communicate this knowledge to interlocutors. Instead intercultural agents may administer subtle cultural cues that are salient to one or both parties in the intercultural interaction to improve the overall intercultural communication competence. In this sense, the software agents, embodied agents, or intelligent user interfaces assist with the process of intercultural communication and serve as part of the intercultural agent community that functions on behalf of users.

Because the community-based systems we design bring diverse users together and facilitate their real-time communication and knowledge sharing, designers need to be more attuned to creating environments that support users' intrinsic motivation for interdependency and participation in a community. We can design intelligent, adaptive environments that utilize cultural footprints to co-create shared narratives and understanding among users about how we do things here and how we behave here. We must begin by designing intelligent environments in which users can leave footprints and which provide users with agent-generated cultural feedback in the form of subtle cues or prompts if we intend to design robust intercultural agents for multicultural interactions.

WHY FOCUS ON MULTICULTURAL INTERACTIONS?

Multiculturalism can be defined as the recognition that several different cultural orientations can co-exist in the same environment and benefit each other. Multicultural interactions are those interactions in which more than one cultural orientation may be operating at the same time. That is, given a particular interaction, several different, yet salient cultural orientations may be operating in each individual. Not only are these orientations diverse, but also the same cultural orientations may not operate in every interaction instance. Hall (1959) indicated that culture hides more than it reveals. That is to say, we can never be absolutely sure which cultural orientations are most salient for individuals in a multicultural setting.

For example, I simultaneously identify with several different groups, each with their own culture. I am a bicultural Latin-American woman in her 30s who is a member of several language communities, works in a large scientific organization, lives both in Western Europe and North America, and is an avid SCUBA diver with a brown belt in Shotokan karate. I use myself as an example to illustrate how any one of us is similarly a member of several different cultural groups at the same time. In any one particular instance, I may identify more strongly with others as a result of a number of cultural or personality variables operating at any given time. In other words, each of us is comprised of multilevel identities that contribute to a salient cultural orientation. Designers of embodied agents and intelligent adaptive systems must be able to engage multilingual, multicultural users on several different cultural levels at very nearly the same time otherwise the environments we create will always lack some richness from the users' points of view. We can engage users on multiple levels by highlighting shared cultural experiences or fostering cross-cultural discovery among users. Cross-cultural discovery is the process in which one willfully seeks more information to understand others and the salient cultural differences that may influence our interactions.

I focus on multicultural interactions in the design of large-scale, agent-based community systems because in such systems we strive to provide the user with a shared sense of community belonging and involvement that should transcend any one individual cultural orientation. Community belonging does not preclude the use of the adaptive properties of intelligent systems to provide culturally appropriate graphical user interfaces or culturally appropriate interactions with embodied agents (see Morel, chap. 8, this volume; de Rosis et al., chap. 4, this volume). However, providing a shared sense of belonging in the design of large-scale, agent-based community systems additionally requires that designers foster among users a spirit of co-existence, co-creation, and collaboration. After all the goal of most community-based systems and social human–computer interaction is to foster learning, knowledge transfer, and communication across various disciplines, individual differences, and diverse user goals.

Designers play to the strengths of multicultural interactions when we design environments that encourage users to engage in cross-cultural discovery with one another. Intercultural agents may play an important role in fostering cross-cultural discovery. Encouraging cross-cultural discovery among individuals, teams, groups, organizations, and nations is a critical step toward facilitating multicultural interactions that transcend any one culture and begin to represent a third culture. According to Brislin and Yoshida (1994), a third culture is neither one culture that subsumes other cultures nor the inequitable combination of a dominant culture and several other minority cultures, but an equitable combination of two or more cultural orientations including one's national culture, ethnicity, organization, gender, age, abilities, and so on. The quality and nature of the users' interactions determine the direction and rate with which a third culture emerges. Most important, a third culture is the dynamic co-creation of meaning in which all interlocutors are equal participants as well as co-owners.

In the next section, I describe a *Designing From the Interaction Out* model (Raybourn, 1999) that has been used for designing multicultural, agent-based interactions from a perspective of supporting intercultural communication and cross-cultural discovery as core interaction goals.

DESIGNING FOR MULTICULTURAL INTERACTIONS FROM THE INTERACTION OUT

A designer I know once indicated in a workshop that before seeing the Designing From the Interaction Out model he had often designed virtual environments from the technology out—that is, he usually focused his attention on acquiring the appropriate technological tools that were needed in the community-based system to support computer-mediated communication. Sometimes in the rush to provide system functionality, the social supports necessary to motivate users to interact with others are overlooked. Unfortunately, technological supports are not enough for engendering rich online communication. Users must be motivated to take that first step toward interaction, which may involve mustering up the courage to contact someone they may not know very well. For example, providing a chat tool in a community-based system may be necessary, but it is not sufficient. Users need contextual social supports for computer-mediated communication. Contextual social supports provided by intercultural agents can help create environments that are lifelike, engaging, adaptive, and fun.

The model in Fig. 12.1 illustrates how through attention to interaction, narrative, place, and emergent culture we can create dynamic multicultural communication patterns in virtual settings and online communities (Raybourn, 1999). Intercultural communication competence serves as a core interaction goal that each of the elements strives to support. Intercultural communication is comprised of several salient elements, among them (a) the type of communication or interaction (interpersonal, group, etc.), (b)

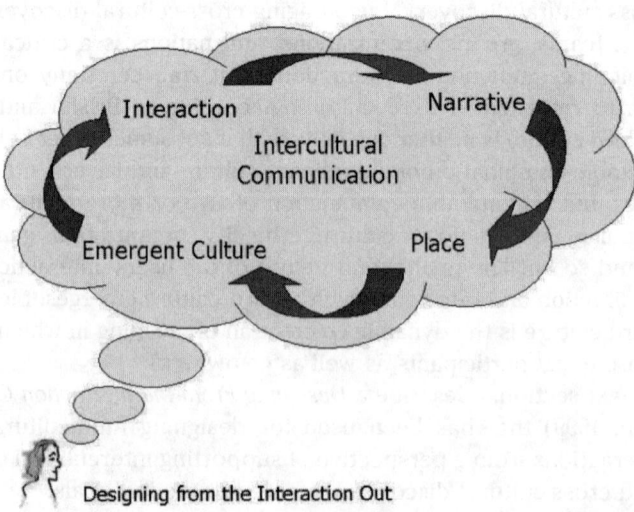

FIG. 12.1. Designing From the Interaction Out Model.

the place or context in which it occurs, (c) the narratives that are co-created and negotiated by the interlocutors, and (d) the culture that emerges from the communication event. Once a designer has considered the design problem in the context of the cycle from interaction to narrative, to place, to emergent culture—then she begins again as emergent culture dynamically spawns new interaction events. Supporting intercultural communication thus remains a core interaction goal that designers and developers may aspire to in the design of more equitable multicultural communities that support the emergence of a third culture that neither belongs to the interlocutors nor to the designers, but instead is a co-creation and artifact of the ongoing dialogue among users, designers, adaptive interface, and intelligent agents.

Interaction

In the interaction phase, a designer specifically considers the types of communication instances that her system will support. Communication is the dynamic process in which two or more interlocutors strive for or engage in shared meaning. Meaning in communication is co-created by interlocutors through the use of a system of mutually agreed upon symbols that are interpreted in a particular context or situation. Therefore, a designer must first consider the possible communication events that can occur in a given context. Earlier in this chapter, I discussed the notion of intercultural agents that facilitated intercultural communication and cross-cultural discovery among users. Engendering trust among users and the spirit of discovery, or curiosity about others, can be accomplished through careful contemplation on the kinds of interactions that one would like to support in the intelligent or adaptive community-based environment.

For example, over a 2-year period, I worked on design issues for an adaptive community-based system that incorporated a WWW-based collaborative virtual environment comprised of intelligent software agents that supported explicit information sharing, chance meetings, and synchronous chat. After conducting a study that involved two different organizations, we realized that the technology lacked the social supports needed to motivate users to engage in cross-cultural discovery, much less take the risk to open a chat with someone in the organization they did not know (Raybourn et al., 2003). Subsequent to the evaluation of organizational cultures, our design question was, "how can we motivate users to take the first steps toward conversing with others whom, according to their profiles, they obviously share common interests with, but do not know well enough to feel comfortable opening a chat in a virtual space?" We concluded that contextual social supports may take the form of cues or signposts in the environment that encourage or motivate the user to ultimately click on the button that opens the chat window, compose a meaningful message, and send it to a complete stranger. Interactive user profiles, team galleries, and virtual tours conducted by embodied agents were among some of the cultural signposts we suggested increasing sociability (Raybourn et al., 2003).

Another way to address cross-cultural discovery among strangers is to engender feelings of trust resulting from informal interpersonal communication. Therefore, raising one's awareness or curiosity about others may motivate increased interpersonal communication in adaptive virtual environments. One way an agent-based community system can motivate a sense of curiosity in users is to provide subtle social prompts or cultural signposts such as having third-party agents introduce them. Because it is difficult to start a conversation with a stranger, the agent could deliver suggestions for conversations based on similarities between users' implicit profiles (Raybourn et al., 2003). We derived these notions by considering the types of interactions we wanted to support—in this case, interactions among strangers in community-based systems. Engendering intercultural communication competence and facilitating cross-cultural discovery among workplace strangers was our core interaction goal, therefore we focused on adding agent-based features to a system that encouraged informal communication among strangers and provided an opportunity to engage in organizational storytelling.

Narrative and Storytelling

In the narrative phase of the Designing From the Interaction Out model (Fig. 12.1), a designer considers the types of narratives among users and intercultural agents that are likely to ensue in the community-based system. Narrative plays a powerful role in virtually all forms of human activity. For example, artificial intelligence has long recognized the power of scripts and other narrative structures in creating and organizing knowledge (Schank & Abelson, 1977; Schank & Morson, 1995). Sengers (2000) discussed the notion of intentional agents as those comprised with narrative properties. Additionally, Laurel (1991) has shown that human–computer interfaces can be improved by paying attention to the narrative structure of the interaction activity. Although a shared graphical environment is not always necessary for groups to establish rapport and trust (Leevers, 2001; Raybourn, 1998a), supporting narrative has been deemed essential (Murray, 1997).

Employees learn about informal norms such as how we do things here and pass along organizational memory by telling myths and stories (Orr, 1996). In the study mentioned in the previous section involving the investigation of two large organizations, we contrasted and compared the work patterns of two distinct, multidisciplinary groups to learn how to best support informal narratives and the exchange of work-related information in adaptive, agent-based community systems (Raybourn et al., 2003). In comparing the data collected from two different organizations, we found that above all both groups used narrative or storytelling to foster a sense of belonging and trust. In considering the narrative phase of the model, we decided to support the users in sharing organizational narratives through storytelling. Our approach resulted in a combination of user-authoring instances and agent-generated narrative support. We offered suggestions that encouraged users to engage in both synchronous and asynchronous informal communication interactions around the routines in which work gets done, author their own individual and organizational stories, and collaboratively author/learn about the informal cultural norms and actions that were characteristic to their community (Raybourn et al., 2003).

Narrative structures and storytelling activities are integral components of one's experience in online communities, MOOs, and adaptive community-based systems. Through narratives and storytelling, interlocutors negotiate power and agency, discover interdependencies, and contribute to the cultural landscape of how things are done here. Intercultural agents can participate in the development of narratives in intelligent or adaptive systems. However, supporting equitable multicultural narratives not only requires careful attention to the flow of power or agency among its coauthors and audience, but also an awareness of one's own cultural biases and the biases inherent in the technologies we design.

Nowhere are the power dynamics of co-creating narrative or storytelling in a virtual environment quite so visible as in a text-based MUD or MOO. The collaborative design of a virtual environment is in many respects the design of a narrative (Clarke & Mitchell, 2001). Users engage in narratives about which objects to build, what meaning to assign them, and what the appropriate social behaviors when interacting with these objects will be. These narratives are often shaped by the constraints of the database, programming language, and possible social interactions supported by a MOO. That is, human behavior in these virtual environments is constrained by both physical (hierarchical nature of the programming code) and social privileges (social power structure) afforded to certain users of online communities (Raybourn, 1998b).

All community-based systems may support hidden power dynamics that privilege some users while disadvantaging others. Therefore, if we choose to meet the challenge of supporting equitable and inclusive multicultural communication with intercultural agents, we need to carefully understand the impact of underlying cultural dimensions such as identity, negotiation, conflict, and trust on agent-based community systems. As we strive to create equitable agent-based interactions and environments, differing cultural values of designers, clients, stakeholders, developers, and users of our technologies make for complications and possibly competing desires or expectations. Nevertheless, we can mitigate frustrations and design intercultural agents for improved multicultural interactions by fostering a climate of cultural sensitivity in our own design teams that continue throughout the entire project cycle (see Mudur, 2001).

Place

In the third phase of the Designing From the Interaction Out model, a designer considers how the dance among interactions and narratives form a multicultural place that engenders cross-cultural discovery. I learned a valuable lesson from my work in text-based MOOs that influenced my approach to interaction design for intercultural agents and adaptive workplace community-based systems. I learned that users appreciate interacting in electronic places that are capable of responding to their actions, therefore providing useful feedback about the state of the virtual world. In a MOO, the user interacts with both other users and the environment. A MOO environment gives the sense of being alive as it engages the user through an interactive textual (physical) space and a communicative, social space.

Therefore, integral to the design of place is allowing users to contribute and develop an ownership for the artifacts in the place as well as the narratives that it supports. Thus, designers should consider engendering a multicultural place in which cultural signposts are not only agent-generated, but also user-generated. We must be allowed to leave our cultural footprints in the intelligent community-based system so that we can feel that the virtual environment is alive. Allowing each person to actively contribute to the development or design of a multicultural place could create more community through a shared, common activity. In other words, users express their identities through visual and textual artifacts. These artifacts become shared signposts that arouse curiosity, spark cross-cultural discovery, and thus engender intercultural communication.

Another design challenge when designing place from the interaction out is acknowledging that a place should be interesting to visit regardless of whether it is populated. That is, we cannot only envision a place in terms of synchronous communication, but should also envision a place where quiet, asynchronous reflection occurs. One of the reasons that text-based MOOs are so successful is because a user can enter the environment on her own and not get bored. Users often enter quiet spaces in MOOs to build objects, read community news, post information, see what others have posted or built, or just wander around. New users may also enter a place when it is unpopulated by others to familiarize themselves with cultural signposts left in the environment.

In summary, designers can create more motivating intelligent or adaptive community-based systems by designing for user fun, curiosity, and cross-cultural discovery. In designing a simulation game in a MOO environment, I learned that adaptive community-based system design and the design of virtual environments in general could engender certain intended or unintended user behaviors and communication (Raybourn, 2001). One user behavior we want to engender in community-based systems is the behavior of returning often to a place to revisit the narratives supported there. Intercultural agents may generate cultural cues that transform spaces into third culture places, or they may participate in the co-creation of the narratives to assist users in the co-creation of rich interaction environments. Intercultural agents may offer subtle motivating cues to the user on how she can contribute to the culture emerging in a place and invite others to participate in the co-creation of a third culture. Agents learn user behaviors by receiving direct feedback from users or through user profiles. In an intelligent, adaptive, community-based system, the intercultural agents themselves can serve as adaptive third culture cues or environmental markers, as in the case of embodied agents that deliver cultural cues about culture change in organizations.

Emergent Culture—Organization Values and Cultural Change

Embodied agents serve as humanlike interfaces to data and information, as well as interaction guides and allies (see Cassell et al., 2000). Embodied agents acting in community-based systems such as intranets or portals to companies' services may also serve as cultural informants who set the tone for informal communications, moderate cultural exchanges, communicate corporate culture, or serve as organizational cultural change agents.

Large multinational organizations need to communicate the cultural values of the organization. Often this responsibility is tasked to groups of individuals on different sites employed in departments such as Human Resources or Corporate Communications, for example. A consultant to the executives of a large multinational information technology (IT) organization once contacted me about the design of a corporate Web-based training program that would be accessed by individuals in 60 countries. The company wanted a solution that was sensitive to diverse cultural orientations and yet would communicate a consistent corporate cultural identity. For many multinational organizations, the prospect of introducing several culturally specific training programs is still too costly in terms of development, human resources, and economics.

Another known issue faced by large organizations is the difficulty in providing opportunities for remote workers such as contractors participating in virtual teams to learn the informal norms and culture that are necessary to maintain productivity and well-being of the team. In the study of two diverse organizations reported in previous sections of the present chapter, we discovered that an embodied agent-guided virtual tour of the organization's physical place and the formal and informal organizational cultural norms could help remote workers feel more like part of the co-located team (Raybourn et al., 2003). A team gallery of interests could also be an informal mechanism for displaying information on team or individual culture. In other words, the team members and their activities could also be showcased in addition to a virtual tour of the place where most of their work was accomplished.

Not only can an embodied agent serve as a virtual tour guide and cultural informant, but also as an organizational ambassador employed by the multinational organization to communicate corporate culture and attend to the virtual teams' needs. Agent ambassadors represent an organization's values and can make the virtual interaction more memorable because they can be given personalities that embody the persona the organization wants to project. These virtual characters can answer questions, guide, impart knowledge or advice, perform tasks, and informally interact with the user.

For example, Morel (chap. 8, this volume), CEO of Cantoche, outlines a number of embodied agent roles such as organizational ambassadors, trainers, assistants, communications directors, and learning companions. Embodied agents designed by Cantoche to perform these roles for multinational organizations have taken the forms of human, animals, and anthropomorphized objects (i.e., Busy Buddy and Qmark). Recall the example provided in his chapter on the development of Qmark for the multinational organization, Microsoft. After the storyboarding and design of several embodied agents, and after much deliberation with the client, the client finally opted for an anthropomorphized question mark as the Windows XP installation agent. Why is this not surprising? Research has shown that, although an embodied agent's appearance is an important design parameter and perceived in-group membership may influence the social attractiveness of embodied agents (Nass et al., 2000), there is no single method of representing a trustworthy virtual character to diverse groups of people (Andersen et al., 2001; Hertzum et al., 2003). Clearly, the task of creating a believable, cross-culturally competent embodied agent (see also de Rosis et al., chap. 4, this volume) for multicultural interactions is neither straightforward nor easily solved. As illustrated by the Microsoft example, when the client conceptualizes culture in terms of national culture, ethnicity, race, or culture-specific verbal and nonverbal behaviors, the task of supporting the sheer number of possible multicultural interactions is daunting.

For these reasons, I believe intercultural agents (of which embodied agents are certainly a part) are best conceptualized as participating with users in multicultural interactions and narratives that dynamically co-

12. DESIGNING INTERCULTURAL AGENTS

create an emergent culture that pertains to all parties involved. In keeping with this theme, I provide an example of an embodied agent whose design—although simple—is also a surprisingly elegant solution to the problem of supporting an emergent culture of organizational change in a multinational organization.

Numix (Fig. 12.2) was developed by Cantoche for the French national TV station, France 2 (Morel, 2003). Numix debuted on the company's new intranet in February 2003. The intranet was developed to disseminate information on the use of new technologies that would subsequently result in changes in the corporate culture and the ways people worked. For example, the France 2 intranet features information about new technologies and training on how to use digital media. The goal of the intranet is to assist employees of France 2 with the transition from current work practices to the adoption of new technologies that require learning new digitalization techniques (Fig. 12.3). The culture of how we do things here was undergoing transition and change as the company introduced new work practices. Change in large organizations is hardly ever easy. Especially for established employees, or those in the sunset of their careers, the adoption of a new technology that changes one's habitual work process is more difficult to accept. Therefore, the design question faced by Cantoche was, what is the best design for an embodied agent that is to help diverse users (age, education, gender, ethnicity, etc.) adopt cultural change?

Numix was designed to serve as a cultural change agent and help individuals understand how the emergence of new technologies benefits their

FIG. 12.2. Numix, cultural change agent.

FIG. 12.3.

work, and also to assist with training on the use of digital media (Broadcast, 2003; Morel, 2003). In discussions with the client, it was determined that Numix should serve as an interactive bridge between the familiar and the new. Therefore, although Numix appears on a new intranet service and is the embodied agent interface to new techniques and new media, Numix was designed to be reminiscent of an older, visually familiar character style that enters the new world of digitalization along with the user. Because the information the user is interacting with is new and unknown, Numix was designed to represent a familiar look that all French users could identify positively with—a look reminiscent of the widely popular French cartoon series of the early 1970s, *Shadocks* (Rouxel, 2000, 2002). The *Shadocks* characters created by Jacques Rouxel were intelligent, provocative, and counterculture. Drawn with simple lines and featuring characters that were neither human nor animal, but a collection of interesting shapes, *Shadocks* appealed to French of all ages and walks of life. Audiences all over France seemed to see themselves and others in the situations and themes encountered and explored in the cartoon series. *Shadocks* remains a part of French culture today. In the case of France 2, the embodied agent Numix serves as the interactional glue or bridge that brings the user closer to foreign materials that are unknown or little understood. Numix is able to engage the multicultural users on the level of being a familiar entity in an unfamiliar setting, although each individual may be identifying with Numix and determining his familiarity for several different culturally bound reasons that may remain unknown.

CONCLUSIONS

When I began working in adaptive agent-based community systems, I sought to develop agents that worked in concert with users to co-create equitable communication environments that facilitated cross-cultural discovery in multinational organizations. My earlier work in designing an online intercultural game provided the impetus for developing the idea to introduce cultural cues or signposts in adaptive agent-based, community systems (Raybourn, 1998a, 2001, 2003a). Using lessons learned from this work, I explored whether user- and agent-generated cultural cues could motivate more frequent informal communication among teams or groups who used a large-scale, adaptive community-based system (Raybourn, 2000). Later we addressed the design challenge of generating signposts in a community-based system to motivate strangers to converse with others with whom they obviously shared common interests, but did not know well enough to feel comfortable opening an informal chat (Raybourn et al., 2003).

Designing agent-based environments that allow users to express their multicultural identities and learn about others is one step toward fostering more meaningful intercultural computer-mediated communication (ICMC) and laying the groundwork for the development of more equitable online communities. By using the term *equitable*, I mean that designers and developers of community-based systems should be held accountable for engendering social inclusivity and ensuring that all users of adaptive community-based systems and interactive software have a fair opportunity to contribute to the communication environment (Raybourn, 2002). We can better foster equitable and inclusive human communication by honoring the contributions and expectations of culturally diverse individuals when interacting with embodied agents or multiuser, agent-based community systems.

As in real life, building a virtual multicultural community that is mindful of inclusivity and social accessibility takes effort and education. In terms of computer-mediated communication and human-computer interaction, this means that designers and developers who are mindful of social accessibility and inclusivity must get involved in the process of supporting multicultural interactions early in the design stages. That is, we should carefully consider our own cultural biases and those communicated by our agent-based software solutions.

Throughout this chapter, I have advocated that in designing intercultural agents for multicultural interactions we should create environments capable of providing all users with a shared sense of community belonging, ownership, and identification with agents or the intelligent user interface that transcends any one particular cultural orientation. Users should feel alive when interacting with our embodied agents or intelligent systems. To achieve this, we must appeal to their multilevel, multicultural identities in each interaction and invite them to participate in the co-creation of narratives and places from which an emergent third culture can grow. I provided concrete examples of how this design approach can be realized through the use of a Designing From the Interaction Out model. Intercultural agents can assist in the process of cross-cultural discovery and third culture formation. Intercultural agents can engage users directly (as in the case of an embodied conversational agent) or indirectly by issuing subtle cultural cues in a community-based system.

Multicultural communication environments need diverse agent solutions. For this reason, I have not advocated a single design solution or a single approach to facilitating cross-cultural discovery. Intercultural agents can interact with users as embodied conversational agents, anthropomorphized agents, narratives, or intelligent user interfaces. Most important, designers should consider the limitations of whichever agent-based solution they choose to implement. Focusing on good interaction design and developing an acute awareness of our own biases (cultural or other) that unwittingly make their way into our designs is a good start.

It is my hope that in this ensuing dialog on agent culture, designing human–agent interactions in a multicultural world, that we designers, researchers, and consumers of these agents hold each other accountable for honing the sensibilities each of us needs to responsibly co-create intelligent, adaptive user interfaces and virtual characters that are truly equitable and multicultural.

REFERENCES

Andersen, V., Hansen, C. B., & Andersen, H. H. K. (2001). Evaluation of agents and study of end-user needs and behaviour for e-commerce: COGITO focus group experiment. Risø Report Number Risø-R-1264(EN), Risø National Laboratory, Roskilde, Denmark. Available at http://www.risoe.dk/rispubl/SYS/ris-r-1264.htm. Consulted September 10, 2003.

Brislin, R., & Yoshida, T. (1994). *Intercultural communication training* Thousand Oaks, CA: Sage.

Broadcast. (2003). Cantoche anime l'Intranet de France 2, p. 12.

Cassell, J., Sullivan, J., Provost, S., & Churchill, E. (Eds.). (2000). *Embodied conversational agents*. Cambridge, MA: MIT Press.

Clarke, A., & Mitchell, G. (2001). Film and the development of interactive narrative. In O. Balet, G. Subsol, & P. Torguet (Eds.), *Virtual storytelling* (ICVS 2001, LNCS 2197, pp. 81–89). Berlin: Springer-Verlag.

Hall, E. T. (1959). *Silent language*. New York: Doubleday.

Hall, E. T. (1976). *Beyond culture*. New York: Doubleday.

Hertzum, M., Andersen, H. H. K., Andersen, V., & Hansen, C. B. (2002). Trust in information sources: Seeking information from people, documents, and virtual agents. *Interacting with Computers, 14*, 575–599.

Laurel, B. (1991). *Computers as theater*. Reading, MA: Addison-Wesley.

Leevers, D. (2001). Collaboration and shared virtual environments—from metaphor to reality. In R. Earnshaw, A. Guedj, A. van Dam, & J. Vince (Eds.), *Frontiers of human-centered computing, online communities, and virtual environments* (pp. 278–298). New York: Springer-Verlag.

Morel, B. (2003). Private conversation with CEO of Cantoche, Benoit Morel. More information available at http://www.cantoche.com/.

Mudur, S. (2001). On the need for cultural representation in interactive systems. In R. Earnshaw, R. Guedj, A. van Dam, & J. Vince (Eds.), *Frontiers of human-centered computing, online communities and virtual environments* (pp. 299–310). New York: Springer.

Murray, J. (1997). *Hamlet on the holodeck: The future of narrative in cyberspace.* Cambridge, MA: MIT Press.

Nass, C., Isbister, K., & Lee, E. (2000). Truth is beauty: Researching embodied conversational agents. In J. Cassell, J. Sullivan, S. Provost, & E. Churchill (Eds.), *Embodied conversational agents* (pp. 374–402). Cambridge, MA: MIT Press.

Orr, J. E. (1996). *Talking about machines: An ethnography of a modern job.* New York: Cornell University Press.

Raybourn, E. M. (1998a). *An intercultural computer-based simulation supporting participant exploration of identity and power in a text-based networked virtual reality: DomeCityMOO.* Unpublished doctoral dissertation, University of New Mexico, Albuquerque.

Raybourn, E. M. (1998b, February 14–17). *The quest for power, popularity, and privilege: Identity construction in a text-based multi-user virtual reality.* Unpublished paper presented at the Western Communication Association, Denver, CO. Available from http://www.cs.unm.edu/~raybourn/. Consulted September 10, 2003.

Raybourn, E. M. (1999, November 14–17). *Designing from the interaction out: Using intercultural communication as a framework to design interactions in collaborative virtual communities.* Workshop presentation given at Group '99 (International Conference on Supporting Group Work), Phoenix, Arizona.

Raybourn, E. M. (2000). *Cultural and organizational cues in CVEs: General principles and their practical introduction to the forum.* British Telecom Research Fellowship Technical Report.

Raybourn, E. M. (2001). Designing an emergent culture of negotiation in collaborative virtual communities: The DomeCityMOO Simulation. In E. Churchill, D. Snowden, & A. Munro (Eds.), *Collaborative virtual environments* (pp. 247–264). New York: Springer.

Raybourn, E. M. (2003a). Design cycle usability and evaluations of an intercultural virtual simulation game for collaborative learning. In C. Ghaoui (Ed.), *Usability evaluation of online learning programs* (pp. 233–253). Hershey, PA: Information Science Publishing.

Raybourn, E. M. (2003b). Toward cultural representation and identification for all in community-based virtual environments. In *Lecture Notes in Computer Science* (Vol. 2615, pp. 219–238). New York: Springer-Verlag.

Raybourn, E. M., Kings, N. J., & Davies, J. (2003). Adding cultural signposts in adaptive community-based environments. *Interacting With Computers: the Interdisciplinary Journal of Human-Computer Interaction* [Special Issue on Intelligent Community-based Systems], *15*(1), 91–107.

Rouxel, J. (2000). *Les shadocks et le désordinateur.* Circonflexe.

Rouxel, J. (2002). *Et les shadocks pompaient.* Le Pouce Et L'index Eds.

Schank, R. C., & Abelson, R. (1977). *Scripts, plans, goals and understanding: An inquiry into human knowledge structures.* Hillsdale, NJ: Lawrence Erlbaum Associates.

Schank, R. C., & Morson, G. S. (1995). *Tell me a story: Narrative and intelligence (Rethinking theory).* Evanston, IL: Northwestern University Press.

Sengers, P. (2000). Narrative intelligence. In K. Dautenhahn (Ed.), *Human cognition and social agent technology, Advances in consciousness series.* Amsterdam: John Benjamins.

Varner, I., & Beaner, L. (1995). *Intercultural communication in the global workplace.* Chicago: Irwin.

Suggested Readings

The following list of recommended readings should be of help if you want to delve more deeply into the issues discussed in this volume. We tried to break it down into thematic domains, but this structure obviously remains somewhat arbitrary. Inside each section, titles are listed by relevance and accessibility so that, for example, "classics" and recent introductory works available as books come first and earlier articles, or those of more specific interest, come later.

EMBODIED AGENTS

Cassell, J., Sullivan, J., Prevost, S., & Churchill, E. (2000). *Embodied conversational agents*. Cambridge, MA: MIT Press.
    Snapshot of that research from around 1998. Probably the best collection of papers on embodied characters so far.

Elliot, C., & Brzezinski, J. (1998, Summer). Autonomous agents as synthetic characters. *AI Magazine*, pp. 13–30.
    A good summary article about much of that work from around the same time period.

Trappl, R., & Petta, P. (Eds.). (1997). *Creating personalities for synthetic actors* (LNAI 1195). Heidelberg, New York: Springer.
    A pioneering book on the domain of agents with personalities.

Bates, J. (1994). The role of emotion in believable agents. *Communications of the ACM, 37*(7), 122–125.

Bates, J., & Loyall, A. B. (1993, June). Real-time control of animated broad agents. *Proceedings of the Fifteenth Annual Conference of the Cognitive Science Society.*
   Two great articles on the philosophy and technology of believable agents.
Blumberg, B. (1994). Action-selection in Hamsterdam: Lessons from ethology. In *From animals to animats: Proceedings of the third international conference on the simulation of adaptive behavior.* Cambridge, MA: MIT Press.
   Classic paper outlining architecture for autonomous agents.
Dautenhahn, K. (Ed.). (2000). *Human cognition and social agent technology.* Amsterdam: John Benjamins.
   Intersects human cognition with social agent technology.
Penny, S. (2000). Agents as artworks and agent design as artistic practice. In K. Dautenhahn (Ed.), *Human cognition and social agent technology.* Amsterdam: John Benjamins.
   Describes what it means to build agents from an artistic, rather than a scientific perspective.
Blair, P. (1995). *Cartoon animation.* Laguna Hills, CA: Walter Foster.
   A master of the craft demonstrates the fundamentals of drawing for the screen, offering tips on two- and four-legged figure construction, body and facial movements, and realistic dialogue, and illustrating concepts such as speed, impact, weight, and recoil.
Brion, P. (1984). *Tex Avery.* Paris: Le Chêne.
   Tex Avery, the creator of Droopy. His cartoon style of slapstick gags is still being imitated today and can help the user to understand the situation better.
Dautenhahn, K., & Nehaniv, C. L. (Eds.). (2002). *Imitation in animals and artifacts.* Cambridge, MA: MIT Press.
Dautenhahn, K., Bond, A., Cañamero, L., & Edmonds, B. (Eds.). (2002). *Socially intelligent agents— Creating relationships with computers and robots.* Dordrecht: Kluwer Academic Publishers.
Trower, T. (1997). Microsoft agent. Washington, DC: Microsoft Corporation. Available: www.microsoft.com/msagent. Last visited September 11, 2003.
   A technology of animated characters designed to enhance, not replace, existing application interfaces such as toolbars and menus.
Yan, H., & Selker, T. (2000). Context-aware office assistant. In Proceedings of the Conference on Intelligent User Interfaces (IUI 2000; pp. 276–279). New York: ACM Press.
   This paper describes the design and implementation of the Office Assistant—an agent that interacts with visitors at the office door and manages the office owner's schedule. The authors claim that rich context information about users is the key to making a flexible and believable interaction, and argue that natural face-to-face conversation is an appropriate metaphor for human–computer interaction.

EMOTIONS AND AFFECTIVE COMPUTING

Picard, R. (1997). *Affective computing.* Cambridge, MA: MIT Press.
   The classical book, a must for anyone willing to work in the domain of affective HCI.
Trappl, R., Petta, P., & Payr, S. (Eds.). (2003). *Emotions in humans and artifacts.* Cambridge, MA: MIT Press.
   Interdisciplinary collection of recent work on emotions in neuroscience, cognitive science, artificial intelligence, and game development.
MacKinnon, N. J. (1994). *Symbolic interactionism as affect control.* Albany: State University of New York Press.
   Presents an overview of the sociological approach to social interaction and additionally offers the most comprehensive book-length presentation of contemporary affect control theory.

SUGGESTED READINGS

Kemper, T. D. (Ed.). (1990). *Research agendas in the sociology of emotions.* Albany: State University of New York Press.
Essays focusing on the sociological view of emotion as a form of communication in social interaction that signals how social relationships are faring.

HUMAN-COMPUTER INTERACTION

Preece, J., Rogers, Y., & Sharp, H. (2002). *Interaction design: Beyond human-computer interaction.* New York: Wiley.
The most recent book on interface design. Updated with consideration of most recent technological solutions (mobile computing and embodied agents).

Dourish, P. (2001). *Where the action is: The foundations of embodied interaction.* Cambridge, MA: MIT Press.
Social dimensions of design for interacting with systems—ubiquity, tangibility, shared awareness, emotions.

Paiva, A. (Ed.). (2000). *Affective interactions. Towards a new generation of computer interfaces* (LNCS 1814). Heidelberg, New York: Springer.
Proceedings of a good workshop in the domain of affective factors in HCI.

Reeves, B., & Nass, C. (1996). *The media equation: How people treat computers, television, and new media like real people and places.* Cambridge: Cambridge University Press.

Maybury, M. T., & Wahlster, W. (Eds.). (1998). *Readings of intelligent user interfaces.* San Mateo, CA: Morgan Kaufman.
A collection of the papers that made the history of intelligent interfaces.

de Rosis, F. (Ed.). (2002). Merging cognition and affect in HCI. *Applied Artificial Intelligence* [Special Issue], *16*, 7–8.

de Rosis, F. (Ed.). (2001–2002). User modeling and adaptation in affective computing. *User-Modeling and User-Adapted Interaction* [Special Issue], *11*(4 & 12), 1.

Hudlicka, E., & McNeese, M. D. (Eds.). (2003). Applications of affective computing in human–computer interaction. *International Journal of Human-Computer Studies* [Special Issue].
Three recent special issues of international journals considering emotional factors in HCI.

Wilks, Y. (Ed.). (1999). *Machine conversations.* Dordrecht: Kluwer Academic Publishers.
Very interesting to those willing to know more about the possibility of entertaining dialogs with a computer today.

Sengers, P. (2003). The engineering of experience. In M. A. Blythe, A. F. Monk, K. Overbeeke, & P. C. Wright (Eds.), *Funology: From usability to enjoyment.* Dordrecht: Kluwer Academic Publishers.
Article looking at the cultural history of algorithmic approaches to human experience and suggesting implications for HCI and ways in which HCI could be rethought.

RETHINKING AGENTS AND AI

Agre, P. E. (1997). *Computation and human experience.* Cambridge: Cambridge University Press.
Analyzes the philosophical background and baggage of historically dominant approaches in AI, and suggests new ways to think about and build agents.

Wise, J. M. (1998). Intelligent agency. *Cultural Studies, 12*(3), 410–428.
Excellent article critically analyzing intelligent agents and our relationship to them.

Mateas, M. (2001). Expressive AI: A hybrid art and science practice. *Leonardo: Journal of the International Society for Arts, Sciences, and Technology, 34*(2), 147–153.

Suggests new ways of thinking about what it means to build AI systems, integrating ideas from an arts perspective.

Mateas, M., & Sengers, P. (Eds.). (2003). *Narrative intelligence. Advances in consciousness series.* Amsterdam: John Benjamins.

Collection of technical and philosophical articles exploring the boundaries between AI and narrative. Many articles examine the relationship between AI technology and culture, and how to design new AI technologies that respond to cultural issues.

Wardrip-Fruin, N., & Moss, B. (2001). The impermanence agent. In M. Eskelinen & R. Koskimaa (Eds.), *Cybertext Yearbook 2001* (Publications of the Research Center for Contemporary Culture #68). Jyväskylä: University of Jyväskylä.

Describes an interesting project to build an agent as a critique of agent-building practice.

Beynon, M., Nehaniv, C. L., & Dautenhahn, K. (Eds.). (2001). *Cognitive technology: Instruments of mind* (LNAI 2117). Heidelberg, New York: Springer.

CULTURAL, SCIENCE, AND TECHNOLOGY STUDIES

Biagioli, M. (Ed.). (1999). *The science studies reader.* New York: Routledge.
A portrait of the field with a focus on recent developments.

Adam, B., & Allan, S. (1995). *Theorizing culture: An interdisciplinary critique after postmodernism.* New York: New York University Press.
Several chapters that touch on technology, nothing specifically on agents.

Fogg, B. J. (2002). *Persuasive technology: Using computers to change what we think and do.* San Mateo: Morgan Kaufmann.

Ritzer, G. (1993). *The McDonaldization of society: An investigation into the changing nature of contemporary social life.* Thousand Oaks: Pine Forge Press.
Sociological classic, easy to read, describing how American (and to a lesser extent) global culture is becoming McDonaldized.

Winner, L. (1985). Do artifacts have politics? In D. MacKenzie & J. Wajcman (Eds.), *The social shaping of technology* (pp. 26–38). Philadelphia: Open University Press.
Classic science studies article that addresses the question of whether technology can be said to have political implications in and of itself (as opposed to in the way in which it is used).

Cowan, R. S. (1985). How the refrigerator got its hum. In D. MacKenzie & J. Wajcman (Eds.), *The social shaping of technology* (pp. 202–218). Philadelphia: Open University Press.
Another classic science studies article. This article analyzes the development of gas versus electrical refrigerators, showing that the technology that eventually becomes adopted is not the best technology, but instead the one that can develop the most market power.

Waldman, S. (1999). The tyranny of choice. In L. B. Glickman (Ed.), *Consumer society in American history* (pp. 359–366). Ithaca: Cornell University Press.
Describes problems with the emphasis on choosing among many options, which is central to both consumer culture and agent architecture.

Tomasello, M. (1999). *The cultural origins of human cognition.* Cambridge, MA: Harvard University Press.

Whiten, A., & Byrne, R. W. (Eds.). (1997). *Machiavellian intelligence: II. Extensions and evaluations.* Cambridge: Cambridge University Press.

Dunbar, R. (1996). *Grooming, gossip and the evolution of language.* Faber & Faber.

McCloud, S. (1994). *Understanding comics: The invisible art.* New York: HarperPerennial. (Original publication 1993 by Kitchen Sink Press.)

INTERCULTURAL COMMUNICATION

Ting-Toomey, S. (1999). *Communicating across cultures.* New York: Guilford.
Hofstede, G. (1997). *Culture and organizations: Software of the mind.* New York: McGraw Hill.
Hall, E. T. (1966). *The hidden dimension.* New York: Anchor Books, Doubleday.
   An "early classic."
Matsumoto, D. (2000). *Culture and psychology. People around the world* (2nd ed.). Belmont, CA: Wadsworth/Thomson Learning.
   A good introduction to cross-cultural psychology not as a special field, but as an approach to mainstream psychology; overview of research results on culture-specific versus universal factors in, among others, personality, emotion, language, nonverbal, and social behavior.
Ekman, P. (1982). *Emotion in the human face* (2nd ed.). Cambridge, MA: Cambridge University Press.
   This book presents an overview of Ekman's pioneering work on the facial expression of emotions.
Osgood, C. E., May, W. H., & Miron, M. S. (1975). *Cross-cultural universals of affective meaning.* Urbana: University of Illinois Press.
   This book—a classic—overviews the most massive quantitative cross-cultural study ever fielded, describing how three dimensions of affective meaning were identified in 25 cultures.
Heise, D. R. (2001). Project magellan: Collecting cross-cultural affective meanings via the Internet. *Electronic Journal of Sociology, 5*(3). http://www.sociology.org/context/vol005.003/mag.html. Last visited March 31, 2003.
   This article presents a practical methodology for getting data needed to implement culturally variable agents.

Contributors

Jan Allbeck is a Systems Programmer in the Center for Human Modeling and Simulation at the University of Pennsylvania. Her responsibilities include managing technical aspects of research projects, supervising students, and designing, coding, and debugging large software systems. She is also a Ph.D. student in the Department of Computer and Information Science at Penn. She is active in the area of autonomous agents and computer animation with more than a dozen publications. Allbeck received a BA degree in Mathematics and a BS degree in Computer Science from Bloomsburg University in 1995. She received an MSE in Computer and Information Science from the University of Pennsylvania in 1997.

Norman I. Badler is a Professor of Computer and Information Science at the University of Pennsylvania and has been on that faculty since 1974. Active in computer graphics since 1968, with more than 200 technical papers, his research focuses on human figure modeling, manipulation, and animation control in real-time three-dimensional graphics. His current research interests include animation via simulation, embodied agent software, human–computer interfaces, and computational connections between language and action. Badler received a BA degree in Creative Studies Mathematics from the University of California at Santa Barbara in 1970, an MSc in Mathematics in 1971, and a Ph.D. in Computer Science in 1975, both from the University of Toronto. He is Co-Editor of the Elsevier journal *Graphical Models*. He was the Cecilia Fitler Moore Department Chair of Computer and Information Science from 1990 to 1994. He directs the Center for Human Modeling and Simulation. Among the Center's achievements are the human modeling software system, Jack, which was the basis for a spin-off company in 1996; the software is now marketed by EDS. He is the Director of the Digital Media Design undergraduate degree program in Computer Science at Penn. Since January 2001, he has also been the Associate Dean for Academic Affairs in the School of Engineering and Applied Science.

Dr. Kerstin Dautenhahn has a background in biology and robotics and is currently Reader in Artificial Intelligence in the Department of Computer Science at University Hertfordshire in England, where she coordinates the Adaptive Systems Research Group. Her main research interests are socially intelligent agents, social robotics, and artificial life. Dr. Dautenhahn has been principal investigator of projects on social robotics and robotic toys in autism therapy. She has edited 10 special journal issues and three books. She is Associate Editor of *Adaptive Behavior* and on the Board of Advisory Editors of the *International Journal of Cognition and Technology: Co-existence, Convergence, Co-evolution (IJCT)*.

Fiorella de Rosis is full Professor of Informatics at the University of Bari. She teaches Human–Computer Interaction at the 2nd year and Intelligent Interfaces at the 5th year of Informatics, University of Bari. She coordinates the Ph.D. School of Informatics and is responsible for the *Intelligent Interface* Research Group in the Department of Informatics of that University. Her interests are in the domain of user modeling and user-adapted interaction, natural language generation, and dialog simulation. More recently, she oriented her interests toward the simulation of affective factors in HCI: She organized several workshops and was the guest editor of two special issues of international journals (*UMUAI* and *AAI*) in this area. More details may be found at the following URL: http://aos2.di.uniba.it:8080/IntInt.html.

Martine Grice is Assistant Professor at the Institute of Phonetics, University of the Saarland. She holds an M.A. and a Ph.D. in Linguistics, and a Habilitation in Phonetics and Phonology. Her main area of research is intonation theory, in particular the structure of tonal representations. She has developed schemes for the database annotation of tonal and junctural phenomena, both for Standard German (GToBI) and a number of varieties of Italian (IToBI). She is also currently investigating the interaction between linguistic and paralinguistic factors determining voice quality and intonation, and the synchronization of speech prosody with visual prosody.

Erich Gstrein is a computer scientist and head of the artificial life group at sysis (www.sysis.at). His work focuses on software architecture and design.

He is the chief architect of the sysis NetLife platform and the client-server platform, which forms the basis of the Flirtboat application.

Barbara Hayes-Roth is a computer scientist, psychologist, author, inventor, and entrepreneur. In 1996, she founded Extempo, Inc., which makes smart, interactive characters, providing automated, personalized two-way communications on the Internet. Dr. Hayes-Roth invented the intellectual property underlying Extempo's technology and products. In February 2000, she was awarded Patent #6031549 for "System and Method of Directed Improvisation by Computer Characters." She has several other patents under review. Dr. Hayes-Roth also teaches and directs research related to interactive characters and other sorts of intelligent agents in the Computer Science Department at Stanford University. Prior to that, she conducted research on human memory and cognition at the Rand Corporation and Bell Laboratories. Dr. Hayes-Roth has published widely and given invited talks at conferences and seminars in the United States and abroad. She holds a Ph.D. in Cognitive Psychology from the University of Michigan and is a Fellow of the American Association for Artificial Intelligence.

Lorna Heaton has an extensive background in cross-cultural communication. She is interested in the technology design process. Her research focuses on organizational change in relation to technology, particularly the communication of different agents across cultural, organizational, and disciplinary boundaries. She has worked in Japan, Africa, and Canada developing and implementing education and training programs in collaboration with local government agencies and community-based groups. Co-author of "The Computerization of Work: A Communication Perspective" (Sage, 2001), Dr. Heaton is currently Assistant Professor in the Communications Department of the University of Montreal and a lead organizer of the biennial CATaC conferences (Cultural Attitudes to Communication and Technology).

David R. Heise, Rudy Professor of Sociology Emeritus at Indiana University, researches the affective foundations of social interaction. Heise received the 1998 Cooley-Mead Award for lifetime contributions from the Social Psychology Section of the American Sociological Association, and he received the 2002 Lifetime Achievement Award from the Sociology of Emotions Section. Additionally, Heise researches issues in quantitative modeling and computer applications in sociology, and he has been editor of *Sociological Methodology* and *Sociological Methods & Research*. He received the 1995 Award for Outstanding Contributions to Computing from the Sociology and Computers Section of the American Sociological Association.

Katherine Isbister (Ph.D., Stanford University, 1998) is a lecturer in Stanford University's Computer Science Department, as well as a consultant in the design of interactive computer characters. Isbister's research includes the study of interactive characters in video games, nonverbal social communication and its applications to agent and robot design, and cross-cultural applications of agents.

Toru Ishida (Doctor of Engineering, Kyoto University, 1989) is a professor of social informatics at Kyoto University and a research professor at NTT Communication Science Laboratories. His research interests include multiagent systems, the human-centered semantic Web, digital cities, and community computing.

Brigitte Krenn is a Senior Scientist at the Austrian Research Institute for Artificial Intelligence. She is a computational linguist. Since 1990 she has worked in a variety of areas ranging from machine translation and robust and corpus-based natural language processing to cognitively adequate language modeling and more recently to animated conversational character technology.

Heidy Maldonado is an international doctoral student at the School of Education at Stanford University. She holds Bachelor and Master of Science degrees in Computer Science, as well as a Master of Arts in Latin American Studies, all from Stanford University. Together with Dr. Hayes-Roth, she has published internationally her research on emotional engagement between humans and rich synthetic characters for learning, embodied onscreen, and as robotic partners or pets. Ms. Maldonado has also published and organized courses at Stanford University on the potential for classroom technology to bridge the digital divide, international technology policy, and Internet entrepreneurship in developing nations.

Benoît Morel is the co-founder and CEO of the French company, Cantoche, created in 1996. Cantoche became a world leader in Embodied Agent technology by creating its own technology, Living Actor™. Formerly a Sound Engineer at Radio France and with numerous private companies, Benoît worked in the video game industry for 10 years producing CGI animation across a variety of formats—notably video games, in-house presentations, interactive shows, Internet Web sites, and, particularly, character animation. Benoît has worked in different countries working on different fields. During an 18-month world tour, he visited more than 60 countries. Today, Benoît consults international companies, research institutes, and universities on embodied agent design and deployment. More information about Benoit and his company Cantoche is available at http://www.cantoche.com.

Hideyuki Nakanishi (Ph.D., Kyoto University, 2001) is an Assistant Professor of Social Informatics at Kyoto University. His research includes virtual environments for communication between people and agents and simulation of social interaction in real worlds.

Clifford Nass (Ph.D., Princeton University, 1986) is a Professor of Communication at Stanford University, with appointments by courtesy in computer science; science, technology, and society; sociology; and symbolic systems (cognitive science). His current research includes social and emotional responses to interactive technologies, voice and avatar-based user interfaces, and human–robot interaction.

Barbara Neumayr is responsible for product development within the sysis NetLife platform and for project management of research projects. As a cofounder of sysis (www.sysis.at), she has been working for sysis on a part-time basis since 1994. Following her studies of business administration and economics, she worked as an assistant professor in both teaching and research at the Vienna University of Economics and Business Administration for 4 years before joining sysis on a full-time basis in 2000.

Sabine Payr (M.A. in modern languages, Ph.D. in linguistics) is author and researcher in e-Learning and media communication. Since 1987, she has been involved in interactive media in training and education, doing research and development in the field of educational technology in higher education, open and distance learning, and tele-learning in vocational training and further education. She has been working for the Austrian Research Institute for Artificial Intelligence since spring 2000 in the framework of the project "An Inquiry into the Cultural Context of the Design and Use of Synthetic Actors," which led to the present book. Currently, she focuses on embodied educational agents. See also http://www.oefai.at/~sabine/.

Catherine Pelachaud is Associate Professor at the IUT of Montreuil, University of Paris 8, France. She received a Ph.D. in Computer Graphics at the University of Pennsylvania, Philadelphia, in 1991. She was then involved in several research projects in the United States, in Italy, and founded by the European Commission. Her research interests include embodied conversational agents, believable agents, human behavior simulation, and multimedia systems. In particular, she is interested in the creation of individual agents, looking at how aspects such as culture, personality, and social role may affect the display of nonverbal behaviors as well as the expressivity of the behaviors.

Isabella Poggi graduated in Philosophy in 1976. She is an Associate Professor of General Psychology and Psychology of Communication at the University Roma Tre. Her main interests and publications concern: Linguistic Education and Teaching Italian as a first Language: Language use, comprehension and production, Metalinguistic and Metacognitive reflection on language, Italian Grammar and Semantics. Pragmatics: Speech Acts, Conversation, Interjections, Theory of Communication, Deception, Persuasion, Political Discourse. Multimodal Communication and Embodied Conversational Agents: gestures, facial expression, gaze, touch, music. Emotions: guilt, shame, humiliation, compassion, enthusiasm.

Dr. Elaine M. Raybourn has a Ph.D. in Intercultural Communication and Human–Computer Interaction. She brings an expertise in understanding culture and communication to the design of interactive software and groupware. In doing so, she collaborates with international organizations including BTexact, British Telecom Advanced Communications Centre in Ipswich, England (1999–2003); Fraunhofer FIT Applied Information Technology Institute in Sankt Augustin, Germany (2001–2003); and INRIA French National Research Institute in Computer Science and Automation located in Paris, France (2003–2004). Her research concerns intelligent community-based systems, intercultural agents, collaborative virtual environments, social-process simulations, games, and interaction design. Current efforts include storytelling in context-aware groupware systems, creating cultural signposts in knowledge-sharing environments, addressing cultural dynamics in agent and avatar behaviors, and designing learning applications and simulations that stimulate intercultural awareness and adaptive thinking. Elaine has been an ERCIM (European Consortium for Research in Informatics and Mathematics) fellow and is currently a member of Sandia National Laboratories and a National Laboratory Professor at the University of New Mexico's Department of Communication & Journalism, Institute for Organizational Communication. See http://www.sandia.gov/media/NewsRel/NR2000/virtual.htm and http://www.cs.unm.edu/~raybourn for more.

Phoebe Sengers is an Assistant Professor in Information Science and Science & Technology Studies at Cornell University, developing applications for everyday computing based on analysis of consumer culture. She graduated from Carnegie Mellon University in 1998 with a self-defined interdisciplinary Ph.D. in Artificial Intelligence and Cultural Theory. In 1998–1999, she was a Fulbright Guest Researcher at the Center for Art and Media Technology (ZKM) in Karlsruhe, Germany. From 1999-2001, she worked at the Media Arts Research Studies group at the GMD Institute for Media Communication in Bonn, Germany.

Robert Trappl is Professor and Head of the Department of Medical Cybernetics and Artificial Intelligence, University of Vienna, Austria. He is director of the Austrian Research Institute for Artificial Intelligence in Vienna, which was founded in 1984. He holds a Ph.D. in psychology (minor: astronomy), a diploma in sociology (Institute for Advanced Studies, Vienna), and is engineer for electrical engineering. He has published more than 150 articles; he is co-author, editor, or co-editor of 29 books, the most recent being "Power, Autonomy, Utopia: New Approaches Toward Complex Systems," Plenum, New York; "Cybernetics and Systems 2002," ASCS, Vienna; "Advanced Topics in Artificial Intelligence," "Creating Personalities for Synthetic Actors," "Multi-Agent Systems and Applications," Springer, Heidelberg/New York; and "Emotions in Humans and Artifacts," MIT Press, 2003. He is Editor-in-Chief of *Applied Artificial Intelligence: An International Journal* and *Cybernetics and Systems: An International Journal*, both published by Taylor & Francis, USA. His main research focus at present is the investigation of the potential contributions of Artificial Intelligence methods to avoid the outbreak of wars or to end them, and the design of emotional personality agents for synthetic actors in films, TV, and interactive media. He has been giving lectures and working as a consultant for national and international companies and organizations (OECD, UNIDO, WHO).

Author Index

A

Abelson, R., 276
Ackoff, R., 22
Agre, P., 17
Albeck, J., 202, 241
Amano, K., 144
Amaya, K., 112
Andersen, H. H. K., 280
André, E., 239
Andrews, S. B., 22
Argyle, M., 85
Arita, K., 29
Arkin, R. C., 37
Aronson, E., 46
Austin, J. L., 39
Axtell, R. E., 108, 112, 114, 158

B

Badler, N., 200, 202, 239, 241
Bagozzi, R. P., 81
Balch, T., 37
Ball, G., 122
Bannon, L., 26
Barnlund, D. C., 33
Barrus, J. W., 245
Bartneck, C., 87
Barton, R. A., 62
Bates, J., 3, 8, 15, 47, 89, 119, 143
Beaner, L., 270
Becheiraz, P., 113
Benford, S., 245
Bequette, J., 144
Berdichevsky, D., 46
Berry, D. C., 96
Bers, M. U., 66
Bickmore, T., 7, 64, 87, 202–203, 236
Bijker, W. E., 23, 25
Bindiganavale, R., 108
Blackmore, S., 61
Blueskies, O., 144
Blumberg, B., 16, 143
Boesch, C., 61
Bourdieu, P., 23
Braitenberg, V., 48, 54
Brand, M., 111
Branham, S., 85, 87
Brave, S., 164
Breazeal, C., 46, 66
Breese, J., 122
Brew, C., 202
Brislin, R., 76, 273

Brown, C. M., 76, 83, 96, 98
Bryant, D., 51
Bumby, K., 48, 49, 54
Burgoon, J. K., 112, 114
Button, G., 22
Byun, M., 112

C

Calhan, C., 130
Calvert, T. W., 112
Cañamero, L., 55
Carley, K. M., 59
Carlson, W. B., 25
Cassell, J., 7, 64, 66, 83–85, 87, 111–112, 114, 162, 202–203, 236, 238–239, 242, 246, 249, 279
Chapman, D., 17
Chi, D., 112
Christian, D., 168
Churchill, E., 238
Cialdini, R., 155–156
Clark, H. H., 236–238, 245, 247
Clarke, A., 277
Clore, G. L., 95, 109, 119, 201
Coakes, E., 22, 25
Cohen, M. F., 240
Colburn, A. R., 200, 234, 236, 240
Coleridge, S. T., 147
Collins, A., 95, 109, 119, 201
Condon, J. C., 79
Cook, M., 85
Cowie, R., 201

D

Davis, A., 151, 159, 165
Dawkins, R., 60
De Carolis, B., 95, 100, 203
De Lannoy, J.-D., 163
de Rosis, F., 186, 273, 280
de Waal, F., 62
Deans, C. P., 23
DeLaet, C., 22
DeLanda, M., 11
DeSanctis, G., 22
Donald, M., 60
Donath, J., 84, 87
Doray, B., 10
Doyle, P., 144, 165–166, 168, 246
Drucker, S. M., 240

Dunbar, R. I., 62, 64
Durig, A., 128

E

Earley, P. C., 23
Eddy, T. J., 52
Edmonds, B., 46
Eibl-Ebesfeldt, I., 82, 110
Ekman, P., 81–82, 85, 110–111, 131, 200–201, 203, 233, 237
Elliot, C., 119, 201, 241
Engeström, Y., 22
Erez, M., 23
Erickson, T., 50
Ewen, R., 151, 154, 164, 170

F

Feyereisen, P., 163
Flores, F., 39
Fogg, B. J., 46
Foner, L. N., 50, 64, 153, 246
Fox, A., 170
Fragaszy, D., 62
Friesen, W., 81–82, 110–111, 201
Frijda, N., 89
Funge, J., 116

G

Garau, M., 236
Gard, 156
Gardner, H., 172
Gash, D. C., 22
Georgeff, M. P., 94
Gibbon, D., 199
Giddens, A., 22
Goldberg, A., 147
Gollob, H. F., 128
Grand, S., 64
Gratch, J., 119
Greenhalgh, C., 245
Grudin, J., 29–30, 40
Gudykunst, W. B., 75

H

Haag, E. van den, 17

Hadley, T., 36
Hagsand, O., 245
Haidt, J., 80
Hales, M., 23
Hall, E. T., 30–31, 76, 235–236, 247, 249, 268–269, 272
Hall, J. A., 82, 85, 107, 111–115, 117
Hall, M. R., 76, 247
Hamm, M., 53–54
Han, J., 245
Hannerz, U., 24
Harnad, S., 58
Hartmann, B., 112
Hayes-Roth, B., 144, 165, 203, 246
Haythornthwaite, C., 22
Heath, C., 22
Heider, F., 47, 53–54
Hertzum, M., 280
Hewes, G. W., 113
Hill, R., 240
Hindmarsh, J., 22
Hofstede, G., 23–24, 76, 81, 87, 95, 103, 118
Höök, C., 204
Hopper, D. I., 204
Huber, M. J., 36
Huffman, M. A., 61
Hughes, T. P., 23
Hutchins, E., 22
Hymowitz, K., 144

I

Ike, S., 140
Immonen, S., 202
Isbister, K., 121, 143, 158, 203
Ishida, T., 158
Ishii, H., 28–29, 30

J

Jack, M., 57
Jackson, M. H., 23
Järvenpää, E., 202
Johanson, B., 170
John, D. P., 90, 103
Johnson, W. L., 114, 240
Johnston, O., 150
Johnstone, K., 237
Jones, C., 150–151, 189

Jones, R. M., 108

K

Keltner, D., 80, 85
Kennedy, N., 5
Kim, Y. Y., 75
King, A. B., 141
Kitayama, S., 159
Knapp, M. L., 82, 85, 107, 111–115
Knodt, E., 172
Kobayashi, M., 29–30
Kozu, J., 151, 159, 165
Kurlander, D., 112

L

LaFrance, M., 114
Laland, K. N., 62
Langton, C. G., 50
Lanier, J., 17
Lasseter, B., 235
Laurel, B., 147, 276
Law, J., 23, 25
Lea, R., 245
Leathers, D. G., 233–234, 236–237
Lee, E. J., 86, 143
Lee, S., 114
Leevers, D., 276
Lester, J., 234, 239–240, 246
Lester, S., 166
Levy, S., 50
Lewis, H., 109, 111
Lloyd-Jones, R., 22
Lord, P., 189
Loyall, A. B., 3, 8, 15, 89
Luff, P., 22
Lukács, G., 9–10

M

Mackenzie, D., 23
MacKinnon, N. J., 129, 139
Madu, C., 22
Maes, P., 15
Magno, C. E., 83
Mahoney, M. S., 11
Markus, H. R., 159

Marsella, S., 55, 119
Mateas, M., 56
Matsuno, T., 139–140
Matsushita, Y., 27
Mayer, R. E., 172
Mayo, C., 114
McBreen, H., 57
McCarthy, J., 204
McCloud, S., 51
McCrae, R., 90, 103
McNeill, D., 83
Middleton, D., 22
Miller, H. G., 22
Miller, P., 151, 159, 165
Mitchell, G., 277
Mitchell, R. W., 53–54
Miyake, N., 28
Moghaddam, F., 151
Montgomery, D. A., 53
Montgomery, D. E., 53
Montigneaux, N., 187
Moon, Y., 202
Moraes, M., 144
Morawetz, C. L., 112
Morel, B., 89, 273, 280, 282
Moreno, R., 172
Mori, M., 45, 51–53
Morris, D., 237
Morson, G. S., 276
Mouer, R., 27
Mudur, S., 278
Müller, J., 239
Murray, J., 166–167, 276

N

Naito, T., 246
Nakanishi, H., 158, 240
Nardi, B. A., 22
Nass, C., 47–49, 86, 89, 121, 143, 150–151, 154, 158, 164, 201–202, 280
Nehaniv, C. L., 46, 56
Neunschwander, E., 46
Newcomb, T., 38
Newell, A., 59
Nichols-English, G., 76, 83, 96, 98
Nielsen, J., 180
Noake, R., 182
Noble, D., 12
Norman, D. A., 46, 56

O

O'Connor, F., 150
O'Neill-Brown, P., 66, 84–85, 87–88
Oatley, K., 53
Oberlander, J., 202
Okabe, R., 118
Okada, K., 31
Orlikowski, W. J., 22
Orr, J. E., 276
Ortony, A., 89–90, 95, 103, 109, 119, 201
Osbeck, L., 151
Osgood, C. E., 129
Ostermann, J., 200

P

Pasquariello, S., 203
Pelachaud, C., 111, 203
Perlin, K., 147
Perreault, S., 151
Persson, P., 47, 204
Picard, R., 164, 168, 170
Pinch, T. J., 23, 25
Poggi, I., 111
Poole, M. S., 22
Porter, R. E., 78
Pufahl, I., 168

R

Raman, K. S., 23
Rao, A. S., 94
Reader, S. M., 62
Reeves, B., 47–49, 143, 148–150, 157, 249
Reichardt, J., 51
Rendell, L., 62
Rhodes, N., 168
Rickel, J., 114, 240
Ricks, D. A., 23
Rimé, B., 53
Rist, T., 239
Ritzer, G., 4, 13–15, 17
Rogers, E. M., 22
Rogers, Y., 22
Rose, C., 111–112
Rouxel, J., 282
Ruttkay, Z., 203

AUTHOR INDEX

S

Samovar, L. A., 78
Scassellati, B., 46–47
Schank, R. C., 276
Schmidt, K., 26
Schneider, A., 129–130
Schröder, M., 201
Searle, J., 39
Sengers, P., 54, 89, 276
Sheridan, E. F., 87
Shneiderman, B., 90
Shoemaker, F., 22
Shuter, R., 114
Siebel, T., 17
Silverman, B. G., 108, 119
Simmel, M., 47, 53–54
Sindermann, C. J., 59
Smith, B., 245
Smith, H. W., 129, 139–141
Smith-Lovin, L., 128–129, 138–139
Stephenson, N., 109
Stewart, E. C., 33
Stigler, J., 167
Stone, M., 162
Strasser, S., 13
Stronks, J. J. S., 201
Sugawara, S., 245
Sugimoto, Y., 27
Swagerman, J., 89
Swartout, W., 108

T

Takeuchi, A., 246
Takeuchi, Y., 57
Taylor, J. R., 40, 42
Thalmann, D., 113
Thomas, F., 150
Ting-Toomey, S., 110–113, 116–119, 233–234, 236–237
Tomasello, M., 46–47, 60
Triandis, H., 111
Tseng, S., 46
Turing, A., 58
Turkle, S., 64

U

Umino, M., 139–140

V

van Gent, R., 168
Varner, I., 270
Vaucanson, J. de, 50
Vilhjálmsson, H., 236
Visalberghi, E., 62

W

Wajcman, J., 23
Wardrip-Fruin, N., 17
Waters, R. C., 245
Watson, R. T., 23
Watt, S. N. K., 52
Weir, B., 130
Weizenbaum, J., 153, 166, 204
Wellman, B., 22
Whitehead, H., 62
Whiten, A., 61
Willis, D., 22
Winner, L., 23
Winograd, T., 39, 170
Wise, J. M., 17

Y

Yan, H., 236
Yoshida, T., 273
Yousef, F. S., 79
Yuill, N., 53
Yum, J. O., 118

Z

Zhao, L., 112
Zong, Y., 201

Subject Index

A

A-B-X system, 38
ACT, *see* Affect Control Theory
Action selection, 15
Affect Control Theory (ACT), 128–134, 138, 140
Affective reasoner (Elliot), 241
Agency, 4
Agent
  animated, 87–90, 94, 144
  appearance, 115, 154–155, 223, 280
  autonomous, 16, 196, 205
  biography (backstory), 154, 157–159
  cultural adaptation, 93–94
  definition, 4, 46, 271
  embodied, 155, 195, 279, 280–283
  embodied conversational (ECA), 46, 64, 198, 200–202, 261
  intercultural, 18, 273, 275–277, 279–280, 283–234
  lifelike, 16, 199–200
  multicultural, 129
  personality, 153–154, 159
  social, 46, 234, 239
  socially intelligent (SIA), 45–46, 50, 54, 57, 60, 63–67, 132, 249, 261

Agent application, 4
Agent architecture, 8, 94
  Hap, 8, 15
  standardization of, 8
Agent behavior, 5, 7, 15, 36, 54, 82,
  consistency of, 89, 102
Agent design, 4, 7, 21, 41, 55–57, 63, 93, 155–157, 238, 273
  culturally specific, 144, 152
Agent functionality, 7
Agent system, *see also* Virtual character
  Agneta and Frida, 203
  Carmen's Bright IDEAS, 55
  EMOTE, 112, 122
  Flirtboat, 206, 209, 214, 224, 226
  Funki Buniz, 168, 170
  GRETA, 94, 99–101
  Living Actor®, 178–179, 183, 191–192, 194
  NetLife, 205–206
  REA, 7, 87, 203
  Victec, 55
  Virtual Theater, 170
Agent, text-based, *see* Chatterbot
Animation markup language, MPML, 201
Anthropomorphization, 52–54, 47, 49
Artificial life, 50

307

Avatar, 108–109, 114, 171, 205–207, 214, 217, 223, 225

B

BDI architecture, 94–95
Behavior, 189
  cultural variation, 139
  nonverbal, 92–93, 132, 202, 233
  rule-driven, 18
  spatial, 114, see also Distance
  touching, 114–115, 202, see also Contact
  verbal, 132, 157, 159
Belief, 95, 102
Belief, Desire, Intention, see BDI
Believability, 35, 45, 51, 79, 89, 102, 143, 147–150, 160, 166, 168, 198, 200
Body language, 31, 109, 111–112, 235

C

Cartoon character, 187, 200
Cartoon series, *Shadocks*, 282
Character, computer-generated, 51, see also Virtual character
Chatterbot
  Cybelle, 204
  Eliza, 153, 204
  Julia, 50, 153, 246
Cognition, socially situated, see Intelligence
Cognitive unit (OCC model), 121
Coherence, 90–92
Communication
  cross-cultural, 144
  definition, 270
  interactional rules, 79–80
  intercultural, 233–234, 242, 274–275, 284
  nonverbal, 29, 31, 34, 46, 83, 100, 109, 203, 234, 238, 241, 249
  organizational, 22–23
  semantic rules, 79–80
Communication style, 79
  person-oriented, 117
  status-oriented, 117
Communication system
  high-context, 117
  low-context, 117
Community system, 269, 272, 275, 279
  agent-based, 276–277, 283

Computer game, 4, 108–109
  Creatures, 64
  Dungeon Keeper, 150
  Grand Theft Auto, 150
  The Sims, 109, 116, 122
  Tomb Raider, 64, 183
Computer-mediated communication (CMC), 270
  intercultural (ICMC), 271, 283
Computer supported cooperative work (CSCW), 21
  ClearBoard, 28–29, 30, 34
  Clearface, 28–29
  Japanese, 26–28, 30, 33–34, 41
  MAJIC, 30–32, 34
  Scandinavian, 26
  TeamWorkStation, 28, 31, 34
Consistency, 90–92, 121
Contact, physical, 238
Cooperativeness, 91–93
Cross-cultural discovery, 276, 278, 284
Cross-cultural portability, 21
Cultural consistency, 88
Cultural cue (signpost), 271, 276, 278–279, 283
Cultural differences, 57
Cultural diversity, 45, 65–66, 88
Cultural doctrine (OCC model), 119
Cultural imperialism, 18
Cultural sensitvity, 88
Cultural universalism, 110
Culture
  animal, 61
  cetaceans, 61
  chimpanzee, 61
  collectivist, 95, 111, 121, 159, 165
  contact, 114
  corporate, see organizational
  definition, 25, 75, 268
  dimensions of (Hofstede), 24, 76, 87, 95, 103, 117–118
  emergent, 274, 279, 281
  evolution of, 60–62, 65
  high-contact, 236
  individualistic, 95, 110, 121, 159, 165
  Internet, 141
  low-contact (noncontact), 114, 236
  national, 23, 25, 246
  occupational, 23–25
  organizational, 23, 25, 275, 279–280
  primate, 45, 60
  transmission of, 61–62

SUBJECT INDEX

transnational, 24
Culture recognition, 261
Cultures
  African American, 236
  American, 82
  Arabic, 139–140, 236
  Austria, 84, 209–211, 214, 217–226
  Belgium, 84
  Brazil, 144, 156–159, 164, 166
  British, 80
  Bulgaria, 84
  China, 141
  Croatia, 209, 214, 217–226
  Eastern, 81
  Euro-American, 236
  Finland, 202
  Germany, 87, 139, 167
  Ireland, 139
  Italy, 88, 96–98
  Japan, 21, 24, 26–27, 29–31, 33–36, 41, 57, 82, 85, 87, 139–141, 158, 167, 236, 238, 240–241, 248, 256–260, 264
  Korea, 86
  Latin, 237
  Malaysia, 87
  North America, 24, 86, 202
  Northern Europe, 92
  Northern Italy, 84
  Scandinavia, 24, 35, 238
  Southern Italy, 80, 84, 92–93
  United Kingdom, 96–98, 209–210, 214, 217–226
  United States, 57, 85, 96, 139, 140, 158, 165–167, 238, 240–241, 248, 256–260, 264
  Venezuela, 144, 158–159, 164, 166
  Western, 81, 85, 87, 88

D

Dialogue generation, 102
Dialogue move, 99
Discourse plan, 92, 97, 101
Display rule, 110, 118, 203
Distance, 200, 202, 235, 242, 249
  conversational, 115
  interaction, 115
  personal, 30
  virtual, 31

E

ECA, *see* Agent, embodied conversational
Embodiment, 35, 249
Emergent behavior, 16
Emotion, 78, 80–81, 89, 130
  basic, 81, 85, 164, 170
  intensity, 81
  models of, 7, 55, 171
Emotion activation, 100–101
Emotion expression, 81–82, 118, 163, 200, 203
  facial, *see* Facial expression
Emotion representation, 201
Emotion theory, 164
Emotional state, 136, 163–164, 201
Empathy, 55, 147, 165
EPA profile, 129, 131–136
  cross-cultural, 138–139
Ethnicity, 86
Evaluation-potency-activity profile, *see* EPA profile
Eye contact, 29, 31, 113, 114, 236, *see also* Gaze

F

Facial expression, 46, 80–82, 84–85, 93, 101, 109, 110–111, 130–131, 136–137, 171, 199–201, 203–204
Facial muscles, 131
Frame of meaning, 25
Frame, technological, 25

G

Gaze, 79–80, 84–85, 100, 113–114, 200, 202, 236, 239
Gaze awareness, 29, 31
Gaze-tracking, 241
Gesture, 28, 31, 79–80, 83, 93, 100–101, 111–112, 162, 199–200, 204, 237, 242
  beat, 83
  coded, 84
  creative, 83–84
  emblematic, 112, 162, 237
  iconic, 83
  metaphoric, 83–84, 162
  symbolic, 83
Globalization, 18, 22–23, 86

Goal, 95, 101–102, 119, 132
  instrumental, 77
  terminal, 77
Greeting, 160–161

H

Habitus, 23
Human-computer interaction, 86–87, 89

I

Identity, sociocultural, 199
Illusion of Life, 47, 150
Industrial Revolution, 5
Intelligence
  behavioral, 195
  narrative, 18, 58–59
  social, 47, 54, 56–57, 59
  socially situated, 46
Intentionality, 47, 53
Interaction
  human-computer, 86–87, 89
  human-human, 246, 263
  multicultural, 273–275
Interaction pattern
  culture-specific, 166
  role-specific, 166
Intercultural contact, 22, see also
    Communication, intercultural
Interface agent, 4, 56, 246
Interface design, 90

K

Knowledge, contextual, 37

L

Laban movement analysis, 112
Learning
  cultural, 60
  imitation, 61
  social, 60
Life-Like Agents Hypothesis, 52, 55
Localization, 144, 151–152, 160, 209–211

M

McDonaldization, 4
Meme, 60, 62
Microsoft® Agent Technology, 35, 88, 157, 178, 280
Modeling, quantitative, 128
MOO, text-based, 277–279
Movie
  101 Dalmatians, 51
  Antz, 50
  Artificial Intelligence, 50
  Bambi, 51
  Final Fantasy, 35, 51, 182
  Jurassic Park, 51
  The Matrix, 51
  Titanic, 182
  Toy Story, 50
MUD, see Multi-user domain
Multiculturalism, 272
Multi-user domain (or dungeon, MUD), 50, 64, 246, 277

N

Narrative, 274, 276–277
Nonverbal behavior, see Behavior, nonverbal
Nonverbal communication, see Communication, nonverbal
Nonverbal vocalization, 115–116
Norm, 78–80, 102, 119
  cultural, 155
  interactional, 200
Normative action, 130, 133

O, P

OCC model (Ortony-Clore-Collins model), 119, 201
Performative, 99, 111
Personality, 118, 201–202, 220, see also
    Agent, personality
  extroversion, 202, 222
  five-factor model, 90, 103
  introversion, 202, 222
  Myers-Briggs Type Indicator, 221–223
Plan, 96
Politeness, 91–92

SUBJECT INDEX

Portability, cross-cultural, 37
Posture, 113, 189, 200–201
Program, Interact, 130–131, 134–136
Project
  MagiCster, 94, 203
  NECA, 227
  Victec, 55
Prosody, 84
Protoculture, see Culture, animal
Proxemics, 117, see also Distance

R

RoboCup, 36
Robot, 4, 8, 48, 54
  Kismet, 66
Robotic toy, 5
  Aibo, 5, 35, 64–65
  Furby, 64
  My Real Baby, 5
Robotics, 46
Role, 166
Role behavior, 157
  teacher, 167
Role dynamics, 157, 166–167
Role expectation, 202

S

SIA, see Agent, socially intelligent
Signal, nonverbal, 101
Skills, intercultural nonverbal, 235, 241, 242
Society
  anonymous (eusocial), 60, 63
  individualized, 60, 63
Speech, 249
  content of, 157
  intonation, 202
Speech act theory, 38–39
Speech recognition, 261
Speech synthesis, 199, 201, 203–204, 250
Status marker, 237
Stereotype, cultural, 85–86, 152
Suspension of disbelief, 47, 143, 147–149
Syntax, 82

T

Task decomposition, 96
Taylorism, 10, 14
Technology, persuasive, 45, 46, 65–66
Technology transfer, 22
Topic knowledge, 157–158, 250
Touch behavior, see Behavior, touching
Turing Test, 58–59
Turn-taking, 166, 200–202, 238

U, V

Usability, 86
User interface, 250
  graphical, 273
User profile, 271
Value, 78–79, 92, 102, 116, 119
Video game, see computer game
Virtual character, see also Agent system, Chatterbot
  10 key qualities, 146, 152–153, 172, 249
  Ananova, 204
  Chase Walker, 204
  Gandalf, 239
  Herman the Bug, 239
  Jack, 148–149
  Jennifer Jones, 165
  Kyra, 144–145, 152–154, 158, 160, 166–167
  Lara Croft, 4, 64, 157, 183
  Mr. Clean, 148
  Pick-up Pete, 207
  REA, 239, 242
  Tigrito, 168–169
Virtual meeting space, 245–246, 263
  Community Place, 245
  CU-SeeMe, 245
  DIVE, 245
  FreeWalk, 247–248, 264
  InterSpace, 245
  Massive, 245
Virtual world, 108
Voice qualities, 115

W

Worldview, 37